# Intermediate Algebra

by Lisa Healey

**Intermediate Algebra**
ISBN: 978-1-943536-30-6
Edition 1.0 Fall 2017
© 2017, Chemeketa Community College. All rights reserved.

**Chemeketa Press**

Chemeketa Press is a nonprofit publishing endeavor at Chemeketa Community College. Working together with faculty, staff, and students, we develop and publish affordable and effective alternatives to commercial textbooks. All proceeds from the sale of this book will be used to develop new textbooks. For more information, please visit chemeketapress.org.

*Publisher:* Tim Rogers
*Managing Editor:* Steve Richardson
*Production Editor:* Brian Mosher
*Manuscript Editors:* Steve Richardson, Matt Schmidgall
*Design Editor:* Ronald Cox IV
*Cover Design:* Ronald Cox IV
*Interior Design:* Ronald Cox IV, Kristi Etzel, Kristen MacDonald
*Layout:* Noah Barrera, Matthew Sanchez, Faith Martinmaas, Emily Evans, Shaun Jaquez, Steve Richardson, Kristi Etzel, Cierra Maher, Candace Johnson

Additional contributions to the design and publication of this textbook come from the students and faculty in the Visual Communications program at Chemeketa.

**Chemeketa Math Faculty**

The development of this text and its accompanying MyOpenMath classroom has benefited from the contributions of many Chemeketa math faculty in addition to the author, including:

Ken Anderson, Benjamin Gort, Kyle Katsinis, Tim Merzenich, Nolan Mitchell, Martin Prather, Keith Schloeman, Rick Rieman, and Toby Wagner

**Text Acknowledgment**

This book was originally developed using materials from *OpenStax College Algebra*, by OpenStax College, which have been made available under a Creative Commons Attribution 4.0 license and may be downloaded for free from cnx.org/contents/9b08c294-057f-4201-9f48-5d6ad992740d.

Printed in the United States of America.

# Contents

**Chapter 1: Graphs and Linear Functions** ............................................................................ 1
    1.1  Qualitative Graphs ............................................................................................... 2
    1.2  Functions ............................................................................................................... 11
    1.3  Finding Equations of Linear Functions ........................................................... 30
    1.4  Using Linear Functions to Model Data ........................................................... 48
    1.5  Function Notation and Making Predictions ................................................... 64

**Chapter 2: Exponential Functions** ....................................................................................... 79
    2.1  Properties of Exponents ..................................................................................... 80
    2.2  Rational Exponents ............................................................................................. 98
    2.3  Exponential Functions ...................................................................................... 105
    2.4  Finding Equations of Exponential Functions .............................................. 119
    2.5  Using Exponential Functions to Model Data .............................................. 127

**Chapter 3: Logarithmic Functions** ..................................................................................... 141
    3.1  Introduction to Logarithmic Functions ......................................................... 142
    3.2  Properties of Logarithms ................................................................................. 151
    3.3  Natural Logarithms .......................................................................................... 170

**Chapter 4: Quadratic Functions** ......................................................................................... 179
    4.1  Expanding and Factoring Polynomials ........................................................ 180
    4.2  Quadratic Functions in Standard Form ....................................................... 205
    4.3  The Square Root Property .............................................................................. 221
    4.4  The Quadratic Formula ................................................................................... 237
    4.5  Modeling with Quadratic Functions ............................................................. 251

**Chapter 5: Further Topics in Algebra** ............................................................................... 263
    5.1  Variation ............................................................................................................. 264
    5.2  Arithmetic Sequences ...................................................................................... 276
    5.3  Geometric Sequences ...................................................................................... 288
    5.4  Dimensional Analysis ...................................................................................... 299

Solutions to Odd-Numbered Exercises ............................................................................... 309

# CHAPTER 1
# Graphs and Linear Functions

Toward the end of the twentieth century, the values of stocks of Internet and technology companies rose dramatically. As a result, the Standard and Poor's stock market average rose as well.

Figure 1 tracks the value of an initial investment of just under $100 over 40 years. It shows an investment that was worth less than $500 until about 1995 skyrocketed up to almost $1500 by the beginning of 2000. That five-year period became known as the "dot-com bubble" because so many Internet startups were formed. The dot-com bubble eventually burst. Many companies grew too fast and then suddenly went out of business. The result caused the sharp decline represented on the graph beginning around the year 2000.

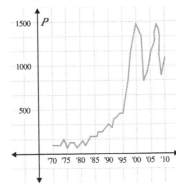

Figure 1.

Notice, as we consider this example, there is a relationship between the year and stock market average. For any year, we choose we can estimate the corresponding value of the stock market average. Analyzing this graph allows us to observe the relationship between the stock market average and years in the past.

In this chapter, we will explore the nature of the relationship between two quantities.

1.1 Qualitative Graphs ................................................................. page 2

1.2 Functions ............................................................................. page 11

1.3 Finding Equations of Linear Functions.................................page 30

1.4 Using Linear Functions to Model Data................................page 48

1.5 Function Notation and Making Predictions.........................page 64

# 1.1 Qualitative Graphs

## Overview

In this section, we will see that, even without using numbers, a graph is a mathematical tool that can describe a wide variety of relationships. For example, there is a relationship between outdoor temperatures over the course of a year and the retail sales of ice cream. We can describe this relationship in a general way using a qualitative graph. As you study this section, you will learn to:

- Read and interpret qualitative graphs
- Identify independent and dependent variables
- Identify and interpret an intercept of a graph
- Identify increasing and decreasing curves
- Sketch qualitative graphs

## A. Reading a Qualitative Graph

Both qualitative and quantitative graphs can have two axes and show the relationship between two variables. We also read both types of graph from left to right — just like a sentence. The difference is that **quantitative graphs** have numerical increments on the axes (scaling and tick marks), while **qualitative graphs** only illustrate the general relationship between two variables.

### Example 1

Use the qualitative graph, Figure 1, and the quantitative graph, Figure 2, to answer the following questions.

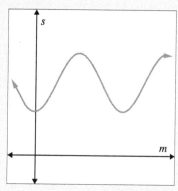

Figure 1. The sale of ice cream at Joe's Café (a qualitative graph).

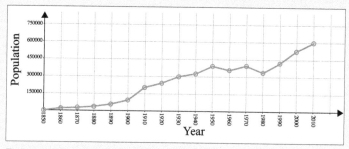

Figure 2. The population of Portland, Oregon (a quantitative graph).

1. What does the qualitative graph tell us about ice cream sales at Joe's Café? Do we know how many servings were sold in June?
2. What does the quantitative graph tell us about the population of Portland, Oregon? What was the population in 1930?

Solutions

1. Ice cream sales are lowest at the beginning and at the end of the year and highest during the middle months. We cannot tell from this graph exactly how many servings are sold in any given month.

2. The population of Portland, Oregon, has been increasing since 1850, except for a slight decrease in the 1950s and 1970s. The population in 1930 was about 300,000.

## B. Independent and Dependent Variables

A qualitative graph is a visual description of the relationship between two variables. The graph tells a "story" about how one quantity is determined or influenced by another quantity. For example, the number of calories one consumes in a week determines the number of pounds one will lose (or gain) that week. Another way to say this is that the change in a person's weight is dependent on the number of calories they consume.

We can assign variables to the quantities in the relationship between calories consumed and weight. Let $c$ be the number of calories consumed in a week and let $w$ be the weight change in pounds of the person who is counting calories. In this example, the quantity of weight change depends on the number of calories consumed, so we call $w$ the **dependent variable**. Because the number of calories consumed determines or influences the weight change, we call $c$ the **independent variable**.

When creating a qualitative graph that depicts the relationship between two variables, the first step is to determine which of the variables is independent and which is dependent. Let's say we want to depict the relationship between $p$, the number of bushels of potatoes produced on an acre of farmland, and $k$, the number of kilograms of fertilizer applied to the acre. We can phrase the relationship two different ways and determine which makes the most sense.

We can say, "The yield of potatoes depends on the amount of fertilizer," or, "The amount of fertilizer depends on the yield of potatoes." It makes more sense to say that $p$, the bushels of potatoes yielded, depends on $k$, the amount of fertilizer used, so $p$ is the dependent variable. The amount of fertilizer used, $k$, is the independent variable because it influences the number of bushels produced.

### Independent and Dependent Variables

In the relationship between two variables, $p$ and $t$, if $p$ depends on $t$, then we call $p$ the **dependent variable** and $t$ the **independent variable**.

## Example 2

Identify the independent variable and the dependent variable for each situation.

1. Let $p$ represent the average price of a home in Salem, Oregon, and let $t$ represent the number of years since 1990.

2. Let $r$ represent the rate in gallons per minute that water is added to a bathtub, and let $m$ be the number of minutes it takes to fill the tub.

**Solutions**

1. We say that the price $p$ *depends on* or *is determined by* the year $t$. It is therefore the dependent variable. We would *not* say the year $t$ depends on the average price of a home $p$. Time is independent of the price. Whether the average price goes up or down, time keeps passing into the future. So we call $t$ the independent variable.

2. The rate of water flow determines how quickly the tub fills, so $r$ is the independent variable. The number of minutes it takes to fill the tub depends on this rate, so $m$ is the dependent variable.

## Practice B

Determine the independent variable and the dependent variable for each situation. Turn the page to check your solutions.

1. Let $m$ be the number of minutes since a cup of hot tea was poured, and let $T$ be the temperature of the tea.

2. Let $g$ be a student's exam score, and let $s$ be the amount of time the student spent studying for the exam.

3. Let $F$ be the outside temperature, and let $c$ be the number of winter coats that a department store sells.

4. Let $v$ be the resale value of a used car, and let $a$ be the age of the car.

## C. Sketching Qualitative Graphs

When graphing, we always represent the independent variable along the horizontal axis, and we always represent the dependent variable along the vertical axis. In Figure 3, for example, we see that the height of a burning candle $h$ is dependent on the number of minutes $m$ since it has been lit. So the independent variable $m$ is represented along the horizontal axis and the dependent variable $h$ is represented along the vertical axis.

In Figure 3, you'll notice that the curve intersects both the horizontal and vertical axes. The point where the curve intersects the vertical axis is called the **vertical intercept**, and the point where the curve intersects the horizontal axis is called the **horizontal intercept**.

## Example 3

The graph in Figure 3 shows the relationship between the height of a burning candle and the number of minutes since it was lit. Interpret the meaning of the intercepts of the graph in Figure 3.

### Solution
The vertical-intercept or $h$-intercept on this graph represents the height of the candle in centimeters when it is first lit, when $m = 0$. The horizontal-intercept or $m$-intercept on this graph represents the time in minutes when the candle has been completely burned, when $h = 0$.

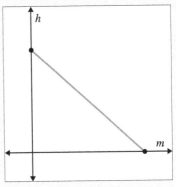

Figure 3. Height of a Burning Candle.

If a curve goes upward from left to right, as in Figures 4a and 4b, the curve is increasing. If the dependent variable increases as the independent variable increases, we sketch a qualitative graph with an **increasing curve**.

If a curve goes downward from left to right, as in Figures 5a and 5b, the curve is decreasing. If the dependent variable decreases as the independent variable increases, we sketch a qualitative graph with a **decreasing curve**.

Notice that in Figures 4 and 5, the independent variable $x$ is represented along the horizontal axis. As we read a graph from left to right, the independent variable is always increasing. The dependent variable $y$ depends on the value of $x$ and may be either increasing or decreasing.

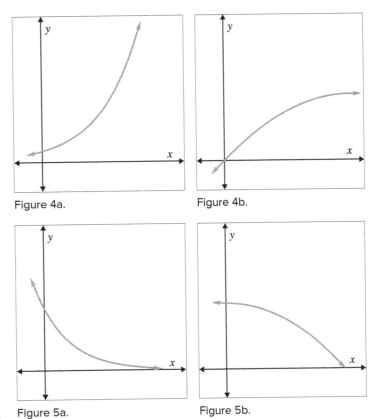

The graph in Figure 3, which represents the height of a burning candle over time, is a decreasing curve because the height of the candle decreases as the number of minutes increases. Although this graph is a straight line, mathematicians still refer to it as a "curve."

Often the dependent variable in a situation will have both intervals of increase and intervals of decrease while the independent variable increases. Sometimes the dependent variable remains constant as the independent variable increases. Examples 4 and 5 involve relationships whose graphs have increasing, decreasing, and constant segments.

### Example 4

A child climbs into a bathtub. After a few minutes of playing around in the water, the child gets out of the tub and pulls the plug so that all of the water drains away. Let $W$ be the water level in the bathtub in inches at $t$ minutes since the child climbed in.

a. Determine the independent variable and the dependent variable.

b. Label the axes of a graph that will describe the relationship between the variables $W$ and $t$.

c. Sketch a qualitative graph that describes the situation taking into consideration any vertical or horizontal intercepts of the graph.

### Solution

a. The water level depends on the number of minutes the child has been in the tub, so $W$ is the dependent variable and $t$ is the independent variable.

b. In Figure 6, we label the horizontal axis with the independent variable $t$, the time in minutes. We label the vertical axis with the dependent variable $W$, the water level in inches.

c. The bathtub is full at the beginning of the situation, when $t = 0$, so we make sure that the vertical-intercept is well above the origin, which is where the axes cross.

When the child climbs into the tub, the water level rises a little, so the curve increases. During the few minutes while the child is playing in the tub, the water level is constant. This is represented by the horizontal segment of the graph. When the child climbs out of the tub, the water level decreases a little, so the curve also decreases.

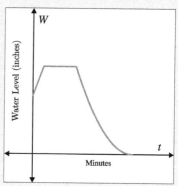

Figure 6. Water level over time in a bathtub.

After the plug is pulled, the water level continues to decrease until the bathtub is drained. The horizontal-intercept represents the time when the water level is zero.

# Example 5

One morning, Paula walked to the bus stop. Once she got there, though, she realized that she had forgotten her backpack. She ran home to get the backpack, and then she ran back to the bus stop to wait for the bus. Let $d$ be Paula's distance from home in meters at $t$, time in seconds.

a. Determine the independent variable and the dependent variable.

b. Label the axes of a graph that describes the relationship between the variables $d$ and $t$.

c. Sketch a qualitative graph that describes the situation, taking into consideration any vertical or horizontal intercepts.

## Solution

a. Paula's distance from home depends on the time since she left, so $d$ is the dependent variable, and $t$ is the independent variable.

b. In Figure 7, we label the horizontal axis with the independent variable $t$, time in seconds. We label the vertical axis with the dependent variable $d$, distance from home in meters.

c. Paula begins at home, so we begin the graph at the origin. This point represents both the vertical intercept and one of the horizontal intercepts because at $t = 0$ seconds, her distance from home is also zero.

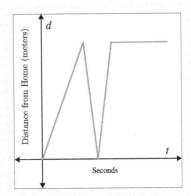

Figure 7. Distance from home on a walk to the bus stop.

Paula walks slowly at first, so the graph increases but not too steeply. When Paula runs home, her distance from home decreases back to zero, so there is a second horizontal intercept. The decreasing segment is steeper than the first segment because it takes less time to run back home from the bus stop than it took to walk there. The next segment of the graph is increasing again and is also relatively steep because Paula runs back to the bus stop. The last segment of the graph is horizontal because Paula's distance from home remains constant while she waits for the bus.

## Practice B — Answers

1. The independent variable is $m$, and the dependent variable is $T$. The temperature of the tea depends on the number of minutes since it was poured.

2. The independent variable is $s$, and the dependent variable is $g$. A student's exam grade depends on the amount of time the student spent studying.

3. The independent variable is $F$, and the dependent variable is $c$. The number of winter coats sold depends on the outside temperature.

4. The independent variable is $a$, and the dependent variable is $v$. The resale value of a used car depends on the age of the car.

# Exercises 1.1

The population of a small town on the Oregon coast is described during the years between 2000 and the present. Let $p$ be the population of the town at $t$ years since 2000. For the following problems, match each of the Figure 8 graphs to the scenario it describes.

1. The population increased steadily.
2. The population decreased steadily.
3. The population increased for 10 years then decreased.
4. The population remained constant.

Alana goes for a 5-kilometer run each morning. Let $d$ be the distance she has run $t$ minutes after she begins. For the following problems, match each of the graphs in Figure 9 to the scenario it describes.

5. She runs at a steady pace the whole time.
6. She increases her speed the whole time.
7. She runs at a steady pace, then stops to rest, then continues at a slower pace.
8. She increases her speed for the first half of the run then decreases her speed.

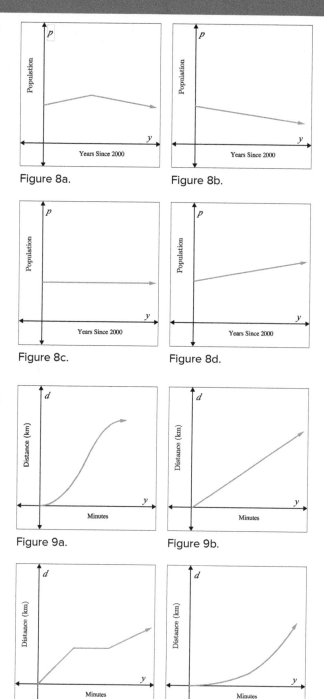

Figure 8a.

Figure 8b.

Figure 8c.

Figure 8d.

Figure 9a.

Figure 9b.

Figure 9c.

Figure 9d.

For the following exercises, identify the independent and dependent variables. Then sketch a qualitative graph that shows the relationship between the variables defined. Consider any vertical or horizontal intercepts. Correct graphs may vary slightly and still accurately represent the given relationship.

9. Let $t$ be the amount of time in minutes that it takes to read a novel with a total of $p$ pages.
10. Let $c$ be the total cost in dollars of $n$ lottery tickets.
11. Let $T$ be the temperature in degrees C of a bowl of hot soup, and let $h$ be the hours it is left uneaten on the dining room table.
12. Let $s$ be the speed (in mph) of a train, and let $t$ be the amount of time in hours that it takes to travel between two cities.
13. Let $h$ be the height in cm of a sunflower, and let $d$ be the days after it was planted as a seed.
14. Let $h$ be the height of baseball in feet and let $t$ be the time in seconds after a baseball bat hits it.

For the following exercises, sketch a qualitative graph that shows the relationship between the variables defined. Consider any vertical or horizontal intercepts. Correct graphs may vary slightly.

15. Rodrigo left home, drove to another city, got gas, and then continued driving to his cousin's house. Let $g$ be the amount of gas in gallons in the gas tank at $t$ minutes after he left home.
16. When Arianna was dieting, she lost weight quickly at first and then more slowly. She was then able to maintain a healthy weight. Let $W$ be her weight in pounds for $m$ months after she began the diet.
17. A plane flies from Portland to Los Angeles. Let $a$ be the altitude in feet at $t$ hours after takeoff.
18. Let $h$ be the height in feet of a rubber ball at $t$ seconds after a child bounces it on the floor. It bounces several times and then stops.
19. Pressure $p$ in pounds per square inch is applied to a volume of gas in a closed container. As the pressure increases, $v$, the volume of gas in cubic cm, decreases.
20. Let $h$ be the height in meters of a hot air balloon at $t$ minutes after it launches. The balloon rises steadily at first, then stays a relatively constant height, and then descends more slowly than it rose.

For the following exercises, let $A$ be the total amount of rain in inches that falls in $t$ hours one afternoon. Sketch a qualitative graph for each of the following scenarios. Correct graphs may vary slightly.

21. The rain fell gently and then stopped. After a while, it began to rain hard.
22. The rain fell harder and harder.
23. The rain fell more and more gently.
24. The rain fell steadily all morning but the sun came out in the afternoon.

It finally stopped raining, so Mario went out for a walk. For the following exercises, let $d$ be his distance from home in meters after leaving for $t$ minutes. Sketch a qualitative graph for each of the following scenarios. Correct graphs may vary slightly.

25. Mario walked quickly until he reached the park, and then he turned and walked slowly home.
26. Mario walked slowly to the park and then turned and ran home.

27. Mario walked slowly at first, realized he forgot his hat, and then ran home. When he set out again, he kept a brisk pace to the park and back.

28. Mario walked steadily to the park, met a friend, and stayed to talk for hours. His friend gave him a ride home in a car.

For the following exercises, write a scenario to match each of the following graphs. Make sure to define the variables $x$ and $y$ in your description.

29.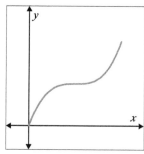

31.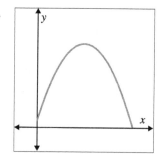

30.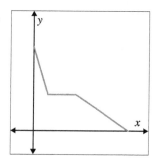

32.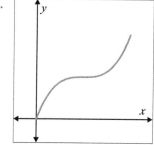

# 1.2 Functions

## Overview

A jetliner changes altitude as the distance increases between it and the starting point of its flight. The weight of a growing child increases with time. In each case, one quantity depends on another. There is a relationship between the two quantities that we can describe, analyze, and use to make predictions. In this section, we continue studying such relationships using quantitative tools. We also introduce the concept of a function.

As you study this section, you will learn to:

- Understand the meanings of relation, domain, range, and function
- Determine whether an equation or table describes a function
- Use the vertical line test to determine whether a graph represents a function
- Write domain and range as inequalities or in interval notation
- Determine a function's domain and range from its graph
- Use the Rule of Four to describe a function in multiple ways

## A. Relations and Functions

In mathematics, a relationship between two variables that change together is called a **relation**. In Section 1.1, we looked at several relationships between pairs of variables and described these relations with qualitative graphs.

A familiar example of a relation is the correspondence between time and height when you toss a ball up into the air. The ball goes up, stops, and falls back down to the ground. As time passes, the height of the ball changes, creating a relationship between the time the ball was in the air and its height. In this relation, time is the independent variable because the height of the ball depends on the amount of time since it was tossed.

There are many kinds of relations. Among the most important relations to mathematicians are functions. A **function** is a relation in which a value of the independent variable specifies a single value of the dependent variable.

For example, when you toss a beach ball into the air, the ball has one and only one corresponding height for each second that passes. We say the height of the ball is *a function of* the amount of time that has passed since it left your hand. Time only moves forward and does not repeat, so each moment of time is unique. However, notice that it's still possible for the beach ball to be at a particular height more than once as it goes up and then comes back down. Knowing the time will tell you the height, but knowing the height won't give you only one time.

When working with functions, we call a specific value of the independent variable an **input value**. An input is the independent, non-repeating quantity. In the case of tossing the beach ball into the air, the input values are measures of time. The value of the dependent variable is called an **output value**. The value of the output depends on the value of the input but may repeat. In the case of tossing a ball in the air, the output values are measures of height.

Although there are many useful relations studied in the field of mathematics, mathematicians distinguish between those relations whose inputs yield a unique output and those that do not. A **function** is a relation that assigns exactly one output value to each input value.

## Function

A **function** is a relation where each value of the input (independent) variable is paired with exactly one value of the output (dependent) variable.

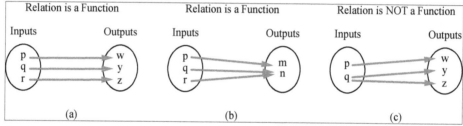

Figure 1. (a) This relationship is a function because each input is associated with a single output. (b) This relationship is also a function. Note that input $q$ and $r$ both give output $n$. (c) This relationship is not a function because input $q$ is associated with two different outputs.

Not all relations are functions. For example, suppose that a baker sells three sizes of chocolate chip cookies — 2-inch, 4-inch, and 6-inch cookies. There is a relationship between the size of the cookie and the number of chocolate chips in the cookie. If the size of the cookie is an input value, then the number of chocolate chips is the output value. However, this relation is not a function because even if the size of the cookie is the same, the number of chocolate chips per cookie will vary slightly. When the input value is the size of the cookie, we are not guaranteed to have one and only one output value. One 4-inch cookie may have 17 chocolate chips, while another 4-inch cookie may have 22 yummy morsels of chocolate.

A good question to ask when we want to determine whether a relation is a function is this: If we choose the same input more than once, are we guaranteed to always get the same output?

With the height of the beach ball over time, the answer is yes. For an input of time, there will be just one height the ball is at that time. This relation is a function. With the cookie sizes and chocolate chips, the input of a cookie size is not guaranteed to always produce the same output because the number of chocolate chips can vary. This relation is not a function.

## Example 1

Determine whether the following relations are functions.

1. To each student ID number, a relation assigns a student birth date.
2. To each birth date, a relation assigns the ID number of the student that has that birth date.
3. To each course name at a college, a relation assigns the number of students enrolled in that course.
4. To a given whole number, a relation assigns another number that is twice its value.

### Solutions

1. The input is a student ID number, and the output is the birth date of the student who has that ID number. This relation is a function because a student with a certain ID number has *only one* birth date.

2. The input is a birth date, and the output is the ID number of a student who has that birth date. Because there can be *more than one* student with the same birth date, this relation is not a function.

3. The input is the name of a college course, and the output is the number of students enrolled in that course. This relation is not a function because there can be *more than one* section of the course taught at the college, and each section may have a different number of students enrolled. We are not guaranteed to have the same output if we repeat the input.

4. The input is a whole number, and the output is a number with twice its value. This relation is a function because even if we repeat the input value, it will have *only one* corresponding output value — the number that is two times the input.

A relation between two variables can also be described by a set of **ordered pairs**. In an ordered pair, the first component represents an input value of the independent variable. The second component represents the corresponding output value of the dependent variable. If $x$ represents the independent variable, and if $y$ represents the dependent variable, then the ordered pair is notated $(x, y)$.

Consider the relation described in number 4 of the previous example. To a given whole number, this relation assigns an output value that is twice the input value. Below are a few randomly selected ordered pairs for this function.

$(1, 2), (2, 4), (3, 6), (14, 28),$ and $(50, 100)$

If we let the variable $x$ represent the input values and the variable $y$ represent the output values, then an equation that describes this relation is $y = 2x$. We determined in the last example that this relation is a function because the input value corresponds to one and only one output value.

We frequently use equations to describe relations and functions. An equation written in terms of the variables $x$ and $y$ represents a large — often infinite — set of ordered pairs concisely. By convention, $x$ represents the independent variable. We can determine whether a relation in the form of an equation is function with a little investigation. First we use the equation to generate a few ordered pairs that are characteristic of the function. Then we determine whether each input will yield one and only one output.

## $y$ Is a Function of $x$

Given a relation in $x$ and $y$, if to each possible value of $x$ there is assigned exactly one value of $y$, then $y$ is said to be a function of $x$.

### Example 2

Consider the relation that can be described by ordered pairs $(x, y)$ such that $y = 3x - 1$. Is this relation a function?

**Solution**

This relation multiplies the input $x$ by 3 and then subtracts 1 to obtain the output $y$. An input for this relation can be any real number because we can do these operations to any real number. Let's try it with a few specific input values: $x = -3$, $x = 0$, and $x = 4.5$.

$y = 3x - 1$
$y = 3(-3) - 1 = -10$   When $x = -3$, $y = -10$
$y = 3(0) - 1 = -1$   When $x = 0$, $y = -1$
$y = 3(4.5) - 1 = 12.5$   When $x = 4.5$, $y = 12.5$

The ordered pairs, $(-3, 10)$, $(0, -1)$, and $(4.5, 12.5)$, are part of the relation. Although we only used a few chosen $x$-values to explore this relation, it seems reasonable to assume that any input will lead to only one output. So the relation $y = 3x - 1$ is a function.

### Example 3

Consider the relation that can be described by ordered pairs $(x, y)$ such that $x = y^2$. In this relation, is $y$ a function of $x$?

**Solution**

This relation takes the input $x$ and matches it with an output $y$, a number that when squared will equal $x$. The inputs for this relation must be real numbers greater than or equal to 0 ($x \geq 0$) because any real number $y$, when squared, is either positive or equal to 0.

Let's try it with a few chosen input values:

When $x = 1$, then $y = 1$ or $y = -1$.
When $x = 4$, then $y = 2$ or $y = -2$.
When $x = 9$, then $y = 3$ or $y = -3$.

So the ordered pairs, (1, 1), (1, –1), (4, 2), (4, –2), (9, 3), (9,–3), and (9, –3) are part of the relation. We can see that these input values each lead to two different output values. This violates the definition of a function, so the relation $x = y^2$ does not describe $y$ as a function of $x$.

Some relations are described by a table of values that list input and output pairs. We can determine whether these relations are functions by checking to see if each input is paired with only one output.

## Example 4

Determine whether the relations described by the following tables are functions.

1.

| $x$ | $y$ |
|---|---|
| 0 | 5 |
| 2 | 15 |
| 4 | 5 |
| 6 | 15 |
| 8 | 5 |
| 10 | 15 |

2.

| $x$ | $y$ |
|---|---|
| –3 | –7 |
| –2 | –1 |
| –1 | 4 |
| 0 | 8 |
| 1 | 11 |
| 2 | 13 |

3.

| $x$ | $y$ |
|---|---|
| –2 | 12 |
| 0 | 9 |
| 2 | 5 |
| –1 | 8 |
| 2 | 7 |
| –4 | 15 |

### Solution

1. Each input value is paired with *only one* output value. Although the output values repeat in the table, none of the input values is paired with more than one output value. This table describes a function.

2. Each input value is paired with *only one* output value. This table describes a function.

3. The input value 2 is paired with two *different* output values, specifically 5 and 7. This table does not describe a function.

## Practice A

Determine if the relations are functions. Explain how you know. When you are done, turn the page to check your solutions.

1. Ordered pairs $(x, y)$ such that $y = \frac{x}{2}$
2. Ordered pairs $(x, y)$ such that $y = \pm x$
3. Ordered pairs $(t, p)$ such that $p = 15t$
4. 

| $x$ | –3 | –2 | –1 | 0 | 1 | 2 | 3 | 4 | 5 |
|---|---|---|---|---|---|---|---|---|---|
| $y$ | 21 | 20 | 14 | 9 | 3 | 8 | 15 | 20 | 26 |

5.

| $t$ | –3 | 0 | 1 | 3 | –1 | –3 |
|---|---|---|---|---|---|---|
| $p$ | 16 | 24 | 38 | 31 | 29 | 13 |

6.

| $x$ | 0 | 5 | 10 | 15 | 20 | 25 |
|---|---|---|---|---|---|---|
| $y$ | 12 | 12 | 12 | 12 | 12 | 12 |

## B. Vertical Line Test

We see in the examples above that we can represent a relation using a verbal description, an equation, or a table. We can also represent a relation using a graph. Graphs display many input-output pairs efficiently. The visual information they provide often makes relationships between variables easier to understand, too. Furthermore, we can use the graph of a relation to determine whether it is a function.

When the input and output values are real numbers, a relation can be represented by a quantitative graph. As with qualitative graphs, the independent variable is plotted on the horizontal axis, which is often the $x$-axis, and the dependent variable is plotted on the vertical axis, which is often the $y$-axis.

One way to determine if a graph represents a function is to imagine a vertical line sweeping across the graph. A vertical line that hits a graph in *exactly one* point pairs that one input value of $x$ with *exactly one* output value of $y$. In that case, the graph represents a function. When a vertical line hits a graph in *two or more* points, it pairs that input value of $x$ with *two or more* output values of $y$. That graph does not represent a function. See Figure 2.

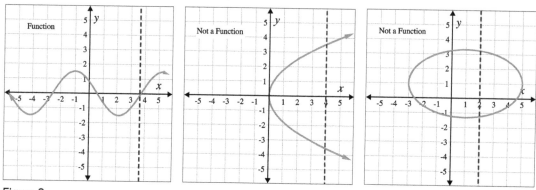

Figure 2.

### Vertical Line Test

A relation is a function if and only if each vertical line intersects the graph of the relation at no more than one point. We call this requirement the **vertical line test**.

### Practice A — Answers

1. Function: Each input is paired with *only one* output.

2. Not a function: Each input is paired with *two* outputs. For example, if $x = 2$, then $y = 2$ and $y = -2$.

3. Function: Each input is paired with *only one* output.

4. Function: Each input is paired with *only one* output.

5. Not a function: The input $t = -3$ is paired with *two* outputs, namely $p = 16$ and $p = 13$.

6. Function: Each input is paired with *only one* output. It's okay that the outputs are the same.

## Example 5

Which of the graphs in Figure 3 represent functions?

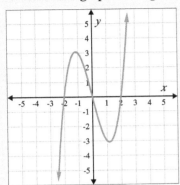 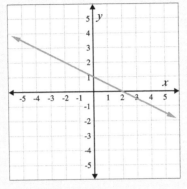 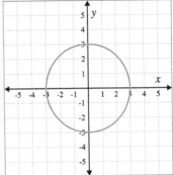

Figure 3.

### Solution

For the first two graphs in Figure 3, notice that any vertical line would pass through only one point. The first two graphs both pass the vertical line test and are therefore both functions. The third graph does not represent a function because at most of the $x$-values a vertical line would intersect the graph at more than one point, as shown in Figure 4.

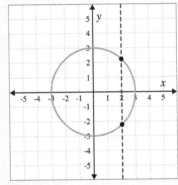

Figure 4.

## Practice B

Do the graphs below represent functions? When you are done, turn the page to check your solutions.

7.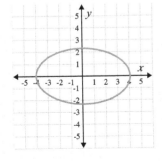

8.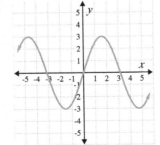

9.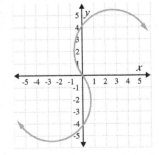

## C. Describing Intervals for Domain and Range

The set consisting of all the input values of a relation is called the **domain** of the relation. The set of input values is associated with the independent variable, which is often the variable $x$.

The set consisting of all the output values of a relation is called the **range** of the relation. The set of output values is associated with the dependent variable, which is often the variable $y$.

### Domain and Range

The **domain** of a relation is the set of all values of the independent variable (input values).
The **range** of a relation is the set of all values of the dependent variable (output values).

Before we learn to depict domain and range symbolically, let's explore the concepts a bit more informally. We'll use some of the functions we've already explored as examples.

### Example 6

Determine the domain and range of the following functions.
1. A ball is tossed in the air. It reaches a height of 20 feet and then lands on the ground after 2 seconds. The height of the ball is a function of the time since it was tossed.
2. To each 7-digit student ID number, a function assigns the student's birth date.
3. To each whole number, a function assigns another whole number that is twice its value.

### Solution

1. The input values for this function are the times that the ball is in the air, so the domain is all values between and including 0 to 2 seconds. The output values for this function are the heights of the ball, so the domain is all values between and including 0 to 20 feet.

2. The input values for this function are 7-digit numbers, so the domain is all 7-digit numbers. The output values for this function are birth dates, so the range is all the dates of the year.

3. Both the input and output values are whole numbers, but the output values will only consist of even whole numbers. Therefore, the domain is all whole numbers, and the range is all even whole numbers.

When the domain and range consist of real numbers in a specified interval, it's convenient to use inequality notation to define the domain and range. Inequalities use values, inequality symbols, and variables to describe a set of numbers.

### Practice B — Answers

7. No
8. Yes
9. No

Here we review the meaning of some inequalities used in mathematics:

$x < a$ means that $x$ is less than $a$.

$x \leq a$ means that $x$ is less than or equal to $a$.

$x > a$ means that $x$ is greater than $a$.

$x \geq a$ means that $x$ is greater than or equal to $a$.

$a < x < b$ means that $x$ is greater than $a$ but less than $b$.

For example, if the domain of a certain function is the set of real numbers less than or equal to 5, then we can represent the domain with the inequality $x \leq 5$. Notice that we use the variable $x$ when describing the domain.

If the range of a function is the set of real numbers greater than $-2$ but also less than 10, we could write the range with the inequality $-2 < y < 10$. Notice we use the variable $y$ when describing the range.

We can also describe domain and range in **interval notation**, which uses values within brackets to describe a set of numbers. In interval notation, we use square brackets [ ] when the set includes the endpoint and parentheses ( ) to indicate that the endpoint is not included. We also use parentheses if the interval is unbounded. For more details, see Figure 4.

If a person has $100 available to spend, for example, the possibilities of how much she or he actually does spend is the interval from 0 to 100, inclusive. In interval notation this is [0, 100]. If we know the person is going to spend at least *some* money, which means that 0 is not a possible value, then the interval is written (0, 100]. If there is somehow — magically — *no* limit at all to how much money this person can spend, including spending nothing, we can use the interval $[0, \infty)$.

If an interval of real numbers is not bounded by a smallest number, then the symbol $-\infty$ is placed on the left side of the interval. For example, to describe the set of numbers less than or equal to 10 using interval notation we write $(-\infty, 10]$.

If an interval of numbers is not bounded by a largest number, then the symbol $\infty$ is placed on the right side of the interval. For example, to describe the set of numbers greater than 25 using interval notation we write $(25, \infty)$.

Here are the conventions of interval notation:

- The smallest number in the interval is written on the left, followed by a comma.
- The largest number in the interval is written on the right.
- Parentheses ( ) signify that an endpoint value is not included or does not exist. On a graph, this is indicated by an open circle or an arrow.
- Brackets [ ] indicate that an endpoint value is included. On a graph, this is indicated by a closed circle (no arrow).

Figure 5 on the next page offers a summary of inequality notation, interval notation, and graphical representations:

# Chapter 1: Graphs and Linear Functions

| Inequality | Interval Notation | Graph on a Number Line | Verbal Description |
|---|---|---|---|
| $a \leq x \leq b$ | $[a, b]$ | | $x$ is between $a$ and $b$, including $a$ and $b$ |
| $a < x \leq b$ | $(a, b]$ | | $x$ is between $a$ and $b$, including $b$ but not $a$ |
| $a \leq x < b$ | $[a, b)$ | | $x$ is between $a$ and $b$, including $a$ but not $b$ |
| $a < x < b$ | $(a, b)$ | | $x$ is between $a$ and $b$, not including $a$ or $b$ |
| $x \geq a$ | $[a, \infty)$ | | $x$ is greater than or equal to $a$ |
| $x > a$ | $(a, \infty)$ | | $x$ is greater than $a$, not including $a$ |
| $x \leq b$ | $(-\infty, b]$ | | $x$ is less than or equal to $b$ |
| $x < b$ | $(-\infty, b)$ | | $x$ is less than $b$, not including $b$ |
| $x \neq a$ | $(-\infty, a) \cup (a, \infty)$ | | all values except for $a$, or $x$ cannot equal $a$ |
| $-\infty < x < \infty$ | $(-\infty, \infty)$ | | all real numbers |

Figure 5.

## Example 7

Use inequalities and interval notation to describe the intervals below.

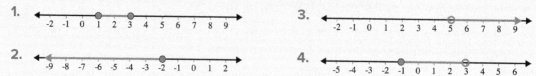

### Solutions

1. The inequality is $1 \leq x \leq 3$, and the interval notation is $[1, 3]$.
2. The inequality is $x > 5$, and the interval notation is $(5, \infty)$.

3. The inequality is $x \leq -2$, and the interval notation is $(-\infty, -2]$.
4. The inequality is $-1 \leq x < 3$, and the interval notation is $[-1, 3)$.

## Example 8

Use inequalities and interval notation to describe the domain and range of the functions. Assume the input variable is $x$ and the output variable is $y$.

1. A function where the inputs and outputs are all real numbers greater than but not equal to zero
2. A function where the inputs are all real numbers, and the outputs are all real numbers between and including $-5$ and $5$
3. A function where the inputs are all real numbers greater than or equal to 0 but less than 33, and the outputs are all real numbers less than or equal to 7.

### Solutions

1. Domain: $x > 0$ or $(0, \infty)$, Range: $y > 0$ or $(0, \infty)$
2. Domain: $-\infty < x < \infty$ or $(-\infty, \infty)$, Range: $-5 \leq y \leq 5$ or $[-5, 5]$
3. Domain: $0 \leq x < 33$ or $[0, 33)$, Range: $y \leq 7$ or $(-\infty, 7]$

## D. Using a Graph to Find the Domain and Range of a Function

Another way to identify the domain and range of functions is by using graphs. Because the domain is the set of input values, the domain of a graph is represented on the horizontal axis. The range is the set of output values, which are represented on the vertical axis. If the input-output pairs of the function continue beyond the portion of the graph we can see, the domain and range may be greater than the visible values.

As you can see in Figure 6, the graph extends horizontally from $-5$ to the right without bound, so the domain is $[-5, \infty)$. The vertical extent of the graph is all values 7 and below, so the range is $(-\infty, 7]$.

Remember when using interval notation, the smaller value is written on the left. For domains that will be the $x$-coordinate of the leftmost point on the graph. For ranges the smaller value will be the $y$-coordinate of the lowest point on the graph.

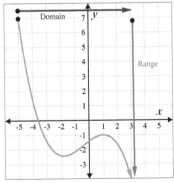

Figure 6.

## Example 9

Find the domain and range of the function whose graph is shown in Figure 6.

### Solution

We can observe that the horizontal extent of the graph is −3 to 1. The closed circles indicate that we should include −3 and 1. So the domain of $f$ is [−3, 1].

The vertical extent of the graph is from −2 to 2, so the range is [−2, 2]. See Figure 7.

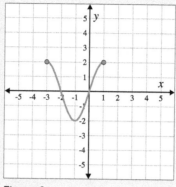

Figure 6.

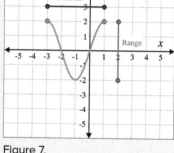

Figure 7.

## Example 10

Find the domain and range of the function whose graph is shown in Figure 8.

### Solution

The input quantity along the horizontal axis is "years," which we represent with the variable $t$ for time. The output quantity is "thousands of barrels of oil per day," which we represent with the variable $b$ for barrels. The graph may continue to the left and right beyond what is viewed, but based on the portion of the graph that is visible, we can determine the domain as $1973 \le t \le 2008$ and the range as approximately $180{,}000 \le b \le 2{,}010{,}000$.

In interval notation, the domain is [1973, 2008], and the range is about [180000, 2010000]. For the domain and the range, we approximate the smallest and largest values since they do not fall exactly on the grid lines.

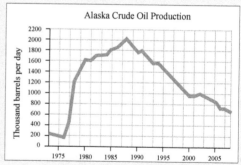

Figure 8. Alaskan crude oil production data from the U.S. Energy Information Administration.

## Practice D

Try the following. When you're done, turn the page to check your solutions.

1. Given Figure 9, identify the domain and range using interval notation.
2. Identity the domain and range of the relations graphed in Figure 10.
3. Which of the graphs in Figure 10 are functions?

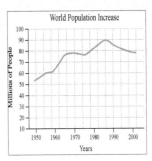

Figure 9.

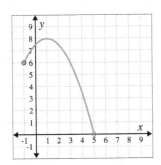

Figure 10a.

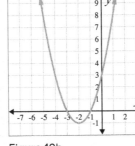

Figure 10b.

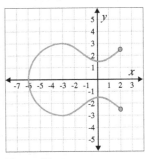

Figure 10c.

## E. Rule of Four for Functions

All of the functions that we will study in this course can be described in four ways: symbolically, graphically, numerically, and verbally. This is known as the **Rule of Four**.

Sometimes one of the four ways to describe a function may be more insightful or useful in a situation than another way. There may also be times when representing a function in multiple ways will be useful. You will benefit by learning to move easily between one way of describing a function and another.

### Rule of Four for Functions

The description of the input-output pairs of a function can be

1. symbolic or algebraic (an equation)
2. verbal (words)
3. graphical (a graph)
4. numeric (a table).

On the next page, you'll find examples of how a simple function can be described by each of the four ways. The table only shows some selected values for the function.

## Rule of Four for Functions — Examples

1. $y = 2x + 1$

2. Multiply the input by 2, then add one to obtain the output.

3.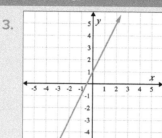

4. 

| $x$ | $y$ |
|---|---|
| −3 | −5 |
| −2 | −3 |
| −1 | −1 |
| 0 | 1 |
| 1 | 3 |
| 2 | 5 |

Example 11 presents a simple function and how it is described by each of the four ways. Notice that the table only shows some selected values for the function.

### Example 11

A function squares the input and then subtracts 3 to obtain the output.

a. Write an equation that matches the function description. Use $x$ for the input variable and $y$ for the output variable.

b. Create a table of values to describe the function. Use $x = -3, -2, -1, 0, 1, 2, 3$ for the input values.

c. Use the table to create a graph of the function.

d. Use the graph to help you determine the domain and range of the function.

### Solution

a. $y = x^2 - 3$

b. The table below shows some possible ordered pairs for the function.

| $x$ | -3 | -2 | -1 | 0 | 1 | 2 | 3 |
|---|---|---|---|---|---|---|---|
| $y$ | 6 | 1 | -2 | -3 | -2 | 1 | 6 |

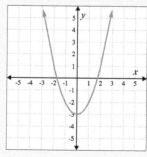

c. In the Figure 11 graph, we connect the points on our graph with a smooth curve since $x$ can be any real number. We put arrows at each end of the curve for the same reason.

Figure 11.

d. Domain: $(-\infty, \infty)$, Range: $[-3, \infty)$

# Example 12

For the function $y = -\frac{1}{2}x + 3$:

a. Use your graphing calculator to make a table of values.

b. Use your graphing calculator to make a graph.

## Solution

a. First, press [Y=] and type the equation in the calculator.

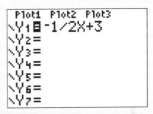

Next, go to the table set-up menu by selecting [2nd] and [WINDOW].

Choose $x = -3$ for a starting value and allow the input variable $x$ to increase in increments of 1.

Notice that the symbol $\Delta$Tbl on the screen, which is read as "delta table," stands for the change in input values. Mathematicians use the Greek letter capital delta ($\Delta$) to stand for the incremental change in a variable.

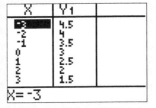

Select Auto for both the independent and dependent variables so that these values will automatically show up in the table. Now look at the table of values by selecting [2nd] and [GRAPH].

We can scroll both up and down through this table of values using the arrow buttons on the calculator.

## Practice D — Answers

10. The domain is [1950, 2002], and the range is approximately [54,000,000; 90,000,000].

11. a. domain: [−1, 5), range: (0, 8]
    b. domain: (−∞, ∞), range: [−1, ∞)
    c. domain: [−6, 2], range: [−3, 3]

12. Figures 10a and 10b represent functions. Figure 10c doesn't pass the vertical line test, so it's not a function.

**b.** We have already entered the equation in our calculator. Now we choose a good viewing window. For this equation we choose the standard viewing window which uses values from −10 to 10 for both the *x* and *y* axes. Select ZOOM and 6 (ZStandard) for a quick way to view the graph.

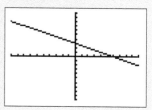

## Exercises 1.2

1. What is the difference between a relation and a function?
2. What is the difference between the input and the output of a function?
3. Why does the vertical line test tell us whether the graph of a relation represents a function?
4. What is the difference between the domain and the range of a function?

For the following exercises, determine whether the relation represents *y* as a function of *x*.

5. $5x + 2y = 10$
6. $3y = x + 6$
7. $x = y^2 + 3$
8. $y = x^2$
9. $y = -2x^2 + 30x$
10. $2x + y^2 = 8$
11. $y = \frac{1}{x}$
12. $x = y^3$

For the following exercises, use the vertical line test to determine which graphs show relations that are functions.

13.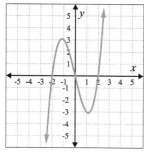

14.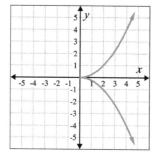

15.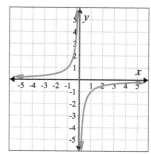

Section 1.2: Functions  27

16.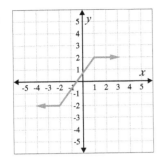

19.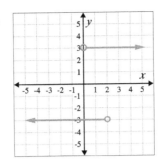

22.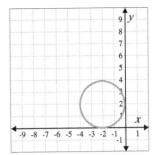

17.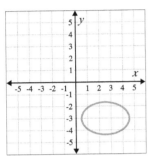

20.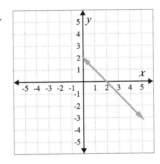

23.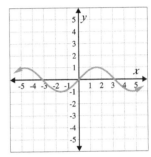

18.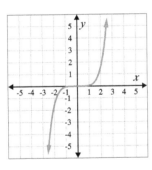

21.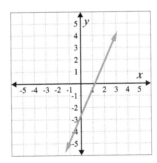

For the following exercises, determine whether the relation represents a function.

24. This relation begins with a person's social security number and pairs it with their date of birth.
25. This relation begins with a date of birth and pairs it with a social security number.
26. This relation assigns a height $h$ to a rocket $t$ seconds after the rocket is launched.

For the following exercises, determine if the relation represented in table form represents $y$ as a function of $x$.

27.
| $x$ | 5 | 10 | 15 | 20 | 25 | 30 |
|---|---|---|---|---|---|---|
| $y$ | 3 | 8 | 14 | 21 | 29 | 38 |

28.
| $x$ | 0 | 3 | 6 | 9 | 12 | 15 |
|---|---|---|---|---|---|---|
| $y$ | 3 | 8 | 14 | 15 | 14 | 8 |

29.
| $x$ | −4 | −2 | −4 | 0 | 2 | 4 |
|---|---|---|---|---|---|---|
| $y$ | −15 | −20 | −10 | −5 | 0 | 5 |

30.
| $x$ | −3 | −2 | −1 | 0 | 1 | 0 |
|---|---|---|---|---|---|---|
| $y$ | 7 | 4 | 1 | −2 | −5 | −8 |

## 28  Chapter 1: Graphs and Linear Functions

For the following exercises, use the graph of the function to determine the function's domain and range. Write both as inequalities and in interval notation.

31.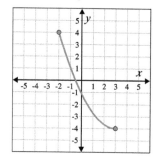

34.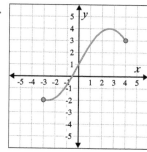

37.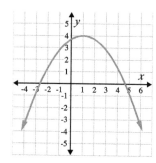

32.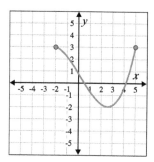

35.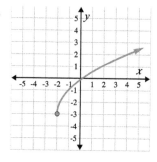

38.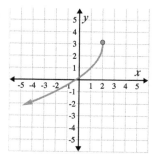

33.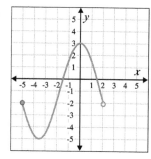

36.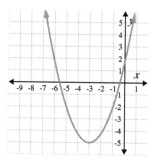

39.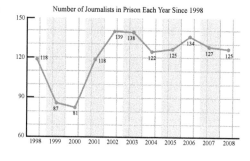

40.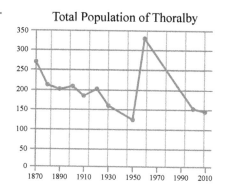

For the following verbally described functions, write the domain and range using interval notation.

41. A function where the inputs are all real numbers between and including 0 and 9.8, and the outputs are all real numbers between and including 0 and 225.

42. A function where the inputs are all real numbers greater than 0 and at most 30, and the outputs are all real numbers greater than 0 but less than 17.6.

43. A linear function where the inputs are all real numbers greater than −12, and the outputs are all real numbers less than −7.

44. A function where the inputs are all real numbers, and the outputs are all real numbers between and including −1 and 1.

For exercises 45 and 46, use your graphing calculator to help you do the following:

a. Write an equation that matches the function description. Use $x$ for the input variable and $y$ for the output variable.

b. Create a table of values to describe this function. Use $x = -3, -2, -1, 0, 1, 2, 3$ for the input values.

c. Sketch a graph of the function.

d. Use the graph to help you determine the domain and range of the function. Remember, the graph on your calculator screen will not show arrows.

45. A function squares the input and then adds 2 to obtain the output.

46. A function multiplies the input by 3 and then subtracts 4 to obtain the output.

For exercises 47 and 48, use your graphing calculator to help you do the following:

a. Write a verbal description that matches the equation.

b. Create a table of values to describe this function. Use $x = -3, -2, -1, 0, 1, 2, 3$ for the input values.

c. Sketch a graph of the function.

d. Use the graph to help you determine the domain and range of the function. Remember, the graph on your calculator screen will not show arrows.

47. $y = -2x + 5$

48. $y = x^3 - 1$

# 1.3 Finding Equations of Linear Functions

## Overview

Imagine placing a plant in the ground one day and finding that it has doubled its height just a few days later. Although it may seem incredible, this can happen with some species of bamboo. These members of the grass family are the fastest-growing plants in the world. One species can grow nearly 1.5 inches every hour. In a twenty-four hour period, this bamboo plant grows about *36 inches*. A constant rate of change, such as the change in height over time of this bamboo plant, is indicative of a linear function.

In Section 1.2, you learned that a function is a relation that assigns to every element in the domain exactly one element in the range. Linear functions are a specific type of function that can be used to model many real-world applications, such as plant growth over time. In this chapter, we will explore linear functions, their graphs, and how to relate them to data.

As you study linear functions, you will learn how to:

- Represent a linear function verbally, algebraically, graphically, and numerically
- Determine whether a linear function is increasing, decreasing, or constant
- Interpret slope as a rate of change
- Build linear models from verbal descriptions
- Build linear models from data in a table
- Interpret the intercepts of a linear model

## A. Representing Linear Functions

Many real world situations exhibit constant change over time. These situations can be represented with a **linear function**, which is a function with a constant rate of change. Consider, for example, the first commercial magnetic levitation (maglev) train in the world, the Shanghai MagLev Train. It carries passengers comfortably for a 40-kilometer trip from the airport to the subway station in only eight minutes.

Suppose a maglev train travels a long distance, maintaining a constant speed of 83 meters per second once it is 250 meters from the station. The function describing the train's distance from the station at a given point in time is a linear function because the speed of the train is a constant rate of change.

The Rule of Four states that we can describe a function in four ways. We will now learn how to do this for linear functions, representing them in verbal form, slope-intercept equation form, tabular form, and graphical form. In the examples to come, we describe the maglev train's motion as a linear function using each of these forms.

## Example 1

For the train problem we considered above, describe the function relationship in *verbal form*.

### Solution
The train's distance from the station is a function of the time in seconds since it was 250 meters from the station. Once the train is 250 meters from the station, it travels at a constant speed of 83 meters per second.

We can also represent a linear function with an algebraic equation. One way to do this is to write an equation in **slope-intercept form**, $y = mx + b$. In the slope intercept form of a line, $x$ is the independent variable and $m$ is the constant rate of change. The graph of a linear function is a straight line whose slope is $m$. The constant $b$ is the initial value of the dependent variable, the value of $y$ when $x = 0$. When we graph a linear function, the $y$-intercept is the point $(0, b)$.

### Slope-Intercept Form of a Line

Linear functions can be written in the **slope-intercept form** of a line

$$y = mx + b$$

where $m$ is the constant rate of change and $b$ is the initial output value. The graph of a linear function is a straight line with slope $m$ and $y$-intercept $(0, b)$.

## Example 2

Using the verbal description from Example 1, write an equation in *slope-intercept form* that represents the motion of the maglev train.

### Solution
Let $x$ represent the time in seconds that the train has been traveling at a constant speed. Let $y$ represent the train's distance from the station in meters.

The constant rate of change $m$ is the speed of the train, 83 meters per second. The initial output value $b$ is the distance from the station, 250 meters. This is the distance the train is from the station when it begins to move at a constant speed.

Because $m = 83$ and $b = 250$, we can write this equation to represent the motion of the train:

$$y = 83x + 250$$

## Example 3

The equation $y = 83x + 250$ represents the motion of the train described in Examples 1 and 2. Represent this linear function in *tabular form* using a table of input and output values. Use the table to verify that the slope, $m = 83$, represents a constant rate of change.

### Solution

First we choose a few input values and calculate the corresponding output values using our equation. We begin with $x = 0, 1, 2,$ and $3$ seconds.

$$y = 83x + 250$$
$$y = 83(0) + 250 = 250 \quad \text{When } x = 0, y = 250$$
$$y = 83(1) + 250 = 333 \quad \text{When } x = 1, y = 333$$
$$y = 83(2) + 250 = 416 \quad \text{When } x = 2, y = 416$$
$$y = 83(3) + 250 = 499 \quad \text{When } x = 3, y = 499$$

Next we create a table that displas these values. See Figure 1.

From the table in Figure 1, we can see that the distance changes by 83 meters for every 1 second increase in time. This verifies that the slope, $m = 83$, represents a constant rate of change, specifically 83 meters per second.

| | 1 second | 1 second | 1 second | |
|---|---|---|---|---|
| $x$ | 0 | 1 | 2 | 3 |
| $y$ | 250 | 333 | 416 | 499 |
| | 83 meters | 83 meters | 83 meters | |

Figure 1.

## Example 4

The equation $y = 83x + 250$ represents the motion of a maglev train as described in the previous examples. Use *graphical form* to show the portion of the function that corresponds to the first 12 seconds that the train is moving at a constant speed.

### Solution

Label the axes with the independent and dependent variables. The $y$-intercept is the point $(0, b)$, so in this case $(0, 250)$. Plot the $y$-intercept and the other ordered pairs from the table in Figure 1.

We can use the equation to find the distance traveled at $x = 12$ seconds.

$$y = 83(12) + 250 = 1246$$

After 12 seconds, the train has traveled 1246 meters, so we plot the point $(12, 1246)$ as well. Now we can connect these points with a straight line and put an arrow on the right end to indicate that the graph continues as the train keeps moving.

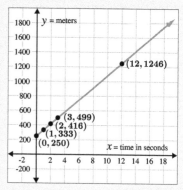

Figure 2

Let's explore the linear function that represents the motion of the maglev train a bit further. The description of the Shanghai maglev train at the beginning of this section states that it can complete a 40-kilometer trip in only 8 minutes. Let's see if our equation, $y = 83x + 250$, supports that statement.

We've been using seconds for our input measurement, so let's convert 8 minutes to seconds by multiplying by 60. 8 minutes = 480 seconds. We now use $x = 480$ as the input value in our equation:

$$y = 83(480) + 250 = 40{,}090$$

In 8 minutes, the train travels 40,090 meters. This is just a bit longer than 40 kilometers because 1 kilometer = 1000 meters so 40 kilometers = 40,000 meters.

In the previous examples, the input cannot be *any* real number. The input represents time *after* a certain moment, so while positive real numbers are possible, negative real numbers are not possible in this situation. Because the trip from the airport to the subway station lasts 480 seconds, it doesn't make sense to use input values larger than 480.

In general, the domain of a linear function is all real numbers, $(-\infty, \infty)$. However, when working with linear functions within the context of a real-world situation, we have to stop and consider any practical limits on the input variable. Any limits would result in a restricted domain, usually called the reasonable domain or the **practical domain**.

The practical domain for the function representing the maglev train consists of non-negative real numbers up to and including 480. In interval notation, this is $[0, 480]$. The corresponding range is the set of output values, distance in meters, covered during this time interval. When the train begins its high-speed journey, $x = 0$ and $y = 250$. When the train ends its journey, $x = 480$ and $y = 40{,}090$. So the corresponding range is $[250, 40090]$.

In Example 4, we gave instructions for graphing the function that represents the train's motion by hand. We can also use our graphing calculator to graph this function. The key to using the calculator is to choose a good viewing window. In Example 5, we use the intervals for the domain and range of the train function to help us select the graphing window.

## Example 5

Use a graphing calculator to graph the function $y = 83x + 250$, representing the journey of a maglev train on an 8-minute (480 seconds) trip.

### Solution

*Step 1:* Type the equation in the [Y=] screen of the calculator. Then press [WINDOW]. If your last viewing window was the standard window (ZoomStandard), the screen should look like this 3. We now need to adjust this window.

```
WINDOW
 Xmin=-10
 Xmax=10
 Xscl=1
 Ymin=-10
 Ymax=10
 Yscl=1
↓Xres=1
```

*Step 2:* As determined above, the domain of this function representing time in seconds is [0, 480], so we let our minimum $x$-value be 0 and our maximum $x$-value be 500. We can also change the $x$-scale to 60. This step isn't crucial, but it makes the tick marks on the $x$-axis occur every 60 seconds rather than every second.

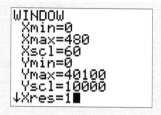

The range of this function representing distance traveled in meters is [250, 40090], so we let our minimum $y$-value be 0 and our maximum $y$-value be 41,000. Notice that we have chosen to go a little above and below the range values. We change the $y$-scale to 10,000.

*Step 3:* Now that we have chosen a good viewing window, we press GRAPH.

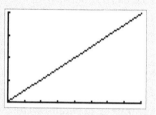

Looking at Example 5, we can write our viewing window concisely this way: $x$: [0, 480, 60] and $y$: [0, 40100, 10000]. We will use this notation for recommended viewing windows throughout the text. For future graphing, we can return to a standard window quickly by pressing ZOOM and choosing 6.

## Practice A

Use the following description of employment at a trucking company to complete the exercises below. Turn the page to check your solutions.

Each week, the DiCicco Brothers Trucking Company pays its drivers $200 plus $18 per hour for every hour on the road. The Department of Transportation (DOT) limits the road time of employed drivers to 60 hours per week. Let $x$ represent the number of hours the employee spends driving in any week. Let $y$ represent the employee's weekly pay in dollars.

1. Write a linear function in slope-intercept form to represent this situation.
2. Make a table of values for inputs $x$ = 0, 10, 20, 30, 40, 50 hours.
3. Use a graphing calculator to graph the function. Make sure to use the values in your table to help you choose a good viewing window.
4. What is the practical domain of this function? What is the corresponding range?

## B. Slope is a Rate of Change

The linear function we used in the previous example increased over time, but not every linear function does. A linear function may be increasing, decreasing, or constant. For an increasing function, as with the train example, the output values increase as the input values increase. The graph of an increasing function has a positive slope. A line with a positive slope slants upward from left to right, as in Figure 6a. For a decreasing function, the slope is negative. The output values decrease as the input values increase. A line with a negative slope slants downward from left to right, as in Figure 3b. If the function is constant, the output values are the same for all input values so the slope is zero. A line with a slope of zero is horizontal, as in Figure 3c.

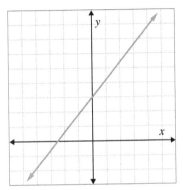

Figure 3a. Increasing linear function.

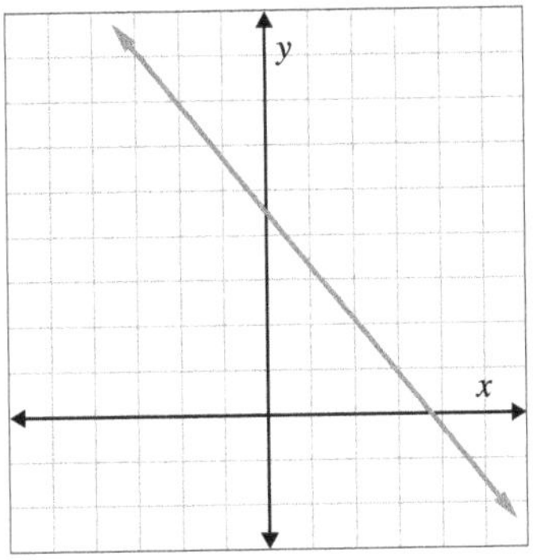

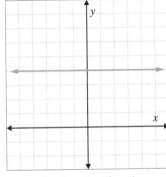

Figure 3b. Decreasing linear function.

Figure 3c. Constant function.

### Using Slope to Determine Increasing and Decreasing Functions

The slope of a linear function determines if it is an increasing linear function, a decreasing linear function, or a constant function.

$y = mx + b$ is an increasing function if $m > 0$.
$y = mx + b$ is a decreasing function if $m < 0$.
$y = mx + b$ is a constant function if $m = 0$.

### Example 6

Suppose that a teenager sends an average of 60 texts per day. For each of the following scenarios, find the linear function that describes the relationship between the input value and the output value. Then determine whether the graph of the function is increasing, decreasing, or constant.

a. The total number of texts a teen sends in a month is considered a function of time in days. The input is the number of days, and output is the total number of texts sent that month.

b. A teen has a limit of 500 texts per month in his or her data plan. The input is the number of days, and output is the total number of texts *remaining* for the month.

c. A teen has an unlimited number of texts in his or her data plan for a cost of $50 per month. The input is the number of days, and output is the total cost of texting each month.

### Solution

a. The function can be represented as $y = 60x$ where $x$ is the number of days. The slope, $m = 60$, which represents the rate of 60 texts per day, is positive, so the function is increasing. This makes sense because the total number of texts for that month increases as the days go by.

b. The function can be represented as $y = -60x + 500$ where $x$ is the number of days. In this case, the slope is negative, so the function is decreasing. This makes sense because the number of texts *remaining* decreases each day, and this function represents the number of texts remaining in the data plan after $x$ days.

c. The cost function can be represented as $y = 50$ because the number of days does not affect the total cost. The slope is 0, so the function is constant.

In some problems involving linear functions, the slope of the line or the constant rate of change is stated in the problem. However, sometimes we need to calculate the slope from ordered pairs of input and output values. If we know just two points on a line, we can use the slope formula to calculate the slope of the line.

### Practice A — Answers

1. $y = 18x + 200$

2. 
| $x$ (hours) | $y$ (dollars) |
|---|---|
| 0 | 200 |
| 10 | 380 |
| 20 | 560 |
| 30 | 740 |
| 40 | 920 |
| 50 | 1100 |

3.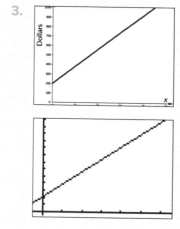

4. The practical domain represents hours on the road, so only zero and positive values make sense. The DOT limits the number of road hours per week to 60, so the domain is [0, 60]. The corresponding range is dollars earned from a minimum of $200 (when $x = 0$) to a maximum of $1280 (when $x = 60$). So the range is [200, 1280].

## Slope of a Line

Given two points $(x_1, y_1)$ and $(x_2, y_2)$, the slope $m$ of a linear function can be calculated as follows:

$$m = \frac{\text{change in output (rise)}}{\text{change in input (run)}} = \frac{\Delta y}{\Delta x} = \frac{y_2 - y_1}{x_2 - x_1}$$

With a real-world linear function, the units for slope are described by $\frac{\text{units for the output}}{\text{units for the input}}$. Think of the units of a slope as "output units per input units."

Examples of units for slope are rates of change such as miles per hour, dollars per day, or pounds per cubic yard. These rates of change may be positive, negative, or zero depending on whether the linear function is increasing, decreasing, or constant.

Given two input-output pairs for a linear function, we can calculate $m$, the slope or rate of change, and interpret it with the following steps:

1. Determine the units for input and output values and write the data given as two points.
2. Calculate the slope according to the formula $m = \frac{y_2 - y_1}{x_2 - x_1}$.
3. Interpret the slope as the change in output values per unit of the input value.

### Example 7

The population of McMinnville, Oregon, increased from 26,695 to 33,131 between 2000 and 2013. Find the slope of the linear function that would represent the population growth if we assume the rate of change was constant from 2000 to 2013.

**Solution**

The input values would be the given years because time is the independent variable. The output values would be the number of people because the population depends on the year. The data can be represented as ordered pairs: (2000, 26695) and (2013, 33131).

We calculate the slope in this way:

$$m = \frac{y_2 - y_1}{x_2 - x_1}$$

$$= \frac{33131 - 26695}{2013 - 2000}$$

$$= \frac{6436}{13}$$

The numerator represents an increase of 6,436 people. The denominator represents a span of 13 years.

$$\approx 495 \frac{\text{people}}{\text{year}}$$

In this situation, it makes sense to round to the nearest whole person (per year).

We interpret the slope to mean that between 2000 and 2013, the population of McMinnville increased by approximately 495 people per year.

## Practice B

Each situation below can be described by a linear function. Find and interpret the slope of the function as a rate of change. Turn the page to check your solutions.

5. The population of Gervais, Oregon increased from 977 people in 1990 to 2,477 people in 2010. Assume the rate of change was constant during those years.

6. A backyard swimming pool is being drained at a constant rate. After 3 hours there are 8450 gallons of water remaining in the pool and after 10 hours there are 5650 gallons.

## C. Finding an Equation of a Linear Model Using Data from Words

We can use linear functions to model real-world situations involving quantities with a constant rate of change — whether or not that rate of change is always perfectly constant. Functions that describe such situations are called **linear models**. As long as we know or can figure out the initial value and the rate of change of a linear model, we can write an equation to represent the function. Then we can use the equation to solve a real-world problem.

When building linear models to solve problems, we use the slope-intercept form: $y = mx + b$. The slope $m$ represents the rate of change in the problem, while the $y$-intercept $b$ represents a starting value.

### Example 8

Marcus currently has 200 songs in his music collection. Every month, he adds 15 new songs. Write a linear model for the number of songs, $y$, in his collection as a function of time, $x$, the number of months. How many songs will Marcus own at the end of one year?

#### Solution

The initial value for this function is 200 because Marcus currently owns 200 songs, so when $x = 0$, $y = 200$, which means that the $y$-intercept is (0, 200) and $b = 200$.

Since Marcus adds 15 new songs to his collection each month, the slope of the line is $m = 15$, representing an increase of 15 songs per month. We can substitute the initial value and the rate of change into the slope-intercept form of a line.

$$y = mx + b$$
$$y = 15x + 200$$

We write the formula $y = 15x + 200$ to model the number of songs in Marcus's music collection after $x$ months. With this formula, we can now predict how many songs Marcus will have at the end of one year (12 months). In other words, we can evaluate the function at $x = 12$.

$$y = 15(12) + 200$$
$$= 180 + 200$$
$$= 380$$

Marcus will have 380 songs in 12 months.

The previous example is simplified by the fact that the data given can directly be interpreted as the slope and the *y*-intercept of the linear model. However, sometimes the data given is in the form of two input-output pairs. In that case, we need to use the steps outlined below to find a linear model.

Given two input-output pairs for a linear model, we can find the slope-intercept form of the line with the following process:

1. Determine the units for input and output values and write the data given as two points.
2. Calculate the slope according to the formula $m = \frac{y_2 - y_1}{x_2 - x_1}$.
3. Using the slope-intercept form of a line $y = mx + b$, substitute your answer from Step 2 in for *m* and substitute the coordinates of one of your points from Step 1 in for values of *x* and *y*.
4. Solve the equation for *b*.
5. Beginning again with slope-intercept form, substitute the values for *m* and *b* into the formula, leaving *x* and *y* as variables.

## Example 9

As an insurance salesperson, Rosa earns a base salary plus commissions on new policies. Therefore, Rosa's weekly income, *y*, depends on the number of new policies, *x*, she sells during the week. Last week she sold 3 new policies and earned $760 for the week. The week before, she sold 5 new policies and earned $920. Find a linear function to model this data. Then interpret the meaning of the slope and *y*-intercept of the equation. Finally, determine Rosa's weekly income if she sells 10 new policies in a week.

### Solution

*Step 1:* The given information gives us two input-output pairs, (3, 760) and (5, 920).

*Step 2:* The real works starts by finding the slope of the linear model.

$$m = \frac{920 - 760}{5 - 3}$$
$$= \frac{160}{2}$$
$$= 80$$

Keeping track of units can help us interpret this quantity. Income increased by $160 when the number of policies increased by 2, so the rate of change is $80 per policy. Therefore, the slope tells us that Rosa earns a commission of $80 for each policy sold each week.

*Step 3:* Substitute one set of coordinates from Step 1 and the value of slope from Step 2.

$y = 80x + b$  Substitute the value of the slope $m = 80$.

$760 = 80(3) + b$  Using one ordered pair $(3, 760)$, we substitute $x = 3$, and $y = 760$.

$760 = 240 + b$  Simplify.

*Step 4:* Now solve for $b$.

$760 = 240 + b$

$\underline{-240 = -240}$  Subtract 240 from both sides of the equation to solve for $b$.

$520 = b$

The value of $b$ is the starting value for the function and represents Rosa's income when $x = 0$, or when no new policies are sold. We can interpret this as Rosa's base salary for the week.

*Step 5:* We can now write the final equation:

$y = 80x + 520$

Our final interpretation is that Rosa's base salary is $520 per week and she earns an additional $80 commission for each policy sold. If Rosa sells 10 new policies in a week, the input variable $x = 10$. Substituting this value in our formula we get:

$y = 80(10) + 520$
$= 800 + 520$
$= 1320$

So if Rosa sells 10 new policies in a week, her salary will be $1320.

## Practice B — Answers

5. The data can be written as ordered pairs: $(1990, 977)$ and $(2010, 2477)$. The slope is

$$m = \frac{2477 - 977}{2010 - 1990}$$
$$= \frac{1500}{20}$$
$$= 75 \text{ people per year}$$

The population of Gervais increased by 75 people per year between 1990 and 2010.

6. The data can be written as ordered pairs: $(3, 8450)$ and $(10, 5650)$. The slope is

$$m = \frac{5650 - 8450}{10 - 3}$$
$$= \frac{-2800}{7}$$
$$= -400 \text{ gallons per hour}$$

The slope is negative and indicates that the water is *draining* out of the pool at a rate of 400 gallons per hour.

## Practice C

Solve the following problems. When you're done, turn the page to check your solutions.

7. Bernardo starts a company in which he incurs a fixed cost of $1,250 per month for the overhead, which includes expenses like his office rent. His production costs are $37.50 per item. Write a linear function in slope-intercept form to model this situation where $y$ is the total cost for $x$ items produced in a given month. What is the company's cost if Bernardo produces 100 items in a month?

8. Walter's Well and Water Company is drilling a well for a customer. On Tuesday, after 5 hours, the drilling equipment has reached a depth of 50 feet, after 8 hours, they have reached a depth of 72.8 feet. Write a linear function in slope-intercept form that models this situation where $y$ is the depth of the well after $x$ hours. Interpret the meaning of the slope and $y$-intercept.

## D. Finding an Equation of a Linear Model Using Data from a Table

At the beginning of this section we saw that a linear function can also be represented by a table of values. If we begin with a table of values representing a constant rate of change, we can then use that table to find the slope and $y$-intercept of a linear model.

### Example 10

The table below relates $y$, the number of dollars in a savings account, to $x$, time in weeks. Use the table to write a linear equation.

| x (weeks) | 0 | 2 | 4 | 6 |
|---|---|---|---|---|
| y (dollars) | 1000 | 1080 | 1160 | 1240 |

### Solution

We see from the table that the initial value for the savings account is 1000, so $b = 1000$. We can tell from looking at the table that the output value increases by 80 each time the input value increases by 2. This means that the rate of change is 80 dollars per 2 weeks, which can be simplified to 40 dollars per week. Because $m = 40$ and $b = 1000$, our equation is

$$y = 40x + 1000$$

If we did not notice the rate of change from the table above, we could still calculate the slope using any two points from the table. For example, using (2, 1080) and (6, 1240), we have:

$$m = \frac{1240 - 1080}{6 - 2}$$
$$= \frac{160}{4}$$
$$= 40$$

The initial value was provided in the table in the previous example, but sometimes it isn't provided. If you see an input of 0 in a table, the initial value $b$ is the corresponding output. If the initial value is not provided, you can find the slope, substitute one ordered pair and the slope into $y = mx + b$, and solve for $b$.

### Example 11

Use the table of values below to write a linear equation.

| x | 2  | 7  | 12 | 17 | 22 |
|---|----|----|----|----|----|
| y | 25 | 17 | 9  | 1  | -7 |

**Solution**

We can tell by looking at the table that the output value decreases by 8 each time the input value increases by 5. We calculate the slope:

$$m = \frac{-8}{5} = -1.6$$

Since we don't know the value of $b$, the output when $x = 0$, we need to find it. Using the slope-intercept form of a line, we substitute $m = -1.6$ and one of the ordered pairs from the table and solve for $b$. We can use any ordered pair from the table. We will use $(2, 25)$.

$$y = mx + b$$
$$y = -1.6x + b \quad \text{Substitute } m = -1.6$$
$$25 = -1.6(2) + b \quad \text{Substitute } x = 2 \text{ and } y = 25$$
$$25 = -3.2 + b \quad \text{Simplify.}$$
$$\underline{+3.2 \quad +3.2} \quad \text{Add 3.2 to both sides of the equation to solve for } b.$$
$$28.2 = b$$

Now we know that the slope $m = -1.6$ and $b = 28.2$, and we can write the equation:

$$y = -1.6x + 28.2$$

### Practice C — Answers

7. The total cost for $x$ items produced in a month is given by $y = 37.50x + 1250$. If Bernardo produces 100 items in a month, his monthly cost is found by substituting 100 for $x$.

$$y = 37.5(100) + 1250$$
$$= 5000$$

Bernardo's monthly cost will be $5,000.

8. The data given can be written as input-output pairs $(5, 50)$ and $(8, 72.8)$. The slope between these two points $m = 7.6$ represents the rate of drilling, specifically 7.6 feet per hour. Solving for the $y$-intercept gives $b = 12$, which represents the depth of the well when the company began drilling on Tuesday. The linear function $y = 7.6x + 12$ models this situation.

## Practice D

Solve the following problems. Then turn the page to check your solutions.

9. A new plant food was introduced to a young tree to test its effect on the height of the tree. Figure 5 shows the height of the tree, in centimeters, for some specific number of months since the measurements began. Write a linear function to model this situation. Let $y$ equal the height of the tree $x$ number of months since the start of the experiment.

| $x$ (months) | 0 | 1 | 2 | 3 | 4 |
|---|---|---|---|---|---|
| $y$ (cm) | 17 | 25.5 | 34 | 42.5 | 51 |

10. Use your equation from the previous problem to estimate the height of the tree after 12 months, assuming it continues to grow at the same rate.

# E. Using and Interpreting the Intercepts of a Linear Model

Some real-world problems provide the $y$-intercept within the context of the problem. This is the initial output value, the $y$ value when $x = 0$. Once an equation for a linear model is known, the $x$-intercept can also be calculated. The $x$-intercept of the function is the value of $x$ when $y = 0$. For linear functions, we find the $x$-intercept by solving the equation $0 = mx + b$.

### Example 12

Hannah plans to pay off a generous, no-interest loan from her parents. Her loan balance is $4,000. She plans to pay $250 per month until her balance is $0. Use this information to do the following:

a. Write an equation to model this situation.

b. Find and interpret the $x$-intercept of the model.

c. Find the practical domain of the function.

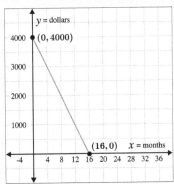

Figure 4. The graph of the function $y = -250x + 4000$.

### Solution

a. We can use the slope-intercept form of a line and the given information to develop a linear model. Let $y$ be the balance on the loan in dollars after $x$ months. The rate of change, or slope, represents a *decrease* in the balance of the loan by 250 dollars per month, so $m = -250$. The initial balance on the loan is $4,000 so $b = 4000$. Our linear model is $y = -250x + 4000$.

b. To find the *x*-intercept, replace *y* with 0, and solve for *x*.

$$0 = -250x + 4000$$
$$\underline{-4000 \qquad \qquad -4000} \quad \text{Subtract 4000 from both sides.}$$
$$-4000 = -250x$$
$$\frac{-4000}{-250} = \frac{-250x}{-250} \quad \text{Divide both sides by } -250.$$
$$16 = x$$
$$x = 16 \qquad \qquad \text{Rewrite the equation with the variable on the left.}$$

The *x*-intercept is (16,0) and represents the number of months it takes Hannah to reach a balance of $0. It will take her 16 months to pay off her loan.

c. The domain of this function refers to the number of months that Hannah pays off her loan. In this case, it doesn't make sense to talk about input values less than zero. We have determined in part b that it will take her 16 months to pay off the loan. So the practical domain is [0, 16].

Figure 4 presents a graph of the linear model for Example 12. The model is a decreasing linear function because the slope is negative and the curve decreases as we read the graph from left to right. Notice the location of the *x*- and *y*-intercepts on the graph.

# Exercises 1.3

1. Terry is skiing down a steep hill. Terry's elevation, *y*, in feet after *x* seconds is given by $y = -70x + 3000$.
   a. Write a complete sentence describing Terry's starting elevation and how it is changing over time.
   b. Make a table of values to represent the function. Let *x* = 0, 10, 20, 30, 40.
   c. Sketch a graph of the function. You can use your graphing calculator to check your work.

2. The amount of water remaining in a water tank, *y*, in gallons after *x* minutes is given by $y = -35x + 1400$.
   a. Write a complete sentence describing the initial amount of water in the tank and how it is changing over time.
   b. Make a table of values to represent the function. Let *x* = 0, 10, 20, 30, 40.
   c. Sketch a graph of the function. You can use your graphing calculator to check your work.

3. A boat that was anchored 25 miles from a marina begins sailing directly away from it at 8 miles per hour.
   a. Write an equation for the distance of the boat *y* from the marina after *x* hours.
   b. Make a table of values to represent the function. Let *x* = 0, 1, 2, 3, 4.
   c. Sketch a graph of the function. You can use your graphing calculator to check your work.

## Section 1.3: Finding Equations of Linear Functions

4. Amanda has $370 in her savings account and she is adding $120 per month.
   a. Write an equation for the balance in the account $y$ after $x$ months.
   b. Make a table of values to represent the function. Let $x = 0, 3, 6, 9, 12$.
   c. Sketch a graph of the function. You can use your graphing calculator to check your work.

For the following exercises, determine whether each function is increasing or decreasing.

5. $y = 14x + 30$
6. $y = 5x + 16$
7. $y = 7 - 2x$
8. $y = 28 - 3x$
9. $y = -17.3x + 41.5$
10. $y = -22.48x + 1.92$
11. $y = \frac{1}{2}x - 3$
12. $y = \frac{1}{4}x - 5$
13. $y = -\frac{2}{5}x - \frac{1}{8}$
14. $y = -\frac{3}{7}x + \frac{1}{7}$

For the following exercises, assume the situation represents a linear function. Find and interpret the slope of the line that models the data.

15. Two months after its grand opening, a new museum has had a total of 489 visitors. After 6 months the total number of visitors to the museum is 1490.
16. A group of hikers sets out a long walk. After 0.25 of an hour, they have gone 1.2 miles. After 0.75 of an hour, they have traveled 3.6 miles.
17. A scuba diver descends to the bottom of a lake. After 30 seconds the diver is 137 feet from the bottom of the lake and after 2 minutes the diver is just 2 feet from the bottom of the lake.
18. Money is dispensed from a retirement account such that after 5 years there is $165,000 left in the account and after 10 years there remains $80,000.

For the following exercises, write the equation of a linear function in slope-intercept form that models the situation. Interpret the meaning of the slope and $y$-intercept in the context of the problem. Then, find the value of the function that satisfies the follow-up question.

19. A plumber charges a flat rate of $120 to make a house call and charges $42.50 per hour for labor. Let $y$ be the total cost of the house call plus labor and let $x$ be the number of hours worked. How much does it cost if the plumber works for 5.5 hours?
20. A hot air balloon is hovering 100 meters above ground and begins to ascend at a rate of 14 meters per second. Let $y$ be the height of the balloon in meters $x$ seconds after it begins to ascend. How high is the hot air balloon after 30 seconds?
21. Grain is pumped out of a full grain silo into a barge at a rate of 13 cubic meters per minute. The silo holds 4500 cubic meters of grain. Let $y$ = cubic meters remaining in the silo after $x$ minutes. How many cubic meters of grain remain in the silo after 4 hours?

### Practice D — Answers

9. $y = 8.5x + 17$
10. $y = 8.5(12) + 17 = 119$, so after 12 months the tree will be 119 cm tall.

22. Stock value per year in a certain company has been steadily decreasing in value by $1.80 per year for the last 10 years. Ten years ago the value of the stock was $39.55. Let $y$ be the value of the stock $x$ years after it began to decrease. What was the value of the stock 7 years after it began to decline?

23. A cable lowers some equipment into a mine shaft. After 5 minutes the equipment is 82 meters from the bottom of the shaft and after 15 minutes the equipment is 46 meters from the bottom of the shaft. Let $y$ = the distance to the bottom of the shaft in meters after $x$ minutes. How far from the bottom of the shaft is the equipment after 20 minutes?

24. Sonny has entered a reading contest at the local library. After reading two books, he has read 330,000 words of text. After 5 books, he has read 825,000 words. Write a linear function to model his progress where $x$ is the number of books read and $y$ is the total number of words read. How many books will he need to read to reach a total of 10,000,000 words?

25. Marco is farming gold coins to buy a new set of armor in an online game. After 3 hours, he has a total of 315 gold coins. After 8, he has 540. Create a linear function in slope-intercept form that models how many gold coins he has, where $x$ is the number of hours he has spent farming and $y$ is his total amount of gold coins. How long will he have to farm in order to have the 855 gold coins he needs?

26. Bianca is analyzing 220 megabits of data from the Mars Rover, *Curiosity*. After 90 minutes, she has 160 megabits left to analyze. Create a linear function where $y$ is the number of megabits remaining to be analyzed and $x$ is the number of minutes that have passed. How long will it take her to analyze all 220 megabits of data?

For the following exercises, find a linear equation that models the data.

27.
| $x$ | 0 | 1 | 2 | 3 | 4 |
|---|---|---|---|---|---|
| $y$ | 5 | -7 | -19 | -31 | -43 |

28.
| $x$ | 0 | 2 | 4 | 6 | 8 |
|---|---|---|---|---|---|
| $y$ | -3 | 7 | 17 | 27 | 37 |

29.
| $x$ | 0 | 5 | 10 | 15 | 20 |
|---|---|---|---|---|---|
| $y$ | -5 | 10 | 25 | 40 | 55 |

30.
| $x$ | 0 | 3 | 6 | 9 | 12 |
|---|---|---|---|---|---|
| $y$ | 184 | 154 | 124 | 94 | 64 |

31.
| $x$ | -2 | 2 | 6 | 10 | 14 |
|---|---|---|---|---|---|
| $y$ | 36 | 30 | 24 | 18 | 12 |

32.
| $x$ | 4 | 6 | 8 | 10 | 12 |
|---|---|---|---|---|---|
| $y$ | 13 | 18 | 23 | 28 | 33 |

33.
| $x$ | 1 | 4 | 7 | 10 | 13 |
|---|---|---|---|---|---|
| $y$ | 7.2 | 10.8 | 14.4 | 18 | 21.6 |

34.
| $x$ | -5 | 5 | 15 | 25 | 35 |
|---|---|---|---|---|---|
| $y$ | 14 | -22 | -58 | -94 | -130 |

For the following exercises, find and interpret the $x$ and $y$-intercepts of each equation.

35. Freddie plays poker with his friends every Friday night. The following equation models the amount of money he has remaining after $x$ rounds of Texas Hold 'Em. $y = -15x + 120$.

36. Sun-Mi has saved some spending money for food and activities while on vacation. The following equation models the amount of money she has remaining after $x$ days. $y = -53x + 490$

For each of the following exercises, answer all four questions.

37. The number of gallons of water remaining in a watering trough $x$ minutes after it begins draining is given by $y = -12x + 350$.
    a. Interpret the slope and $y$-intercept of this equation in the context of the problem.
    b. How many gallons remain in the trough after 14 hours?
    c. When will there be only 100 gallons remaining in the trough?
    d. How long does it take for the trough to be empty (0 gallons)?

38. A hiker descends from a mountain top. Her altitude above sea level in feet $x$ hours after she begins her descent is given by $y = -260x + 1400$.
    a. Interpret the slope and $y$-intercept of this equation in the context of the problem.
    b. What is the hiker's altitude after 2.5 hours?
    c. When will the hiker be at an altitude of 555 feet?
    d. How long does it take the hiker to descend to sea level (0 feet)?

# 1.4 Using Linear Functions to Model Data

## Overview

A professor wants to identify trends among final exam scores. His class has a mixture of students, so he wonders if there is any meaningful relationship between their ages and the final exam scores. One way for him to analyze the scores is by creating a diagram that relates the age of each student to the exam score received.

In this section, we will examine one such diagram known as a scatter plot. Models such as scatter plots can be extremely useful for analyzing relationships and making predictions based on those relationships. As you study this section, you will learn how to:

- Draw and interpret scatter plots
- Distinguish between linear and nonlinear relations
- Find a line of best fit by hand
- Use a graphing utility to find a linear regression
- Fit a regression line to a set of data and use the linear model to make predictions
- Interpret the intercepts of a model
- Determine when model breakdown occurs

## A. Scatter Plots and Linear Models

A *scatter plot* or scattergram is a graph of plotted points. A scatter plot can give us useful information about the relationship between two sets of data.

If the scattergram shows a relationship that is linear or nearly linear, we can write a linear equation to model the relationship. We can then use it to help us make predictions and draw conclusions.

Of course, not all relationships can be represented by linear models. The professor mentioned in this section's overview, who wants to see if a student's age is meaningfully related to their final exam score, can create a scatter plot of the data. See Figure 1.

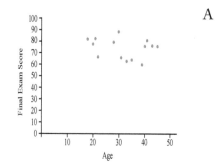

Figure 1. Scatter plot of final exam score vs. age.

You'll notice this scatter plot does not indicate a linear relationship. In fact, the points don't appear to follow any trend. In other words, there doesn't appear to be a meaningful relationship between the age of the student and the score on the final exam.

To determine whether a set of data is linearly related or nearly linearly related, we look at the scatter plot of the ordered pairs. If a set of data has a positive linear relationship, it can be modeled with a line having a positive slope. If a set of data has a negative linear relationship, it can be modeled with a line having a negative slope. We can see both of these relationships in Figure 2. We can also see that not all data can be modeled with a linear function.

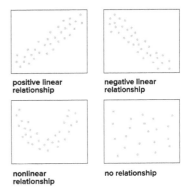

Figure 2. Scatter plots may or may not show a nearly linear relationship.

## Example 1

The table below shows the number of chirps made by a cricket during a 15-second interval, for several different air temperatures, in degrees Fahrenheit. Make a scatter plot of this data, and determine whether the data appears to be linearly related.

| Temperature (°F) | 52 | 53 | 57 | 61 | 66 | 68 | 70.5 | 72 | 73.5 | 80.5 |
|---|---|---|---|---|---|---|---|---|---|---|
| Chirps (in 15-second interval) | 18.5 | 23 | 21.5 | 27 | 33 | 31 | 35 | 35 | 37 | 44 |

Figure 3. Air temperatures versus cricket chirps.

### Solution

Recall that data in a table like this represent input-output pairs, which in turn can be written as ordered pairs. Temperature is the independent variable in this example because the number of cricket chirps may *depend upon* the temperature but will not *determine* temperature.

From the table in Figure 3, we see that when the temperature was 53°F, the number of cricket chirps was 23. This can be written as the ordered pair (53, 23).

When the data is plotted, as in Figure 4, we can see from the trend in the data that the number of chirps increases as the temperature increases. The trend appears to be roughly linear, with a positive slope, though certainly not perfectly so.

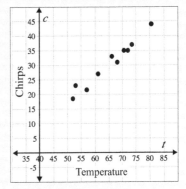

Figure 4. Scatter plot of temperature (°F) vs. cricket chirps (in 15 second intervals).

## Creating Scatter Plots on a Graphing Calculator

We can also use our graphing calculators to make a scatter plot. We'll do this with the date from Example 1.

*Step 1:* Press [STAT] and [ENTER] to access Lists 1 and 2, $L_1$ and $L_2$. To clear previous data, arrow up to the list title and press [CLEAR] and [ENTER].

*Step 2:* Enter the input values into L1 and the output values into L2.

*Step 3:* Access the StatPlot above the [Y=] button. Press [ENTER] to select Plot 1 and [ENTER] again to turn the plot on. Make sure the other icons on this screen match the selections in in this screen image.

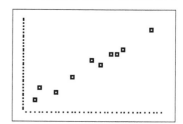

*Step 4:* To view the scatter plot, Figure 7, select [ZOOM] and 9 for a viewing window that matches the data.

## B. Approximating Lines of Best Fit

Once we recognize that a set of data is nearly linear, we can try to find the equation of a linear function that "best fits" the description of the data. One way to approximate our linear function is to sketch the line that seems to come closest to most data points while following the trend of the data. We call this a trend line or a **line of best fit**. See Figure 8 for an example.

To find the equation of our line, we can follow the steps outlined in Section 1.3 for writing an equation in slope-intercept form using two points. However, not all data points are created equal! Make sure to choose two points that are on or near your trend line.

## Example 2

Use the data in the table in Figure 3 from Example 1 and the trend line drawn in Figure 5 to find the equation in slope-intercept form of a linear model that fits the data. Interpret the slope of the line in the context of the problem.

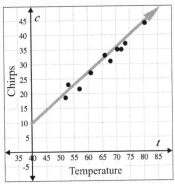

### Solution

Two data points appear to be on or near the line that's drawn in Figure 5. Checking our data table, we find the coordinates of these points are (61, 27) and (73.5, 37). Now we calculate the slope of the line:

$$m = \frac{37 - 27}{73.5 - 61}$$
$$= \frac{10}{12.5}$$
$$= 0.8$$

Figure 5. A line of best fit follows the trend of the data.

The output units are chirps, while the input units are °F. We interpret the slope to mean there is an increase of 0.8 (almost 1) chirp per increase of 1°F. We can now solve for the $y$-intercept $(0, b)$.

| | |
|---|---|
| $y = 0.8x + b$ | Substitute the value of the slope, $m = 0.8$ |
| $27 = 0.8(61) + b$ | Using one ordered pair (61, 27), substitute $x = 61$, and $y = 27$. |
| $27 = 48.8 + b$ | Simplify. |
| $\underline{-48.8 \quad -48.8}$ | Subtract 48.8 from both sides of the equation to solve for $b$. |
| $-21.8 = b$ | |

Substituting the values we found for $m$ and $b$ into $y = mx + b$, we write the linear model $y = 0.8x - 21.8$.

If you created a scatter plot on your calculator, you can enter this equation into Y1 to verify that the line we found is a good model for the data. See Figure 6. This linear equation can now be used to approximate answers to various questions we might ask about the trend.

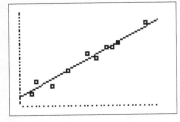

Figure 6.

## Example 3

Using the equation of the trend line found in Example 2, $y = 0.8x - 21$, predict the number of cricket chirps in a 15-second interval when the temperature is 78°F.

### Solution

Substituting $x = 78$ into the equation $y = 0.8x - 21.8$, we obtain

$$y = 0.8(78) - 21.8$$
$$= 40.6$$

So we estimate that a cricket will chirp about 40.6 times in a 15-second interval when the outside temperature is 78°F.

If you entered the equation from Examples 2 and 3 into your graphing calculator, you can also find a predicted number of chirps using your [TRACE] button. Make sure you are tracing on the equation in Y1 rather than on the scatter plot. Use the up and down arrows to toggle between the data in the scatter plot and points on the line. Look in the upper left corner of the screen to see whether you are tracing the plot or the equation. See Figures 7 and 8.

When the calculator is in [TRACE] mode, the left and right arrows move the cursor along the curve showing various ordered pairs. However, we can find the corresponding output for a specific input by just typing the input value, as long as the value is within the viewing window.

In Example 3, we are asked to predict the number of chirps when the temperature is 78°F. In trace mode we just type 78 and press [ENTER] to find the corresponding number of chirps. See Figure 9. At 78° a cricket will chirp about 40.6 times in a 15-second interval.

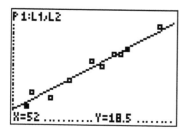

Figure 7. Tracing on the scatter plot.

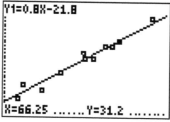

Figure 8. Tracing on the equation.

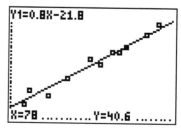

Figure 9. Finding a specific output.

## C. Finding Linear Regression Equations

While eyeballing a line works reasonably well, there are statistical techniques for fitting a line to data that minimize the differences between the line and data values. One such technique is called **least squares regression** and can be computed by many graphing calculators, spreadsheet software, statistical software, and many web-based calculators. Least squares regression is one way to define the line that best fits the data. We will refer to this method as **linear regression**.

Before we use our graphing calculators to find a linear regression equation, it's a good idea to graph the scatter plot of the data first. This way we can determine if a linear model makes sense in a given situation. Besides, entering the data into Lists 1 and 2 on the calculator is a required step for finding a linear regression line.

### Finding a Linear Regression Line on a Graphing Calculator

To find a linear regression on a graphing calculator, follow these steps:

*Step 1:* Enter the input values in $L_1$ and the output values in $L_2$.

*Step 2:* From the home screen, return to [STAT] and arrow right to [CALC]. Then arrow down to select linear regression, LinReg, and press [ENTER].

To calculate the linear regression press [ENTER] again. For newer models, your screen should match this screen image and you will need to press [ENTER] several times until the `Calculate` instruction is selected.

*Step 3:* Use the displayed values of a and b to write down the linear equation, rounding these values to a reasonable number of decimal places. Notice the calculator uses *a* for the slope rather than *m*. This screen image shows a sample linear regression.

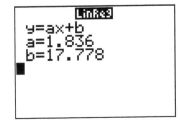

*Step 4:* Methods for viewing the equation on the graphing screen vary by calculator edition, but you can always enter the equation by hand in $Y_1$.

*Step 5:* To see both the regression equation and the scatter plot of the data, make sure Plot 1 is turned on and that you have chosen a good viewing window ([ZOOM] 9).

## Example 4

Use your graphing calculator to find the linear regression line using the cricket-chirp data in Figure 3 from Example 1. Use this equation to predict the number of chirps when the temperature is 78°F.

### Solution

Enter the input values (temperature) in List 1 ($L_1$). Then enter the output values (chirps) in List 2 ($L_2$).

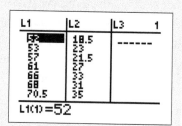

On a graphing utility, select Linear Regression, LinReg. Press press ENTER to calculate.

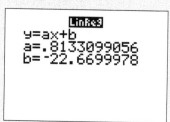

After rounding the *a* and *b* values, we obtain this equation: $y = 0.828x - 23.586$

Notice that this line is quite similar to the equation we calculated by hand in Example 2, but it fits the data better. Notice also that using *this* equation will change our prediction for the number of chirps in 15 seconds at 78°F to

$$y = 0.828(78) - 23.586$$
$$= 40.998$$
$$\approx 41 \text{ chirps}$$

This prediction is just slightly greater than our prediction of 40.6 chirps in Example 2.

Figure 10 shows the graph of the scatter plot with the least squares regression line.

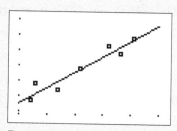

Figure 10.

## Practice C

The data in this table record the speed of a truck in miles per hour and its fuel efficiency in miles per gallon when traveling at that speed. Use this data to answer the questions below. Check your solutions on the following page.

| $x$ mph | $y$ mpg |
|---|---|
| 20 | 38 |
| 33 | 35 |
| 35 | 34 |
| 40 | 33 |
| 48 | 32 |
| 60 | 25 |
| 65 | 22 |

1. Use a graphing calculator to draw a scatter plot of the data. Describe the trend of these points.

2. Use the calculator to find a linear regression equation to model the data. Round $a$ and $b$ to three decimal places.

3. Use the regression equation to estimate the fuel efficiency when the speed of the truck is 55 mph.

## D. Using a Linear Model to Makes Estimates and Predictions

When working with a linear model, we must often evaluate the linear model at a given input. We did this in Examples 3 and 4 when we predicted the rate of cricket chirps at 78°F. We used our equation to calculate that the output would be approximately 41 chirps in a 15-second interval.

While the data for most examples does not fall perfectly on the line, the equation is our best guess as to how the relationship will behave outside of the values for which we have data. We use a process known as **interpolation** when we predict a value *inside* the domain and range of the data. We use the process of **extrapolation** when we predict a value *outside* the domain and range of the data.

Figure 11 compares the two processes for the cricket-chirp data addressed in previous examples. We can see that interpolation will occur if we used our model to predict chirps when the values for temperature are between 52°F and 80.5°F. Extrapolation will occur if we used our model to predict chirps when the values for temperature are less than 52°F or greater than 80.5°F.

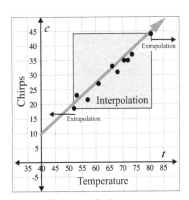

Figure 11. Interpolation occurs within the domain and range of the provided data. Extrapolation occurs outside the domain and range.

## Example 5

Using the cricket data from Example 1, which appears again below, and our regression model, $y = 0.828x - 23.586$, answer the following questions.

| Temperature (°F) | 52 | 53 | 57 | 61 | 66 | 68 | 70.5 | 72 | 73.5 | 80.5 |
|---|---|---|---|---|---|---|---|---|---|---|
| Chirps (in 15-second interval) | 18.5 | 23 | 21.5 | 27 | 33 | 31 | 35 | 35 | 37 | 44 |

a. Will predicting the number of chirps crickets will make during a 15-second interval when the outside temperature is 55°F be interpolation or extrapolation? Make the prediction, and discuss whether it is reasonable.

b. Will predicting the rate of cricket-chirps at 40°F be interpolation or extrapolation? Make the prediction, and discuss whether it is reasonable.

### Solution

a. The temperatures in the data are between 52°F and 80.5°F. A prediction at 55°F is inside the domain of our data, so it will be interpolation. Using our model, we find:

$$y = 0.828(55) - 23.586 = 21.954 \approx 22$$

We predict that crickets will chirp 22 times in a 15-minute interval. Based on the data we have, this seems reasonable.

b. Predicting the number of chirps at 40°F is extrapolation because 40 is outside the domain of our data. Using our model, we find:

$$y = 0.828(40) - 23.586 = 9.534 \approx 9.5$$

Our model predicts the crickets would chirp 9.5 times in 15 seconds.

While this might be possible, we have no reason to believe our model is valid outside the domain and range. In fact, crickets generally stop chirping when the temperature is below 50 degrees.

### Practice C — Answers

1. The data appear to be approximately linearly related. As the speed of the truck increases the fuel efficiency decreases. This indicates a linear function with a negative slope could model the data.

2. $y = -0.351x + 46.364$

3. $y = -0.351(55) + 46.364 \approx 27.1$ At a speed of 55mph, the fuel efficiency of the truck will be approximately 27 miles per gallon.

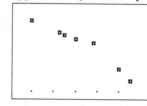

Another way to use a linear model is to set the equation of the model equal to a specified output. If we then solve the equation, we can answer a question that asks us to find an input that will predict the specified output. For example, if we are outside on a warm evening and count 25 chirps in 15 seconds, we can use our regression model to predict the temperature.

$$y = 0.828x - 23.586$$
$$25 = 0.828x - 23.586 \quad \text{25 chirps is an output, so } y = 25$$
$$\underline{+23.586 \qquad\qquad +23.586} \quad \text{To solve for } x, \text{ add 23.586 to both sides.}$$
$$48.586 = 0.828x$$
$$\frac{48.586}{0.828} = \frac{0.828x}{0.828} \qquad \text{Divide both sides by 0.828.}$$
$$58.7 \approx x$$

A count of 25 chirps in 15 seconds predicts that the temperature is approximately 58.7°F. This is an example of interpolation because the number of chirps is within the range of our data.

### Example 6

As the table below shows, gasoline consumption in the U. S. has been increasing. T

| Year | 1994 | 1995 | 1996 | 1997 | 1998 | 1999 | 2000 | 2001 | 2002 | 2003 | 2004 |
|---|---|---|---|---|---|---|---|---|---|---|---|
| Consumption (billions of gallons) | 113 | 116 | 118 | 119 | 123 | 125 | 126 | 128 | 131 | 133 | 136 |

Using the data from this table, do the following:

a. Determine whether the trend is linear, and if so, find a model for the data.
b. Interpret the slope and $y$-intercept of the model in the context of the problem.
c. Use the model to predict the consumption in 2008.
d. Use the model to predict when the consumption will reach 175 billion gallons.

### Solution

In a problem that uses years like this, it's useful to define the independent variable with respect to a starting year. This ensures that the input values will be smaller, more manageable numbers. So, let $x$ be the number of years since 1994. Then $x = 0$ corresponds to 1994, $x = 1$ corresponds to 1995, and so on. Let $y$ be the gas consumption in the U.S. in billions of gallons.

We can use the table above for our input-output pairs in List 1 and List 2 on the graphing calculator.

| $x$ (years since 1994) | 0 | 1 | 2 | 3 | 4 | 5 | 6 | 7 | 8 | 9 | 10 |
|---|---|---|---|---|---|---|---|---|---|---|---|
| $y$ (billions of gallons) | 113 | 116 | 118 | 119 | 123 | 125 | 126 | 128 | 131 | 133 | 136 |

a. We create a scatter plot using a calculator and see that the data are nearly linear with a positive slope. We then calculate the linear regression line and graph it along with our scatter plot as shown in Figure 12. After rounding the values of $a$ and $b$, our regression equation is $y = 2.209x + 113.318$.

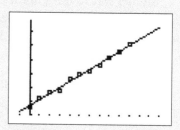

Figure 12. Scatter plot of years vs. consumption.

b. The slope indicates that gasoline consumption in the U.S. has been increasing at an average rate of 2.209 billion gallons per year. The $y$-intercept indicates that gas consumption in the initial year, 1994 for our data set, was approximately 113.3 billion gallons.

c. For the year 2008, $x = 2008 - 1994 = 14$. We calculate that $y = 2.209(14) + 113.318 = 144.244$.
  The model thus predicts 144.244 billion gallons of gasoline consumption in 2008. This prediction is an extrapolation since the year 2008 is outside of the domain of the data set.

d. To predict when the consumption will reach 175 billion gallons, substitute 175 for $y$ in our equation and solve for $x$.

$$175 = 2.209x + 113.318$$
$$\underline{-113.318 \qquad\quad -113.318} \quad \text{Subtract 113.318 from both sides.}$$
$$61.682 = 2.209x$$
$$\frac{61.682}{2.209} = \frac{2.208x}{2.209} \quad \text{Divide both sides by 2.209.}$$
$$27.923 \approx x$$

Our solution, $x \approx 28$ years, corresponds to the year 1994 + 28 = 2022. So we predict that gas consumption in the U.S. will reach 175 billion gallons in 2022. This prediction is also an extrapolation.

## Practice Set D

Use the model in Example 6 to answer the questions below. When youre are done, turn the page to check your solutions.

4. Predict the gas consumption in 2017. Is this an interpolation or an extrapolation?
5. Predict the year the gas consumption will reach 200 billion gallons, assuming the model is still relevant that far into the future.

# E. Intercepts of a Model and Model Breakdown

There's a big difference between making predictions inside the domain and range of values for which we have data and making predictions outside that domain and range. Predicting a value outside the domain and range has its limitations.

In Example 6, the U.S. gas consumption problem, the further we move beyond the domain of the provided data — that is, the further into the future we predict — the less sure we can be about our predictions. The rate of gas consumption *could* increase or decrease dramatically, but it might also remain constant because of the increase of electric or hybrid vehicles, the use of alternative energy, or other changes that would affect gasoline consumption. When we extrapolate much beyond our data set, we lose confidence in our prediction.

**Model breakdown** occurs when a model no longer applies beyond a certain point because a value of the input or output variable is obviously wrong based on common knowledge.

Figure 13 presents a linear model that represents $y$ as the percentage of U.S. homes with Internet access for $x$ years since 1990. As we know, that percentage has increased over the years. However, the input values, years, are limited by the fact that *no one* had Internet access from their home before 1990. The $y$-intercept, when $x = 0$, indicates a practical limit on our domain.

The $x$-intercept, when $y = 0$, indicates that 0 percent of homes have Internet access. Clearly, there can't be a negative percent of homes with access. The output values, percentages of homes, can only be between 0 and 100%, representing a practical limit on our range.

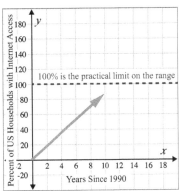

Figure 13.

## Example 7

Students in a college dormitory carried out an experiment on hours of TV watched per day ($x$) and number of sit-ups a person can do ($y$). After collecting data and plotting a scattergram, the students could see that there was a linear relationship between the two activities. They then ran a linear regression resulting in the equation $y = -1.341x + 32.234$.

a. Determine the $y$-intercept of the regression model and interpret its meaning in the context of the problem.

b. Use the regression model to find the $x$-intercept. Interpret its meaning in the context of the problem. Discuss the potential for model breakdown.

c. Determine a practical domain for this function. Justify your answer.

### Solution

a. The $y$-intercept is (0, 32.234). It means that a student in the dorm experiment who watched no TV could do about 32 sit-ups.

b. Setting the equation equal to 0 and solving for $x$, we have

$$0 = -1.341x + 32.234$$
$$\underline{-32.234 \qquad\qquad -32.234}$$
$$-32.234 = -1.341x$$
$$\frac{-32.234}{-1.341} = \frac{-1.341x}{-1.341}$$
$$24.037 \approx x$$

The $x$-intercept is thus approximately $(24, 0)$, which means that a student who watches TV for 24 hours per day will not be able to do any sit-ups. This situation represents model breakdown because it's probable that no one can watch TV for 24 hours a day, however hard they might try.

c. Determining the practical domain is partly subjective and correct answers might vary. One reasonable practical domain could be $[0, 16]$, representing from 0 to 16 hours of TV time. It's impossible to have a negative number of hours watching TV, and it is unlikely that students would watch TV for more than 16 hours per day. They do have classes to attend.

## Exercises 1.4

1. What is interpolation when using a linear model?
2. What is extrapolation when using a linear model?
3. Explain the difference between a positive and a negative linear relationship.
4. What does it mean to say that model breakdown has occurred?

For the following exercises, draw a scatter plot for the data provided. Does the data appear to be linearly related? If so, is it a positive or negative linear relationship?

5.

| $x$ | 0  | 1  | 3  | 4  | 7  | 8 | 10 |
|-----|----|----|----|----|----|---|----|
| $y$ | 22 | 23 | 19 | 15 | 11 | 6 | 5  |

6.

| $x$ | 1  | 2  | 3  | 4  | 5   | 6   | 7   |
|-----|----|----|----|----|-----|-----|-----|
| $y$ | 46 | 50 | 59 | 75 | 100 | 136 | 185 |

7.

| $x$ | 100 | 190 | 250 | 310 | 380 | 430 | 450 |
|-----|-----|-----|-----|-----|-----|-----|-----|
| $y$ | 12  | 28  | 25  | 43  | 60  | 73  | 75  |

8.

| $x$ | 1 | 3 | 5  | 6  | 7  | 9   | 11  |
|-----|---|---|----|----|----|-----|-----|
| $y$ | 1 | 9 | 28 | 57 | 65 | 125 | 216 |

# Section 1.4: Using Linear Equations to Model Data

9. The town of Midgar increased in population from 1990 to 2010. For the following data, draw a scatter plot and estimate a line of best fit. Choose two points on or near your line and use them to write a linear model for the data. Let $y$ be the population $x$ years since 1990. Use your model to predict the population in 2018. Does this prediction involve interpolation or extrapolation?

| Year | 1990 | 1995 | 2000 | 2005 | 2010 |
|---|---|---|---|---|---|
| Population | 11,500 | 12,100 | 12,700 | 13,000 | 13,750 |

10. Vincent is defrosting frozen vegetables in the microwave. For the following data, draw a scatter plot and estimate a line of best fit. Choose two points on or near your line and use them to write a linear model for the data. Use your model to predict when the temperature would reach 48°F. Does this prediction involve interpolation or extrapolation?

| Time (seconds) | 46 | 50 | 54 | 58 | 62 |
|---|---|---|---|---|---|
| Temperature (°F) | 26 | 29 | 31 | 35 | 40 |

For the following exercises, match each scatterplot with one of the four specified relationships in Figure 14.

11. Nearly linear with a positive slope
12. Nearly linear with a negative slope
13. No relationship
14. Far from linear with a possible negative slope.

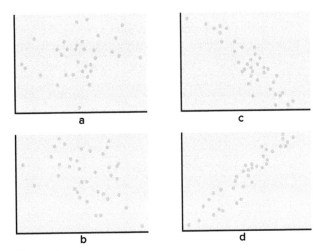

Figure 14.

15. A regression was run to determine whether there is a relationship between $x$, years since 1990, and $y$, the number of people in millions in the U.S. labor force. The results of the regression are shown below. Assuming the trend continues, predict the number of people in the labor force in 2012 and in 2020.

$y = ax + b$
$a = 1.7$
$b = 124$

## Practice D — Answers

4. According to the model, gas consumption in 2017 ($x = 23$) is 164.125 billion gallons. This number was found through extrapolation.

5. According to the model, gas consumption will reach 200 billion gallons when $x \approx 39.240 \approx 39$. This corresponds to the year 2033.

16. A regression was run to determine whether there is a relationship between $x$, the diameter of a tree in inches, and $y$, the tree's age in years. The results of the regression are shown below. Use this to predict the age of a tree with diameter 10 inches and a tree with diameter 15 inches.

    $y = ax + b$
    $a = 6.301$
    $b = -1.044$

17. The U.S. Census Bureau tracks the percentage of persons 25 years or older who are college graduates. The table below provides several years worth of data. Use your graphing calculator to make a scatter plot of the data and to find a linear regression equation to model the data. Assuming the trend continues, in what year will the percentage exceed 35%?

    | Year | 1990 | 1992 | 1994 | 1996 | 1998 | 2000 | 2002 | 2004 | 2006 | 2008 |
    |---|---|---|---|---|---|---|---|---|---|---|
    | Graduates (%) | 21.3 | 21.4 | 22.2 | 23.6 | 24.4 | 25.6 | 26.7 | 27.7 | 28 | 29.4 |

18. The table below provides the U.S. import of wine (in hectoliters) for several years. Let $y$ be the number of hectoliters imported $x$ years after 1990. Find a linear regression equation to model the data. Assuming the trend continues, use your model to predict when imports will exceed 12,000 hectoliters.

    | Year | 1992 | 1994 | 1996 | 1998 | 2000 | 2002 | 2004 | 2006 | 2008 | 2009 |
    |---|---|---|---|---|---|---|---|---|---|---|
    | Imports | 2665 | 2688 | 3565 | 4129 | 4584 | 5655 | 6549 | 7950 | 8487 | 9462 |

19. The table below shows the year and the number of people unemployed in the city of Waldorf over several years. Let $y$ be the number of people who are unemployed $x$ years after 1990. Find a linear regression equation and, assuming the trend continues, use it to predict the number unemployed in 2016.

    | Year | 1990 | 1992 | 1994 | 1996 | 1998 | 2000 | 2002 | 2004 | 2006 | 2008 |
    |---|---|---|---|---|---|---|---|---|---|---|
    | Number of Unemployed | 750 | 670 | 650 | 605 | 550 | 510 | 460 | 420 | 380 | 320 |

20. The table below shows a small sample of heights and weights of women aged 30 to 39. Find a linear regression equation to model this data. Let $y$ be the weight of a woman who is $x$ inches tall. Use your model to predict the height of a 120-pound woman.

    | Height (inches) | 62 | 67 | 68 | 70 | 71 | 74 | 75 |
    |---|---|---|---|---|---|---|---|
    | Weight (pounds) | 123.1 | 138.1 | 134.3 | 151 | 145.5 | 164.2 | 176.4 |

21. The Oregon Department of Education is conducting a review of mathematics education over the last 25 years. Using the table below, determine if there is a linear trend between $x$, the number of years since 1990, and $y$, the percentage of students who took Algebra 2. Find a linear regression equation to model this data and use it to predict the percentage of high school students taking Algebra 2 in 2020.

    | Year | 1992 | 1997 | 2000 | 2004 | 2008 | 2010 | 2015 |
    |---|---|---|---|---|---|---|---|
    | Students (%) | 39.9 | 49.0 | 58.8 | 61.5 | 61.7 | 67.6 | 70.3 |

22. Veronica is looking through records in her hometown of Silver Creek. She is comparing $y$, the average age of mothers at the time of the birth of their first child, with the year in which the births took place. Let $x$ be the number of years since 1980. Using the data from the table below, determine if there is a linear trend. If there is a trend, find a linear regression equation and use it to predict the average age of a first-time mother in the year 2020.

| Year | 1980 | 1983 | 1985 | 1988 | 1990 | 1993 | 1995 | 1998 | 2000 | 2003 |
|---|---|---|---|---|---|---|---|---|---|---|
| Age of Mother | 19 | 22 | 21 | 23 | 23 | 24 | 22 | 23 | 21 | 25 |

23. The data below shows both the number of calories and the number of grams of fat in seven sandwiches at a McDonald's restaurant. Find a linear regression equation to model the data and use it to predict the number of grams of fat in a sandwich with 620 calories.

| Sandwich Name | Calories | Fat (grams) |
|---|---|---|
| Big Mac | 550 | 29 |
| Quarter Pounder w/ cheese | 520 | 26 |
| Double quarter-pounder w/ cheese | 750 | 42 |
| Hamburger | 250 | 9 |
| Hamburger w/ cheese | 300 | 12 |
| Double Cheeseburger | 440 | 23 |
| McDouble | 390 | 19 |

# 1.5 Function Notation and Making Predictions

## Overview

Once we determine that a relationship is a function, we want to display and define the function so that we can easily understand and use it. As we have already seen, there are various ways of representing functions. Standard function notation is one representation that makes working with functions easier. In this section, we introduce a new notation for functions commonly used by mathematicians, scientists, and computer programmers. You will learn to:

- Use function notation to represent functions
- Find inputs and solve for outputs using function notation
- Use function notation with graphs and tables
- Interpret models described with function notation
- Make predictions from models described with function notation
- Find and interpret intercepts, domain, and range of a model

## A. Function Notation

When using an equation, graph, table, or words to represent a function, it is very useful to name the function. The letters $f$, $g$, and $h$ are often used to name functions although any letter can be used.

Once we name a function, say $f$, and if we are using $x$ and $y$ as input and output variables, we can use the symbol $f(x)$ to represent the output $y$. We refer to "$f(x)$" as **function notation** and we read the expression as "$f$ of $x$". We use parentheses to contain the function input — in this case $x$. Note that in this context, the parentheses *do not* indicate multiplication.

Since $f(x)$ is another name for the output, $y$, we can write $y = f(x)$ which nicely communicates that $y$ is a function of $x$.

### Function Notation

The notation $y = f(x)$ defines a function named $f$. The letter $x$ represents the input value, or independent variable. The letter $y$, or $f(x)$, represents the output value, or dependent variable.

When we know an input value and want to determine the corresponding output value for a function, we **evaluate** the function. Evaluating will always produce one result because each input value of a function corresponds to exactly one output value.

When we have a function in algebraic form, it's usually easy to evaluate the function. We replace the input variable with a given value and then simplify according to the correct order of operations.

## Section 1.5: Function Notation and Making Predictions

For example, we can evaluate the function $f(x) = 5 - 3x^2$ by squaring the input value, multiplying by 3, and then subtracting the product from 5. To evaluate this function when $x = 2$, we write:

$f(x) = 5 - 3x^2$     Equation of $f$.
$f(2) = 5 - 3(2)^2$     Replace $x$ on both sides with the input value 2.
$\phantom{f(2)} = 5 - 3(4)$     Simplify.
$\phantom{f(2)} = 5 - 12$
$\phantom{f(2)} = -7$

The equation $f(2) = -7$ means the input $x = 2$ leads to the output $y = -7$. This equation is of the form

$f(\text{input}) = \text{output}$

This is the essential meaning of function notation.

### Example 1

For the functions $f(x) = 6x - 5$ and $g(x) = x^3 + 1$, evaluate the functions at $x = 3$, $x = -4$, and $x = 2.59$.

**Solution**

First, we replace the $x$ in the functions with each specified value, starting with $f$ and $x = 3$.

$f(3) = 6(3) - 5$     Substitute 3 for $x$.
$\phantom{f(3)} = 18 - 5$     Multiply.
$\phantom{f(3)} = 6$     Subtract.

Now with $x = -4$:

$f(-4) = 6(-4) - 5$     Substitute $-4$ for $x$.
$\phantom{f(-4)} = -29$     Simplify.

And with $x = 2.59$:

$f(2.59) = 6(2.59) - 5$     Substitute 2.59 for $x$.
$\phantom{f(2.59)} = 10.54$     Simplify.

Next we replace the $x$ in $g$ with $x = 3$:

$g(3) = (3)^3 + 1$     Substitute 3 for $x$.
$\phantom{g(3)} = 27 + 1$     Perform exponentiation by raising 3 to the third power.
$\phantom{g(3)} = 28$     Add.

Now with $x = -4$:

$g(-4) = (-4)^3 + 1$     Substitute $-4$ for $x$.
$\phantom{g(-4)} = -63$     Simplify.

And with $x = 2.59$:

$g(2.59) = (2.59)^3 + 1$     Substitute 2.59 for $x$.

$\phantom{g(2.59)} \approx 18.374$     Simplify.

We define some functions with a variable in more than one place, such as with $f(x) = x^2 + 3x - 4$. In this case, when we evaluate the function for a specific input value, we must substitute that value everywhere we find an $x$.

We can also give an algebraic expression as the input to a function. For example, $f(a + 5)$ means to "evaluate the function at $a + 5$."

## Example 2

For $f(x) = x^2 + 3x - 4$, find the following:

a. $f(-2)$     b. $f(a)$     c. $f(a + 5)$

### Solution

a. Because the input value is a number, $-2$, we just use the correct order of operations to simplify.

$f(-2) = (-2)^2 + 3(-2) - 4$     Substitute $-2$ for $x$.

$\phantom{f(-2)} = -6$

b. In this case, the input value is a letter, and we cannot simplify the answer.

$f(a) = (a)^2 + 3(a) - 4$     Substitute $a$ for $x$.

$\phantom{f(a)} = a^2 + 3a - 4$

c. In this case, the input is an algebraic expression, $a + 5$, so we must use the distributive property to simplify.

$f(a + 5) = (a + 5)^2 + 3(a + 5) - 4$     Substitute $a + 5$ for $x$.

$\phantom{f(a + 5)} = (a + 5)(a + 5) + 3(a + 5) - 4$     Expand.

$\phantom{f(a + 5)} = a^2 + 10a + 25 + 3a + 15 - 4$     Use the distributive property.

$\phantom{f(a + 5)} = a^2 + 13a + 36$     Combine like terms.

In Example 2, we see that when an input is a number for an algebraic function, the output will be a number. However, when the input is a variable expression, the output will be a variable expression.

Sometimes we know an output value and want to determine the input value or values that produce that output value. In these cases, we set the output equal to the function's formula and solve for the input.

## Example 3

Given the function $f(x) = \frac{1}{2}x + 9$, do the following:

a. Find $f(10)$.
b. Find $x$ when $f(x) = -6$.

**Solution**

a. $f(10)$

$\quad f(10) = \frac{1}{2}(10) + 9 \quad$ Substitute 10 for $x$.

$\qquad\quad = 14 \qquad\qquad$ Simplify.

b. Substitute $-6$ for $f(x)$ and solve for $x$. Notice that we set the output value on the right side of the equation, but it can be set on the left side as well.

$\quad \frac{1}{2}x + 9 = -6 \qquad$ Substitute $-6$ for $f(x)$.

$\qquad \underline{-9 \quad\; -9} \qquad$ Subtract 9 from both sides.

$\qquad \frac{1}{2}x = -15$

$\quad 2 \cdot \frac{1}{2}x = 2(-15) \qquad$ Multiply both sides by 2 (the reciprocal of $\frac{1}{2}$).

$\qquad\quad x = -30$

When solving for an input value, given an output value, we sometimes get more than one solution. This is because for some functions different input values can produce the same output value.

## Example 4

Given the function $f(x) = x^2 - 3$, do the following:

a. Find $f(5)$ and $f(-5)$.
b. Solve $f(x) = 97$.

**Solution**

a. Notice that for this function, the two given inputs produce the same output.

$\quad f(5) = (5)^2 - 3 \qquad$ Substitute 5 for $x$.

$\qquad\; = 22 \qquad\qquad$ Simplify.

And:

$\quad f(-5) = (-5)^2 - 3 \qquad$ Substitute $-5$ for $x$.

$\qquad\;\; = 22 \qquad\qquad$ Simplify.

b. We substitute 97 for $f(x)$ and solve for $x$.

$$x^2 - 3 = 97 \quad \text{Substitute 97 for } f(x).$$
$$\underline{+3 \quad\quad +3} \quad \text{Add 3 to both sides.}$$
$$x^2 = 100$$
$$x = \pm 10 \quad 10^2 = 100 \text{ and } (-10)^2 = 100.$$

Both inputs, $x = 10$ and $x = -10$, produce the same output: $f(x) = 97$.

## Practice A

When you're finished with these problems, turn the page and check your solutions.

1. Given the function $f(x) = \sqrt{x - 4}$, evaluate the function at $x = 40$ and $x = 15$.
2. Given the function $g(x) = 5x^2 + x$, find $g(3.4)$, $g(-2)$, and $g(a - 1)$.
3. Given the function $h(x) = 8 - 2.5x$, find $x$ when $h(x) = -10.5$.

## B. Using Function Notation with Graphs and Tables

### Functions in Tabular Form

As we saw in Section 1.2, we can represent functions in tables. Let's now apply function notation to the process of determining input and output values from a table. When a function is represented by a table, the process for evaluating the function at a given input value is as follows:

1. Find the given input in the row (or column) of input values.
2. Identify the corresponding output value paired with that input value.

To determine the input or inputs that resulted in a given output, do the following:

1. Find the given output values in the row (or column) of output values.
2. Identify the input value(s) corresponding to the given output value.

### Example 5

Use the table below to find the values asked for.

| $t$ | 2 | 5 | 9 | 12 | 17 | 21 | 25 |
|---|---|---|---|---|---|---|---|
| $f(t)$ | 11 | 13 | 17 | 26 | 18 | 15 | 11 |

a. Find $f(2)$.
b. Find $f(17)$.
c. Find $t$ when $f(t) = 26$.
d. Find $t$ when $f(t) = 11$.

## Solution

a. The value in parentheses always represents the input value, in this case $t = 2$. The corresponding output value in the table is 11. So $f(2) = 11$.

b. The number 17 occurs in both the input and output columns, so be careful here! In this case, the input value $t = 17$. The corresponding output value in the table is 18. So $f(17) = 18$.

c. This time we are given the *output* value and asked to solve for the input value. We find 26 in the second column of the table. The corresponding input value is 12. So $t = 12$.

d. Again, we are given the *output* value. Notice that the number 11 occurs *twice* in the output column. There are two corresponding inputs, 2 and 25. So $t = 2$ and $t = 25$.

## Finding Function Values from a Graph

For functions represented by a graph, the function notation $y = f(x)$ means that the point $(x, y)$ is on the graph of $f$. For example, if for some function $f$, $f(7) = 11$, then the point $(7, 11)$ is on its graph.

Evaluating a function requires finding the corresponding output value for a given input value. When the function is represented by a graph, we first locate the point on the graph that has the given input value. We then read off the $y$-coordinate of the point. Finding the input value of a function equation using a graph requires finding all instances of the given output value on the graph and observing the corresponding input value(s).

### Example 6

Given the graph in Figure 1, do the following:

a. Evaluate $f(2)$.
b. Find $x$ when $f(x) = 4$.

### Solution

a. To evaluate $f(2)$, locate the point on the curve where $x = 2$, and then read the $y$-coordinate of that point. The point has coordinates $(2, 1)$, so $f(2) = 1$. See Figure 2.

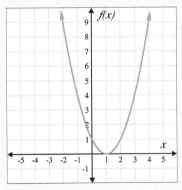

Figure 1.

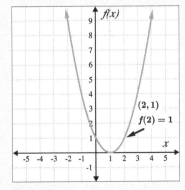

Figure 2.

b. To solve $f(x) = 4$, find the output value 4 on the vertical axis. Moving horizontally along the line $y = 4$, we notice that there are two points of the curve with output value 4: $(-1, 4)$ and $(3, 4)$. So the two solutions are $x = -1$ or $x = 3$. See Figure 3.

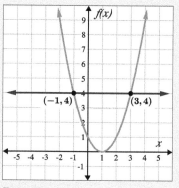

Figure 3.

## Practice B

Now it's your turn. When you are done, turn the page and check your solutions.

4. Use the table to find the values of function h.

| t    | 0 | 5  | 10 | 15 | 20 | 25 | 30 |
|------|---|----|----|----|----|----|----|
| h(t) | 0 | 12 | 15 | 20 | 25 | 12 | 3  |

a. Find $h(15)$.

b. Find $h(30)$.

c. Find $t$ when $h(t) = 20$.

d. Find $t$ when $h(t) = 12$.

5. Use the graph to find the values asked for.

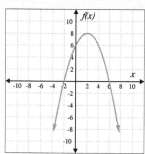

a. Find $f(0)$.

b. Find $f(5)$.

c. Find $x$ when $f(x) = 8$.

d. Find $x$ when $f(x) = 0$.

## C. Making Predictions from Models Described with Function Notation

So far, we've mostly used our favorite variables, $x$ and $y$, to name the input-output pairs of a function. When working with a real world problem, it's common to choose letters that will help us remember what the variables represent in the problem. For example, to represent a relation where "height is a function of age," we start by identifying the descriptive variables $h$ for height and $a$ for age.

| | |
|---|---|
| $h$ is $f$ of $a$ | We name the function $f$. Height is a function of age. |
| $h = f(a)$ | We use parentheses to indicate the function input. |
| $f(a)$ | The expression is read as "$f$ of $a$." |

The notation $h = f(a)$ shows us that $h$ depends on $a$. The value $a$ must be put into the function $h$ to get a result. Remember, the parentheses indicate that age is an input to the function — they do *not* indicate multiplication.

Notice that the inputs to a function do not have to be numbers. Function inputs can be names of people, labels of geometric objects, or any other element that determines some kind of output. However, most of the functions that we will work with in this book will have numbers as inputs and outputs.

### Practice A — Answers

1. $f(40) = \sqrt{40-4} = 6$ and $f(15) = \sqrt{15-4} \approx 3.317$
2. $g(3.4) = 5(3.4)^2 + (3.4) = 61.2$ and $g(-2) = 5(-2)^2 + (-2) = 5(4) - 2 = 18$ and

| | |
|---|---|
| $g(a-1) = 5(a-1)^2 + (a-1)$ | Substitute $a-1$ for $x$. |
| $= 5(a-1)(a-1) + a - 1$ | Expand. |
| $= 5(a^2 - 2a + 1) + a - 1$ | Use the distributive property to multiply the binomials. |
| $= 5a^2 - 10a + 5 + a - 1$ | Multiply by 5 according to the distributive property. |
| $= 5a^2 - 9a + 4$ | Combine like terms. |

3. In the equation, replace $g(x)$ with $-10.5$ and solve for $x$.

| | |
|---|---|
| $8 - 2.5x = -10.5$ | Substitute $-10.5$ for $g(x)$. |
| $-2.5x = -18.5$ | Subtract 8 from both sides. |
| $x = 7.4$ | Divide both sides by $-2.5$. |

### Example 7

Use function notation to represent a function whose input, $m$, is the name of a month and whose output, $d$, is the number of days in that month.

**Solution**

The number of days in a month is a function of the name of the month, so if we name the function $f$, we write:

$$\text{days} = f(\text{month}) \text{ or } d = f(m)$$

The name of the month is the input to a "rule" that yields the number of days in that month as the output. For example, $f(\text{March}) = 31$ because March has 31 days and $f(\text{June}) = 30$ because June has 30 days. The notation $d = f(m)$ reminds us that the number of days, $d$ (the output), is dependent on the name of the month, $m$ (the input).

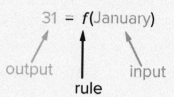

Figure 4.

### Example 8

A function, $N = f(x)$, gives $N$, the number of police officers in Salem, in $x$, a given year. What does $f(2016) = 300$ represent?

**Solution**

When we read $f(2016) = 300$, we see that the input year is 2016. The value for the output, the number of police officers $N$, is 300. Remember, $N = f(x)$. The statement $f(2016) = 300$ tells us that in the year 2016, there were 300 police officers in Salem.

Now let's combine the skills we've learned in this section and apply them to the following problem.

### Example 9

The average cost of in-state tuition at community colleges has been increasing approximately linearly over the last two decades. A function modeling this is given by $C = f(t) = 320t + 3030$, where $C$ is the cost at $t$ years since the year 2000. Given this information, do the following:

a. Find $f(0)$ and interpret its meaning in the context of the problem.

b. Find $f(20)$ and interpret its meaning in the context of the problem.

c. Find $t$ when $f(t) = 11{,}350$ and interpret the solution in the context of the problem.

**Practice B — Answers**

4. a. $h(15) = 20$; b. $h(30) = 3$; c. $t = 15$; d. $t = 5$ and $t = 25$
5. a. $f(0) = 6$; b. $f(5) = 3.5$ (This is a bit of a guess because the curve doesn't fall exactly on the intersection of grid lines when $x = 5$.); c. $x = 2$; d. $x = -2$ and $x = 6$

## Solution

a. $f(0) = 320(0) + 3030 = 3030$

The input value 0 represents years since 2000, while the output value 3030 represents cost. So our interpretation is that in the year 2000, the average cost of in-state tuition at community colleges was $3030.

b. $f(20) = 320(20) + 3030 = 9430$

We predict that in the year 2020, the average cost of in-state tuition at community colleges will be $9430.

c. Because we are given an output value and asked to find an input value, we substitute 11,350 for $f(t)$ and solve for $t$.

| | |
|---|---|
| $320t + 3030 = 11{,}350$ | Substitute 11,350 for $f(t)$. |
| $320t = 8320$ | Subtract 3030 from both sides. |
| $t = 26$ | Divide both sides by 320. |

We predict the cost of in-state tuition and fees at public, four-year universities will be $11,350 in the year 2026.

### Practice C

Now it's your turn to use function notation with the following problems. When you are done, turn the page and check your solutions.

6. Use function notation to express $w$, the weight of a pig in pounds, as a function of $d$, its age in days.

7. The function $L = f(g)$ gives the $L$, length of a hanging spring in centimeters, when $g$, a weight of grams, is added to the spring. Interpret $f(100) = 15$.

8. $P$, the pressure in pounds per square inch (psi) on a scuba diver, depends upon $d$, her depth below the water surface in feet. This relationship may be modeled by the equation, $P = f(d) = 0.434d + 14.696$. Given that information, do the following:

   a. Find and interpret $f(20)$.      b. Find $d$ when $f(d) = 35$.

## D. Finding Intercepts of a Model Using Function Notation

Let's look at vertical and horizontal intercepts of a graph of a function when its equation is given in function notation — this time with variables other than our favorite variables $x$ and $y$.

For example, if a function is described by $q = f(n)$, then $n$ is the independent variable and is described along the horizontal axis. The $n$-intercept is the point $(a, 0)$ where the function intersects the horizontal axis. The $q$-intercept is the point $(0, b)$ where the function intersects the vertical axis. Note that $a$ and $b$ represent numerical values, not variables.

It can help to sketch a graph of the function to see how this works. Figure 5 is a possible graph of $q = f(n)$.

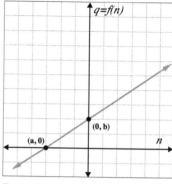

Figure 5.

### Example 10

The U.S. market share of Ford vehicles has been decreasing in recent years. Suppose that the function $M = f(t) = -1.2t + 35.3$ models this situation where $M$ is the market share in percents at $t$ years since 1990.

a. Find the $M$-intercept. Interpret its meaning in the context of the problem.

b. Find and interpret $M = f(21)$.

c. Find the $t$-intercept. Interpret its meaning in the context of the problem.

**Solution**

a. The $M$-intercept is the vertical intercept and occurs when $t = 0$.

$$f(0) = -1.2(0) + 35.3 = 35.3$$

This means that in 1990, Ford had a U.S. market share of 35.3%. In other words, 35.3% of the vehicles sold in the U.S. in 1990 were Fords.

b. $f(21) = -1.2(21) + 35.3 = 10.1$

In 2011, 21 years after 1990, Ford's U.S. market share was only 10.1%.

c. The $t$-intercept is the horizontal intercept and occurs when $M = 0$. In the equation replace $f(t)$ with 0 and solve for $t$.

$-1.2t + 35.3 = 0$  Substitute 0 for $f(t)$.

$\underline{\phantom{-1.2t}\; -35.3 \quad -35.3}$  Subtract 35.3 from both sides.

$-1.2t = -35.3$

$\frac{-1.2t}{-1.2} = \frac{-35.3}{-1.2}$  Divide both sides by $-1.2$.

$t \approx 29.4$

We predict that in 2019, 29 years after 1990, Ford's market share will decrease to 0%. However, this solution is most likely an example of **model breakdown**. In all likelihood, Ford will still be making and selling vehicles in the U.S. in 2019.

A graphical representation of a function helps us visualize the relationship between the variables, and that visualization helps us interpret applications.

Figure 7 gives us the graph of the linear function representing Ford's market share over time. The solutions to Example 10 are labeled on the graph. The $m$-intercept shows that in 1990 ($t = 0$) Ford's market share was 35.3%. By the year 2011, it had decreased to 10.1%. If the trend continues, it looks like Ford's market share will be zero in 2019, which is likely a breakdown of the applicability of this linear model.

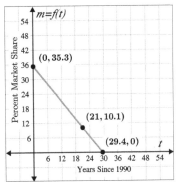

Figure 7. The graph of y = -1.2x + 35.3

### Practice D

Eduardo and Hector plan to hike a section of the Pacific Crest Trail at a pace of about 15 miles per day. The function $R = f(d) = -15d + 212$ models the number of miles *remaining* after $d$ days on the trail. Given this information, do the following. Then turn the page to check your solutions.

9. Find and interpret $f(6)$.

10. Find and interpret the $R$-intercept.

11. Find and interpret the $d$-intercept.

### Practice C — Answers

6. $w = f(d)$

7. When a 100 gram weight is added to the spring, the length of the spring will be 15 centimeters.

8. a. $P = f(20) = 0.434(20) + 14.696 = 23.376$; When the diver is 20 feet below the water surface there are 23.376 psi of pressure on the diver.
   b. In the equation replace $f(d)$ with 35 and solve for $d$.

$$0.434d + 14.696 = 35$$
$$-14.696 \quad -14.696$$
$$0.434d = 20.304$$
$$\frac{0.434d}{0.434} = \frac{20.304}{0.434}$$
$$t \approx 46.783$$

We predict the pressure on the diver will be 35 psi when she is approximately 46.783 feet below the surface.

# Exercises 1.5

For Exercises, 1 – 4 evaluate the function $f$ at the indicated values $f(-3), f(2),$ and $f(a + 3)$.

1. $f(x) = 2x - 5$
2. $f(x) = \sqrt{2-x} + 5$
3. $f(x) = 3x^2 + 2x - 1$
4. $f(x) = \frac{6x-1}{5x+2}$

5. Given the function $k(t) = 2t - 1$:
   a. Evaluate $k(2)$.
   b. Find $t$ when $k(t) = 7$.

6. Given the function $f(x) = 8 - 3x$:
   a. Evaluate $f(-2)$.
   b. Find $x$ when $f(x) = -1$.

7. Given the function $p(c) = c^2 + 6$:
   a. Evaluate $p(-3)$.
   b. Find $c$ when $p(c) = 31$.

8. Given the function $f(x) = 3 - 3x^2$:
   a. Evaluate $f(5)$.
   b. Find $x$ when $f(x) = 0$.

9. Given the function $f(x) = \sqrt{x} + 2$:
   a. Evaluate $f(9)$.
   b. Find $x$ when $f(x) = 4$.

10. Given the function $r(x) = \frac{x^3}{2}$:
    a. Evaluate $r(-1)$.
    b. Find $x$ when $r(x) = 32$.

11. Given the following graph:
    a. Evaluate $f(-1)$.
    b. Find $x$ when $f(x) = 3$.

    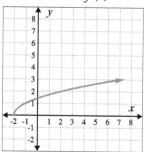

12. Given the following graph:
    a. Evaluate $f(0)$.
    b. Find $x$ when $f(x) = -3$.

    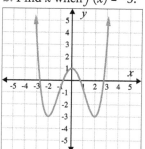

13. Given the following graph:
    a. Evaluate $f(4)$.
    b. Find $x$ when $f(x) = 1$.

    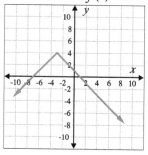

14. Given the following graph:
    a. Evaluate $f(6)$.
    b. Find $x$ when $f(x) = 21$.

    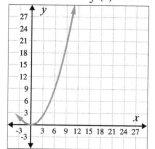

## Practice D — Answers

9. $f(6) = -15(6) + 212 = 122$
   This means that after 6 days, Eduardo and Hector will have 122 miles remaining for their trip.

10. The $R$-intercept, or vertical intercept occurs where $d = 0$.
    $f(0) = -15(0) + 212 = 212$
    This means that at the beginning of the hike, they have 212 miles ahead of them.

11. Replace $R$ with 0 and solve for $d$.
    $-15d + 212 = 0 \rightarrow d = 14.333$.
    This means that after a little more than 14 days, Eduardo and Hector will be done with the hike. There will be zero miles remaining.

For the following exercises, use the function $f$ represented in each table.

15.

| $x$ | 0 | 1 | 2 | 3 | 4 | 5 | 6 | 7 | 8 | 9 |
|---|---|---|---|---|---|---|---|---|---|---|
| $f(x)$ | 74 | 28 | 1 | 53 | 56 | 3 | 36 | 45 | 14 | 47 |

a. Evaluate $f(3)$.
b. Find $x$ when $f(x) = 1$.

16.

| $x$ | 0 | 1 | 2 | 3 | 4 | 5 | 6 | 7 | 8 | 9 |
|---|---|---|---|---|---|---|---|---|---|---|
| $f(x)$ | 0 | 1 | 1 | 2 | 3 | 5 | 8 | 13 | 21 | 34 |

a. Evaluate $f(8)$
b. Find $x$ when $f(x) = 1$

For the following functions, find $f(-2)$, $f(-1)$, $f(0)$, and $f(1)$.

17. $f(x) = 4 - 2x$
18. $f(x) = 8 - 3x$
19. $f(x) = x^2 - 7x + 3$
20. $f(x) = 3 + \sqrt{x+3}$
21. $f(x) = \frac{x-2}{x+3}$
22. $f(x) = x^3$

23. $G$, the amount of garbage produced by a city with population $p$, is given by $G = f(p)$. $G$ is measured in tons per week, and $p$ is measured in thousands of people.

   a. The town of Tola has a population of 40,000 and produces 13 tons of garbage each week. Use function notation to express this information in terms of the function $f$.
   b. Explain the meaning of the statement $f(5) = 2$.

24. $D$, the number of cubic yards of dirt needed to cover a garden with area $a$ square feet, is given by $D = g(a)$.

   a. A garden with area 5000 square feet requires 50 cubic yards of dirt. Use function notation to express this information in terms of the function $g$.
   b. Explain the meaning of the statement $g(100) = 1$.

25. Let $f(t)$ be the number of ducks in a lake $t$ years after 1990. Explain the meaning of each statement.
   a. $f(5) = 30$
   b. $f(10) = 40$

26. Let $h(t)$ be the height above ground in feet of a rocket $t$ seconds after launching. Explain the meaning of each statement.
   a. $h(1) = 200$
   b. $h(2) = 350$

27. The weight of a newborn baby, Arlo, is 7.5 pounds. Arlo gains 1.5 pounds per month during his first year. Given this scenario, do the following:
   a. Use function notation to write an equation for a function $f$ that models $W$, Arlo's weight in pounds, after $t$ months.
   b. Find and interpret the $W$-intercept of the graph of the function.
   c. Find and interpret the $t$-intercept of the graph of the function.
   d. Find and interpret $f(6)$.
   e. Predict when Arlo will weigh 24 pounds.

28. At a local community college, the number of students afflicted with the common cold during winter term has steadily decreased by about 150 students per year since the year 2000. In 2000, 6300 students had a cold during winter term. Given this scenario, do the following:
   a. Use function notation to write an equation for a function $f$ that models $C$, the number of students who had a cold, $t$ years since 2000.
   b. Find and interpret the $C$-intercept of the graph of the function.
   c. Find and interpret the $t$-intercept of the graph of the function.
   d. Find and interpret $f(16)$.
   e. Assuming this trend continues, predict when the number of students who catch a cold during winter term will be 2700.

# CHAPTER 2
# Exponential Functions

Focus in on a square centimeter of your skin. Look closer. Closer still. If you could look closely enough, you would see hundreds of thousands of microscopic organisms. They are bacteria, and they're not only on your skin, but in your mouth, nose, and even your intestines. In fact, the bacterial cells in your body at any given moment outnumber your own cells. However, while some bacteria can cause illness, many are healthy and even essential to the body.

Bacteria commonly reproduce through a process called binary fission, during which one bacterial cell splits into two. When conditions are right, bacteria can reproduce very quickly. Unlike humans and other complex organisms, the time required to form a new generation of bacteria is often a matter of minutes or hours, as opposed to days or years.

For simplicity's sake, suppose we begin with a culture of one bacterial cell that can divide every hour. The table below shows the number of bacterial cells at the end of each subsequent hour. We see that the single bacterial cell leads to over one thousand bacterial cells in just ten hours! And if we were to extrapolate the table to twenty-four hours, we would have over 16 million cells!

| *Hour* | 0 | 1 | 2 | 3 | 4 | 5 | 6 | 7 | 8 | 9 | 10 |
|---|---|---|---|---|---|---|---|---|---|---|---|
| *Bacteria* | 1 | 2 | 4 | 8 | 16 | 32 | 64 | 128 | 256 | 512 | 1024 |

In this chapter, we will explore exponential functions, which can be used for modeling growth patterns such as those found in bacteria. Exponential functions have numerous other real-world applications in the fields of finance, archeology, and medicine to name a few. Here are the topics we'll cover in this chapter:

2.1  Properties of Exponents ................................................................. page 80

2.2 Rational Exponents ....................................................................... page 98

2.3 Exponential Functions .................................................................. page 105

2.4 Finding Equations of Exponential Functions ................................. page 119

2.5 Using Exponential Functions to Model Data ................................. page 127

# 2.1 Properties of Exponents

## Overview

Mathematicians, scientists, and economists commonly encounter very large and very small numbers, but it may not be obvious how common such figures are in everyday life. For example, a digital video camera records many tiny picture elements, called pixels, to form a single image, or frame. A camera might record frames in 2,048 pixels by 1,536 pixels, and each pixel might contain 48 bits of information pertaining to the color that pixel should represent. If the camera shoots video at 24 frames per second, the number of bits of information in a one-hour video is extremely large.

We can compute the maximum number of bits of information in our example above. Using a calculator, we enter 2048 × 1536 × 48 × 24 × 3600 and press [ENTER]. The calculator displays 1.304596316E13. What does this mean? The "E13" portion of the result represents the exponent 13 of ten, so there are a maximum of approximately $1.3 \times 10^{13}$ bits of data in that one-hour film.

In this section, we review rules of exponents first and then apply them to calculations involving very large or small numbers. As you study this section, you will learn how to:

- Use the five basic properties of exponents
- Use the zero exponent rule of exponents
- Use negative integer exponents
- Simplify exponential expressions
- Use scientific notation

## A. Definition of an Exponent

We use exponents as a kind of shorthand for repeated multiplication. If we multiply the number 5 times itself 6 times, we can write $5 \cdot 5 \cdot 5 \cdot 5 \cdot 5 \cdot 5$, but it's much shorter to write $5^6$. In this example, we call 5 the **base**, and we call 6 the **exponent**. When written as $5^6$, the meaning is that 5 is a factor 6 times. We call $5^6$ a **power** of 5, namely the sixth power of 5.

In general, for any natural number $n$, the notation $b^n$ indicates that $b$ is a factor $n$ times.

### Exponents

For any natural number $n$,
$$b^n = b \cdot b \cdot b \cdot \ldots \cdot b \text{ where } b \text{ is a factor } n \text{ times.}$$
We refer to $b^n$ as "$b$ raised to the $n$th power." We call $b$ the **base** and $n$ the **exponent**.

For most powers of a base, we use our calculators to find the value. For $5^6$, type the base 5 and then use the exponent button ⌐^⌐ before typing 6. In Figure 1, we see that $5^6 = 15625$.

Notice that $b^1$ stands for one factor of $b$, so $b^1 = b$. We typically don't write the exponent in this case. Two other powers have specific names. We refer to $b^2$ as "$b$ squared" and $b^3$ as "$b$ cubed." These nicknames come from geometry formulas for the area of a square and the volume of a cube.

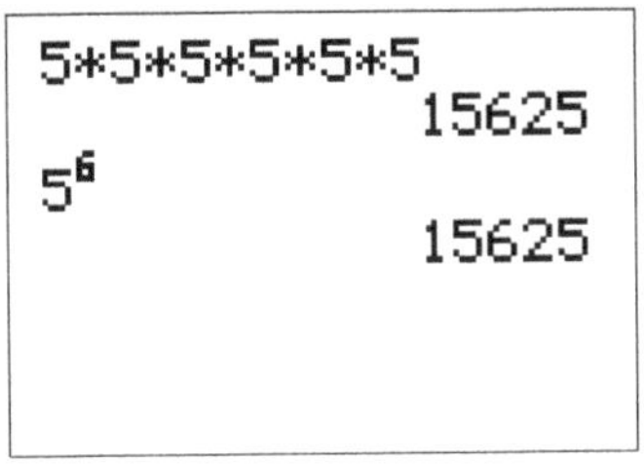

Figure 1. Calculating powers on the calculator.

It's common to confuse the two expressions $-3^4$ and $(-3)^4$. The first expression indicates that we should raise 3 to the 4th power and then find its opposite.

$$-3^4 = -(3^4) = -(3 \cdot 3 \cdot 3 \cdot 3) = -81$$

While the second expression indicates that we should raise $-3$ to the 4th power.

$$(-3)^4 = (-3)(-3)(-3)(-3) = 81$$

We can use a calculator to check both computations, as you see in Figure 2.

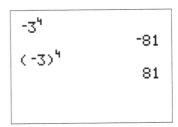

Figure 2. Compute $-3^4$ and $(-3)^4$.

## B. Properties of Exponents

### The Product Property of Exponents

Consider the product $x^3 \cdot x^4$. Both terms have the same base, $x$, but they are raised to different exponents. We expand each expression, and then rewrite the resulting expression.

$$x^3 \cdot x^4 = \underbrace{x \cdot x \cdot x}_{\text{3 factors}} \cdot \underbrace{x \cdot x \cdot x \cdot x}_{\text{4 factors}}$$

$$= \underbrace{x \cdot x \cdot x \cdot x \cdot x \cdot x \cdot x}_{\text{7 factors}}$$

$$= x^7$$

The result is that $x^3 \cdot x^4$ equals $x^{3+4}$, which equals $x^7$.

Notice that the exponent of the product is the sum of the exponents of the factors. In other words, when multiplying exponential expressions with the same base, we write the result with the common base and add the exponents. This is the product property of exponents and can be stated in general terms as

$$b^m \cdot b^n = b^{m+n}$$

## The Quotient Property of Exponents

In a similar way to the product property, we can simplify an expression such as $\frac{y^m}{y^n}$, where $m > n$. Consider the example $\frac{y^9}{y^5}$. Perform the division by canceling common factors.

$$\frac{y^9}{y^5} = \frac{y \cdot y \cdot y \cdot y \cdot y \cdot y \cdot y \cdot y \cdot y}{y \cdot y \cdot y \cdot y \cdot y} \qquad \text{Definition of exponent.}$$

$$= \frac{\cancel{y} \cdot \cancel{y} \cdot \cancel{y} \cdot \cancel{y} \cdot \cancel{y} \cdot y \cdot y \cdot y \cdot y}{\cancel{y} \cdot \cancel{y} \cdot \cancel{y} \cdot \cancel{y} \cdot \cancel{y}} \qquad \text{Divide: } \frac{y}{y} = 1$$

$$= \frac{y \cdot y \cdot y \cdot y}{1} \qquad \text{Simplify.}$$

$$= y^4$$

Notice that the exponent of the quotient is the difference between the exponents of the divisor and dividend. So the shortcut would look like this:

$$\frac{y^9}{y^5} = y^{9-5} = y^4$$

In other words, when dividing exponential expressions with the same base, we write the result with the common base and subtract the exponents. This is the quotient property of exponents, which can be stated in general terms as

$$\frac{b^m}{b^n} = b^{m-n}, \text{ where } m > n \text{ and } b \neq 0$$

For the time being, we must be aware of the condition $m > n$. Otherwise, the difference $m - n$ could be zero or negative. We'll explore those possibilities soon.

There are three other basic properties of exponents. These rules can be demonstrated by expanding and simplifying each expression as we did with the product and quotient properties. The five basic properties of exponents are listed below.

### Five Properties of Exponents

For any real numbers $b$ and $c$ and natural numbers $m$ and $n$,

| | |
|---|---|
| $b^m \cdot b^n = b^{m+n}$ | Product property of exponents |
| $\frac{b^m}{b^n} = b^{m-n}, b \neq 0, m > n$ | Quotient property of exponents |
| $(bc)^n = b^n c^n$ | Raising a product to a power |
| $\left(\frac{b}{c}\right)^n = \frac{b^n}{c^n}, c \neq 0$ | Raising a quotient to a power |
| $(b^m)^n = b^{mn}$ | Raising a power to a power |

Section 2.1: Properties of Exponents

## Example 1

Use the product and quotient properties to simplify the following expressions.

1. $b^5 \cdot b^3$
2. $(-2)^5 \cdot (-2)$
3. $x^2 \cdot x^5 \cdot x^3$
4. $\dfrac{b^{18}}{b^{11}}$
5. $\dfrac{7^{11}}{7^9}$
6. $\dfrac{24\,a^6}{3\,a^5}$

### Solutions

1. $b^5 \cdot b^3 = b^{3+5} = b^8$

2. $(-2)^5 \cdot (-2) = (-2)^5 (-2)^1 = (-2)^6 = 64$

We can extend the product property of exponents to simplify a product of three factors having the same base. Simply add the exponents of all three factors:

3. $x^2 \cdot x^5 \cdot x^3 = x^{2+5+3} = x^{10}$

4. $\dfrac{b^{18}}{b^{11}} = b^{18-11} = b^7$

5. $\dfrac{7^{11}}{7^9} = 7^{11-9} = 7^2 = 49$

6. $\dfrac{24\,a^6}{3\,a^5} = \dfrac{24}{3} a^{6-5} = 8a$

We can use all five properties of exponents in a variety of combinations to simplify more complex expressions. At first, it's wise to proceed methodically using one property at a time. With practice, you may find that you're able to combine some of the steps.

## Example 2

Simplify the following expressions.

1. $(2\,b^3\,c^6)^4$
2. $(5\,b\,c^5)(9\,b^4\,c^7)$
3. $\dfrac{4\,b^8\,c^9}{20\,b^7\,c^5}$
4. $\left(\dfrac{3\,b^5\,c^4}{b^3\,c\,d^{10}}\right)^3$

### Solutions

1. $(2\,b^3\,c^6)^4$

$\begin{aligned}(2\,b^3\,c^6)^4 &= 2^4(b^3)^4(c^6)^4 \\ &= 16\,b^{12}\,c^{24}\end{aligned}$   Raise factors to the fourth power: $(bc)^n = b^n c^n$
Simplify the numerical part. Multiply exponents:
$(b^m)^n = b^{mn}$

2. $(5\,b\,c^5)(9\,b^4\,c^7)$

$(5\,b\,c^5)(9\,b^4\,c^7) = (5 \cdot 9)(b\,b^4)(c^5\,c^7)$   Rearrange factors.

$\qquad\qquad\qquad = 45\,b^5\,c^{12}$   Simplify the numerical part. Add exponents: $b^m\,b^n = b^{m+n}$

3. $\dfrac{4b^8 c^9}{20 b^7 c^5}$

$\dfrac{4b^8 c^9}{20 b^7 c^5} = \dfrac{1}{5} b^1 c^4$  Simplify numeric fraction. Subtract exponents: $\dfrac{b^m}{b^n} = b^{m-n}$

$= \dfrac{bc^4}{5}$  Simplify.

4. $\left( \dfrac{3 b^5 c^4}{b^3 c d^{10}} \right)^3$

$\left( \dfrac{3 b^5 c^4}{b^3 c d^{10}} \right)^3 = \left( \dfrac{3 b^2 c^3}{d^{10}} \right)^3$  Subtract exponents: $\dfrac{b^m}{b^n} = b^{m-n}$

$= \dfrac{(3 b^2 c^3)^3}{(d^{10})^3}$  Raise numerator and denominator to the third power: $\left( \dfrac{b}{c} \right)^n = \dfrac{b^n}{c^n}$

$= \dfrac{3^3 (b^2)^3 (c^3)^3}{(d^{10})^3}$  Raise factors to the third power: $(bc)^n = b^n c^n$

$= \dfrac{27 b^6 c^9}{d^{30}}$  Multiply exponents: $(b^m)^n = b^{mn}$

## Practice B

Simplify the following expressions. Turn the page to check your work.

1. $t^3 \cdot t^6 \cdot t^5$

2. $(6 a^6 b^3)(4 a^9 b^2)$

3. $\dfrac{33 a^9 b^5}{3 a^2 b^2}$

4. $(4 x^2 y^7)^5$

5. $\left( \dfrac{2 x^3 y^{13}}{10 x y^{10}} \right)^4$

## C. Zero Exponents and Negative Exponents

### The Zero Exponent Rule of Exponents

Let's return to the quotient rule. We started with the condition $m > n$ so that the difference $m - n$ would not be zero or negative. What would happen if $m = n$? Let's simplify the expression $\dfrac{b^8}{b^8}$ as an example. If we simplify this expression using the quotient property, we have this:

$\dfrac{b^8}{b^8} = b^{8-8} = b^0$

Let's view the expression $\dfrac{b^8}{b^8}$ in a different way. Remember that any number divided by itself is equal to 1. That means:

$\dfrac{b^8}{b^8} = 1$

If we equate the two answers, the result is $b^0 = 1$. This is true for any nonzero real number or any variable representing such a number. In other words, any nonzero base raised to the zero power equals 1.

## The Zero Exponent Rule

For any nonzero real number $b$, $b^0 = 1$.

### Example 3

Use the zero exponent rule to simplify each expression.

1. $\dfrac{a^0 b^4 c^5}{c^5}$
2. $(7a^3 b^0)^2$
3. $9 + a^0 + b^0$

**Solution**

1. $\dfrac{a^0 b^4 c^5}{c^5} = a^0 b^4 c^0 = (1) b^4 (1) = b^4$

   Recall that any number times 1 equals itself ($1a = a$) and that we do not need to write the number 1 when it is a coefficient.

2. $(7a^3 b^0)^2 = (7a^3 \cdot 1)^2 = (7a^3)^2 = 49 a^6$

   Notice in the last step we are squaring the 7 as well as the variable term because it is in parentheses.

3. $9 + a^0 + b^0 = 9 + 1 + 1 = 11$

   Be careful! Adding 1 to a term is not the same as multiplying by 1.

### The Negative Integer Rule of Exponents

Another useful result occurs if we relax the condition that $m > n$ in the quotient property even further and let $m < n$. For example, let's simplify $\dfrac{b^3}{b^5}$. Begin by using the definition of an exponent.

$$\dfrac{b^3}{b^5} = \dfrac{b \cdot b \cdot b}{b \cdot b \cdot b \cdot b \cdot b} \qquad \text{Definition of exponent.}$$

$$= \dfrac{\cancel{b} \cdot \cancel{b} \cdot \cancel{b}}{\cancel{b} \cdot \cancel{b} \cdot \cancel{b} \cdot b \cdot b} \qquad \text{Divide: } \dfrac{b}{b} = 1$$

$$= \dfrac{1}{b \cdot b} \qquad \text{Simplify.}$$

$$= \dfrac{1}{b^2}$$

If we simplify the original expression using the quotient property, we have

$$\dfrac{b^3}{b^5} = b^{3-5} = b^{-2}$$

Putting the answers together, we have $b^{-2} = \dfrac{1}{b^2}$. This relationship is true for any nonzero base raised to an integer exponent. That is to say, for any nonzero base raised to a negative exponent we can write its reciprocal and change the sign of the exponent. Here is the general rule:

$$b^{-n} = \dfrac{1}{b^n}$$

Here are a couple more examples of this rule:

$b^{-6} = \frac{1}{b^6}$ and $2^{-3} = \frac{1}{2^3} = \frac{1}{8}$

For the quotient $\frac{b^m}{b^n}$, when $m < n$, we can use the negative integer rule of exponents to simplify the expression to its reciprocal because the difference $m - n$ is negative. Here's another example:

$$\frac{b^6}{b^{11}} = b^{-5} = \frac{1}{b^5}$$

Next, we start with an expression that has a negative exponent on a nonzero base in the denominator of a fraction, such as $\frac{1}{b^{-4}}$. We can rewrite this expression as follows:

$\frac{1}{b^{-4}} = 1 \div b^{-4}$  A fraction represents division.

$\phantom{\frac{1}{b^{-4}}} = 1 \div \frac{1}{b^4}$  Rewrite so the exponent is positive.

$\phantom{\frac{1}{b^{-4}}} = 1 \cdot \frac{b^4}{1}$  Dividing by a fraction is equivalent to multiplying by its reciprocal.

$\phantom{\frac{1}{b^{-4}}} = b^4$  Simplify.

This equivalency applies to all similar expressions, so for any nonzero base raised to a negative exponent, we can rewrite the expression with a positive exponent by writing its reciprocal and changing the sign of the exponent. This is the general rule:

$$\frac{1}{b^{-n}} = b^n$$

### Rules for Negative Integer Exponents

For any nonzero real number $b$ and natural number $n$,

$$b^{-n} = \frac{1}{b^n} \quad \text{and} \quad \frac{1}{b^{-n}} = b^n.$$

Students often find it easiest to think of the rules for negative exponents this way: A factor with a negative exponent becomes the same factor with a positive exponent if we move it across a fraction bar — from numerator to denominator or vice versa. Here's an example that combines these rules in one step to simplify an expression:

$$\frac{a^{-8}}{b^{-5}} = \frac{b^5}{a^8}$$

### Practice B — Answers

1. $t^3 \cdot t^6 \cdot t^5 = t^{14}$

2. $(6 a^6 b^3)(4 a^9 b^2) = (6 \cdot 4) a^{6+9} b^{3+2} = 24 a^{15} b^5$

3. $\frac{33 a^9 b^5}{3 a^2 b^2} = \frac{33}{3} a^{9-2} b^{5-2} = 11 a^7 b^3$

4. $(4 x^2 y^7)^5 = 4^5 (x^2)^5 (y^7)^5 = 1024 x^{10} y^{35}$

5. $\left(\frac{2 x^3 y^{13}}{10 x y^{10}}\right)^4 = \left(\frac{1 x^2 y^3}{5}\right)^4 = \frac{(x^2)^4 (y^3)^4}{5^4} = \frac{x^8 y^{12}}{625}$

## Example 4

Simplify and write each expression with positive exponents.

1. $\dfrac{c^3}{c^{10}}$
2. $b^2 \cdot b^{-8}$
3. $\dfrac{4a^{-5}}{b^{-9}}$
4. $\dfrac{-3a^8 b^{-7}}{c^{-1}}$

### Solutions

1. We use the quotient property in the first step and the negative exponent rule in the last step.

$$\dfrac{c^3}{c^{10}} = c^{3-10} = c^{-7} = \dfrac{1}{c^7}$$

2. We use the product property in the first step and the negative exponent rule in the last step.

$$b^2 \cdot b^{-8} = b^{2+8} = b^{-6} = \dfrac{1}{b^6}$$

3. A factor with a negative exponent becomes the same factor with a positive exponent if it's moved across the fraction bar.

$$\dfrac{4a^{-5}}{b^{-9}} = \dfrac{4b^9}{a^5}$$

4. Be careful! The coefficient, −3, is a negative *number*, not a negative exponent.

$$\dfrac{-3a^8 b^{-7}}{c^{-1}} = \dfrac{-3a^8 c}{b^7}$$

## Practice C

Simplify each expression and rewrite with positive exponents. Turn the page to check your work.

6. $(a^3 b c^0)(b^9 c^2)$
7. $a^0 b^8 + b^0$
8. $\dfrac{2b^4 c^3}{b^7 c^3}$
9. $\dfrac{-9a^{13} c^5}{a^9 c^{12}}$
10. $\dfrac{b^{-9} c^2}{a^7 d^{-4}}$

## D. Simplifying Expressions Involving Exponents

We have seen that the exponential expression $b^n$ is defined when $n$ is a natural number, 0, or the negative of a natural number. That means that $b^n$ is defined for any integer $n$, and the five basic exponent properties introduced at the beginning of this section hold for any integer $n$. Furthermore, we can combine any of these properties with the rules for zero and negative exponents to simplify complex expressions.

### Simplifying Expressions Involving Exponents

An expression involving exponents is simplified if

- It includes no parentheses.
- Each variable or constant appears as a base as few times as possible.
- Each numerical expression has been calculated, and each numerical fraction has been written in lowest terms.
- Each exponent is positive.

### Example 5

Simplify each expression.

1. $(6m^2n^{-1})^3$
2. $17^5 \cdot 17^{-4} \cdot 17^{-3}$
3. $\left(\dfrac{u^{-1}v}{v^{-1}}\right)^2$
4. $(-2a^3b^{-1})(5a^{-2}b^2)$
5. $(x^2y^3)^4(x^2y^3)^{-4}$
6. $\dfrac{(3w^2)^5}{(6w^{-2})^2}$

**Solutions**

1. $(6m^2n^{-1})^3$

$$\begin{aligned}(6m^2n^{-1})^3 &= (6)^3(m^2)^3(n^{-1})^3 &&\text{Raise a product to a power.}\\ &= 6^3 m^{2\cdot3} n^{-1\cdot3} &&\text{Raise a power to a power.}\\ &= 216 m^6 n^{-3} &&\text{Simplify.}\\ &= \dfrac{216 m^6}{n^3} &&\text{Apply the negative exponent rule.}\end{aligned}$$

### Practice C — Answers

6. $(a^3bc^0)(b^9c^2) = a^3b^{10}c^2$

7. $a^0b^8 + b^0 = b^8 + 1$

8. $\dfrac{2b^4c^3}{b^7c^3} = \dfrac{2}{b^3}$

9. $\dfrac{-9a^{13}c^5}{a^9c^{12}} = \dfrac{-9a^4}{c^7}$

10. $\dfrac{b^{-9}c^2}{a^7d^{-4}} = \dfrac{c^2d^4}{a^7b^9}$

2. $17^5 \cdot 17^{-4} \cdot 17^{-3}$

$$\begin{aligned}17^5 \cdot 17^{-4} \cdot 17^{-3} &= 17^{5-4-3} &&\text{Apply the product property.}\\ &= 17^{-2} &&\text{Simplify.}\\ &= \frac{1}{17^2} &&\text{Apply the negative exponent rule.}\\ &= \frac{1}{289} &&\text{Simplify.}\end{aligned}$$

3. $\left(\dfrac{u^{-1}v}{v^{-1}}\right)^2$

$$\begin{aligned}\left(\frac{u^{-1}v}{v^{-1}}\right)^2 &= \frac{(u^{-1}v)^2}{(v^{-1})^2} &&\text{Raise a quotient to a power.}\\ &= \frac{u^{-2}v^2}{v^{-2}} &&\text{Raise a product to a power.}\\ &= u^{-2}v^{2-(-2)} &&\text{Apply the quotient property.}\\ &= u^{-2}v^4 &&\text{Simplify.}\\ &= \frac{v^4}{u^2} &&\text{Apply the negative exponent rule.}\end{aligned}$$

4. $(-2a^3b^{-1})(5a^{-2}b^2)$

$$\begin{aligned}(-2a^3b^{-1})(5a^{-2}b^2) &= -2 \cdot 5 \cdot a^3 \cdot a^{-2} \cdot b^{-1} \cdot b^2 &&\text{Rearrange terms.}\\ &= -10 \cdot a^{3-2} \cdot b^{-1+2} &&\text{Apply the product property.}\\ &= -10ab &&\text{Simplify.}\end{aligned}$$

5. $(x^2y^3)^4(x^2y^3)^{-4}$

$$\begin{aligned}(x^2y^3)^4(x^2y^3)^{-4} &= (x^2y^3)^{4-4} &&\text{Apply the product property.}\\ &= (x^2y^3)^0 &&\text{Simplify.}\\ &= 1 &&\text{Apply the zero exponent rule.}\end{aligned}$$

6. $\dfrac{(3w^2)^5}{(6w^{-2})^2}$

$$\begin{aligned}\frac{(3w^2)^5}{(6w^{-2})^2} &= \frac{(3)^5 \cdot (w^2)^5}{(6)^2 \cdot (w^{-2})^2} &&\text{Raise a product to a power.}\\ &= \frac{3^5 w^{2 \cdot 5}}{6^2 w^{-2 \cdot 2}} &&\text{Raise a power to a power.}\\ &= \frac{243 w^{10}}{36 w^{-4}} &&\text{Simplify.}\\ &= \frac{27 w^{10-(-4)}}{4} &&\text{Apply the quotient property. Reduce the fraction.}\\ &= \frac{27 w^{14}}{4} &&\text{Simplify.}\end{aligned}$$

## Practice D

Simplify each expression. Turn the page to check your work.

11. $(2uv^{-2})^{-3}$

12. $x^8 \cdot x^{-12} \cdot x$

13. $\left(\dfrac{a^2 b^{-3}}{b^{-1}}\right)^2$

14. $(9r^{-5}s^3)(3r^6 s^{-4})$

15. $\left(\dfrac{4}{9}tw^{-2}\right)^{-3}\left(\dfrac{4}{9}tw^{-2}\right)^3$

16. $\dfrac{(2h^2 k)^4}{(7h^{-1}k^2)^2}$

## E. Scientific Notation

We use exponents on a base of ten to describe numbers in **scientific notation**. Scientific notation is a shorthand form of writing numbers whose absolute values are very large or very small. It is commonly used by scientists, mathematicians, and engineers in part because it can simplify certain arithmetic operations with numbers that are too big or too small to be conveniently written in standard decimal form.

At the beginning of this section, for example, we calculated that there can be as many as 13,046,000,000,000 of bits of information in a digital image. Using scientific notation, we write the number like this:

$1.3046 \times 10^{13}$

An example of a very small number is the mass of a dust particle, about 0.000000000753 kg. We write the number in scientific notation like this:

$7.53 \times 10^{-10}$

Here are a few more examples of numbers written in scientific notation:

The speed of light is approximately 300,000,000 meters per second: $3.0 \times 10^8$

The width of a human hair is about 0.00005 m: $5.0 \times 10^{-5}$

The radius of an electron is nearly 0.00000000000047 m: $4.7 \times 10^{-13}$

The mass of the sun is roughly 1,988,000,000,000,000,000,000,000,000,000 kg: $1.988 \times 10^{30}$

This last example gives us a good idea of why we might want a way to describe a very large number without having to write all those zeros!

In scientific notation all numbers are written in the form $a \times 10^n$, where the exponent $n$ is an integer and $a$ is a real number whose absolute value is greater than or equal to 1 but less than 10.

## Scientific Notation

A number is written in **scientific notation** if it is written in the form $a \times 10^n$, where $1 \leq |a| < 10$ and $n$ is an integer.

## Converting from Standard Notation to Scientific Notation

To write a number in scientific notation, we need to move the decimal point to the right of the first non-zero digit in the number so that we write the digits as a decimal number between 1 and 10. Count the number of places, $n$, that you move the decimal point. You then multiply the decimal number by 10 raised to a power of $n$. The sign of $n$ (positive or negative) depends on whether you moved the decimal point of the number to the right or to the left.

For example, consider the number 2,780,000. This is a whole number so the decimal point is understood to be to the right of the digit in the ones column:

2,780,000 = 2,780,000.0

We move the decimal point until it is to the right of the first nonzero digit. We get 2.78, which is between 1 and 10.

2,780,000.0 → 2.780000
           6 places

In scientific notation, we have $2.78 \times 10^n$. We move the decimal point of 2.78 six places to the right to get 2,780,000, so $n = 6$, and the scientific notation is:

$2.78 \times 10^6$

Now let's see how to write a very small number in scientific notation. An example we can use is the radius of an electron, 0.00000000000047 m. We move the decimal point until it's to the right of the first nonzero digit. We get 4.7, which is between 1 and 10.

0.00000000000047 → 0000000000004.7
                   13 places

In scientific notation, we have $4.7 \times 10^n$. We move the decimal point of 4.7 thirteen places to the left to get 0.00000000000047, so $n = -13$, and the scientific notation is:

$4.7 \times 10^{-13}$

## Example 6

Write each number in scientific notation.

1. Distance to Andromeda Galaxy from Earth: 24,000,000,000,000,000,000,000m
2. Diameter of Andromeda Galaxy: 1,300,000,000,000,000,000,000m
3. Number of stars in Andromeda Galaxy: 1,000,000,000,000
4. Diameter of electron: 0.00000000000094m
5. Probability of being struck by lightning in any single year: 0.00000143

**Solutions**

1. Move the decimal point 22 places:

    $24{,}000{,}000{,}000{,}000{,}000{,}000{,}000 = 2.4 \times 10^{22}$

2. Move the decimal point 21 places:

    $1{,}300{,}000{,}000{,}000{,}000{,}000{,}000 = 1.3 \times 10^{21}$

3. Move the decimal point 12 places:

    $1{,}000{,}000{,}000{,}000 = 1 \times 10^{12}$

4. Move the decimal point 13 places:

    $0.00000000000094 = 9.4 \times 10^{-13}$

5. Move the decimal point 6 places:

    $0.00000143 = 1.43 \times 10^{-6}$

Notice that if a number is greater than 1, then when it is written is in scientific notation, the exponent of 10 is positive. We see this in numbers 1–3 of Example 6. If a number is less than 1, then when it is written in scientific notation, the exponent of 10 is negative. We see this in numbers 4 and 5 of Example 6.

## Converting from Scientific Notation to Standard Notation

We can convert a number in scientific notation, $a \times 10^n$, to standard notation. If the exponent $n$ is positive, we multiply $a$ by 10 $n$ times, so we move the decimal point $n$ places to the right. If $n$ is negative, we divide $a$ by 10 $n$ times, so we move the decimal $n$ places to the left. This is demonstrated in Example 7.

### Example 7

Convert each number in scientific notation to standard notation.

1. $3.547 \times 10^{14}$
2. $-2 \times 10^{6}$
3. $7.91 \times 10^{-7}$
4. $-8.05 \times 10^{-12}$

### Practice D — Answers

11. $\dfrac{v^6}{8u^3}$
12. $\dfrac{1}{x^3}$
13. $\dfrac{a^4}{b^4}$
14. $\dfrac{27r}{s}$
15. $1$
16. $\dfrac{16h^{10}}{49}$

## Solutions

1. Move the decimal point 14 places:

   $3.547 \times 10^{14} = 354{,}700{,}000{,}000{,}000$

2. Move the decimal point 6 places:

   $-2 \times 10^6 = -2{,}000{,}000$

3. Move the decimal point 7 places:

   $7.91 \times 10^{-7} = 0.000000791$

4. Move the decimal point 12 places:

   $-8.05 \times 10^{-12} = -0.00000000000805$

## Using Scientific Notation in Applications

Calculating with large or small numbers is much easier to do if they are written in scientific notation rather than standard notation. We can use our calculators to perform such operations. One reason a calculator will express numbers in scientific notation is so the numbers fit on the screen.

To represent $5.58 \times 10^{19}$, for example, most calculators use the notation 5.58 E 19, where E indicates that the following number is the exponent on base 10. Calculators represent $7.612 \times 10^{-35}$ as 7.612E–35. See Figure 3.

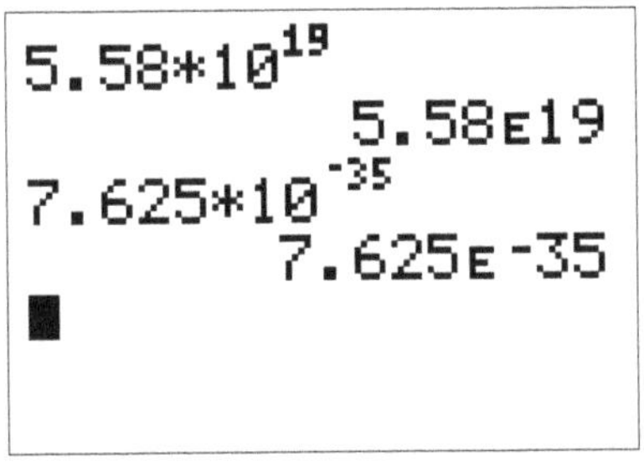

Figure 3. Representing $5.58 \times 10^{19}$ and $7.612 \times 10^{-35}$.

### Example 8

A tiny rectangular space inside a computer chip is measured as $3.56 \times 10^{-5}$ mm wide, $2.8 \times 10^{-4}$ mm long, and $2.75 \times 10^{-6}$ mm high. Use your calculator to find out the volume.

### Solution

The volume of a rectangular space can be computed by multiplying the length by the width and the height. So the volume of the space inside the computer chip is

$(0.56 \times 10^{-5})(2.8 \times 10^{-4})(2.75 \times 10^{-6})$

We use our calculator to multiply these values. Figure 4 shows two different ways this might be done depending on your calculator model.

The volume of the space inside the computer chip is $2.7412 \times 10^{-14}$ cubic mm.

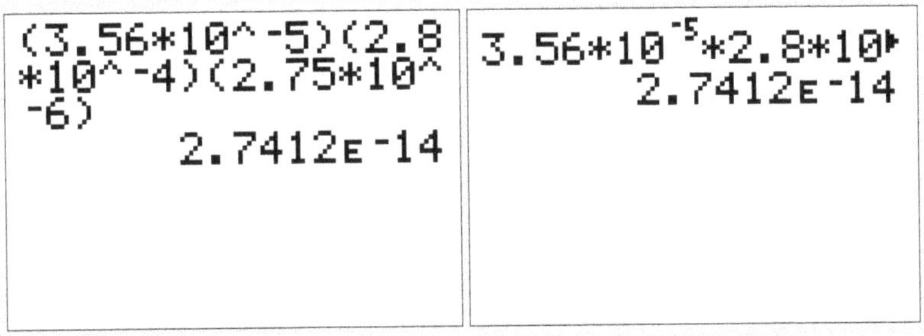

Figure 4. Calculating the product of numbers in scientific notation.

There are many instances in the sciences where one may need to multiply or divide two or more factors that are written in scientific notation. When using a calculator to divide by one or more numbers in scientific notation, it's important to use parentheses to ensure we divide by all the factors. This is demonstrated in Example 9.

### Example 9

Use a calculator to perform the following operations. Write the answer in scientific notation.

1. $(7.4 \times 10^{19}) \div (1.35 \times 10^{8})$
2. $\dfrac{(9.33 \times 10^{7})(3.06 \times 10^{13})}{6.4 \times 10^{29}}$
3. $\dfrac{1.88 \times 10^{26}}{(-4.7 \times 10^{5})(2.35 \times 10^{9})}$

### Solutions

1. When dividing by a number in scientific notation, we are dividing by *two* factors, the decimal number *and* the power of 10. Make sure to use parentheses around both when dividing on a calculator. See Figure 5. Notice that we don't need parentheses around the first number, but it would still be correct to use them.

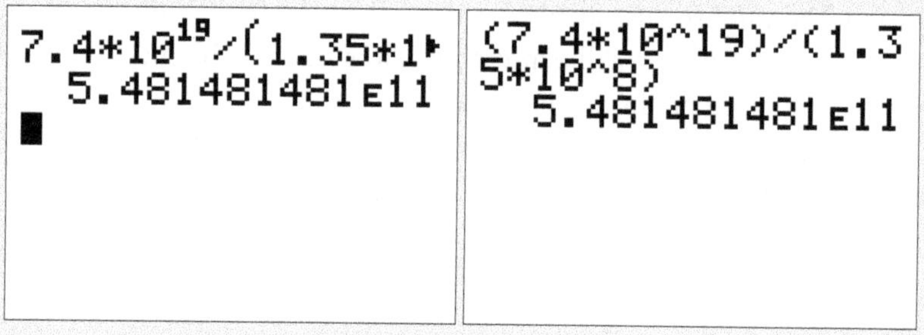

Figure 5. Dividing by a number in scientific notation.

$(7.4 \times 10^{19}) \div (1.35 \times 10^{8}) \approx 5.481 \times 10^{11}$

2. Using a calculator, the parentheses in the denominator are essential, but those in the numerator are not. See Figure 6.

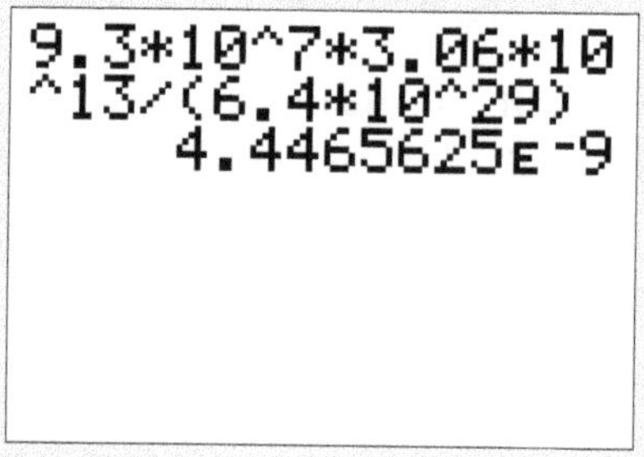

$\dfrac{(9.33 \times 10^{7})(3.06 \times 10^{13})}{6.4 \times 10^{29}} \approx 4.461 \times 10^{-9}$

Figure 6.

3. For this operation, we don't need parentheses between the two expressions in the denominator but it is essential to use them around *all four factors* in the denominator. See Figure 7.

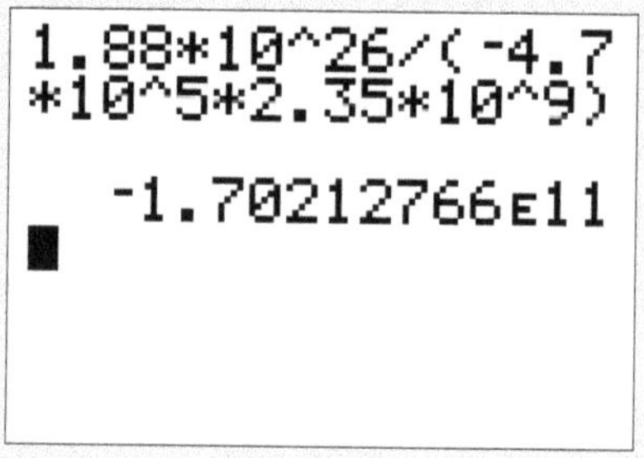

$\dfrac{1.88 \times 10^{26}}{(-4.7 \times 10^{5})(2.35 \times 10^{9})} \approx -1.702 \times 10^{11}$

Figure 7.

## Practice E

Write each number in scientific notation.

17. U.S. national debt per taxpayer (June 2017): $166,000

18. World population (June 2017): 7,509,000,000

19. World gross national income (April 2014): $85,500,000,000,000

20. Time for light to travel 1 m: 0.00000000334 s

21. Probability of winning a lottery by matching 6 of 49 possible numbers: 0.0000000715

Convert each number in scientific notation to standard notation.

22. $7.03 \times 10^5$
23. $-8.16 \times 10^{11}$
24. $-3.9 \times 10^{-13}$
25. $8 \times 10^{-6}$

Perform the operations and write the answer in scientific notation.

26. $(-7.5 \times 10^8)(1.13 \times 10^{-2})$
27. $(1.24 \times 10^{11}) \div (1.55 \times 10^{18})$
28. $(3.72 \times 10^9)(8 \times 10^3)$
29. $(9.933 \times 10^{23}) \div (-2.31 \times 10^{17})$

When you're finished, turn the page to check your answers.

## Exercises 2.1

1. Is $2^3$ the same as $3^2$? Explain.
2. When can you add two exponents?
3. What is the purpose of scientific notation?
4. Explain what a negative exponent does.

For the following exercises, simplify the given expression.

5. $9^2$
6. $15^{-2}$
7. $3^2 \times 3^3$
8. $4^4$
9. $(2^2)^{-2}$
10. $(5-8)^0$
11. $11^3 \div 11^4$
12. $6^5 \cdot 6^{-7}$
13. $(8^0)^2$
14. $5^{-2} \div 5^2$

For the following exercises, write each expression with a single base. Do not simplify further.

15. $4^2 \cdot 4^3 \div 4^{-4}$
16. $\dfrac{6^{12}}{6^9}$
17. $(12^3 \cdot 12)^{10}$
18. $10^6 \div (10^{10})^{-2}$
19. $7^{-6} \cdot 7^{-3}$
20. $(3^3 \div 3^4)^5$

For the following exercises, simplify the given expression. Write answers with positive exponents.

21. $\dfrac{a^3 a^2}{a}$

22. $\dfrac{mn^2}{m^{-2}}$

23. $(b^3 c^4)^2$

24. $\left(\dfrac{x^{-3}}{y^2}\right)^{-5}$

25. $ab^2 \div d^{-3}$

26. $(w^0 x^5)^{-1}$

27. $\dfrac{m^4}{n^0}$

28. $y^{-4}(y^2)^2$

29. $\dfrac{p^{-4} q^2}{p^2 q^{-3}}$

30. $(kw)^2$

31. $(y^7)^3 \div x^{14}$

32. $\left(\dfrac{a}{2^3}\right)^2$

33. $5^2 m \div 5^0 m$

34. $\dfrac{(16\sqrt{x})^2}{y^{-1}}$

35. $\dfrac{2^3}{(3a)^{-2}}$

36. $(ma^6)^2 \dfrac{1}{m^3 a^2}$

37. $(b^{-3} c)^3$

38. $(x^2 y^{13} \div y^0)^2$

For the following exercises, simplify the given expression. Write answers with positive exponents.

39. $\left(\dfrac{3^2}{a^3}\right)^{-2} \left(\dfrac{a^4}{2^2}\right)^2$

40. $(6^2 - 24)^2 \div \left(\dfrac{x}{y}\right)^{-5}$

41. $\dfrac{m^2 n^3}{a^2 c^{-3}} \cdot \dfrac{a^{-7} n^{-2}}{m^2 c^4}$

42. $\left(\dfrac{x^6 y^3}{x^3 y^{-3}} \cdot \dfrac{y^{-7}}{x^{-3}}\right)^{10}$

43. $\left(\dfrac{(ab^2 c)^{-3}}{b^{-3}}\right)^2$

For the following exercises, express the number in scientific notation.

44. 148,000,000

45. 0.0000314

For the following two exercises, convert each number in scientific notation to standard notation.

46. $9.8 \times 10^{-9}$

47. $1.6 \times 10^{10}$

48. To reach escape velocity from Earth, the Millennium Falcon must travel at the rate of $2.2 \times 10^6$ ft/min. Rewrite the rate in standard notation.

49. A dime is the thinnest coin in U.S. currency. A dime's thickness measures $2.2 \times 10^{-6}$ m. Rewrite the number in standard notation.

## Practice E — Answers

17. $1.66 \times 10^5$
18. $7.509 \times 10^9$
19. $8.55 \times 10^{13}$
20. $3.34 \times 10^{-9}$
21. $7.15 \times 10^{-8}$
22. 703,000
23. –816,000,000,000
24. –0.00000000000039
25. 0.000008
26. –8.475 × $10^6$
27. $8 \times 10^{-8}$
28. $2.976 \times 10^{13}$
29. $-4.3 \times 10^6$

50. The average distance between Earth and the Sun is 92,960,000 mi. Rewrite the distance using scientific notation.
51. A terabyte is made of approximately 1,099,500,000,000 bytes. Rewrite in scientific notation.
52. The Gross Domestic Product (GDP) for the United States in the first quarter of 2014 was $1.71496 \times 10^{13}$. Rewrite the GDP in standard notation.
53. One picometer is approximately $3.397 \times 10^{-11}$ in. Rewrite this length using standard notation.
54. The value of the services sector of the U.S. economy in the first quarter of 2012 was $10,633.6 billion. Rewrite this amount in scientific notation.
55. Avogadro's constant is used to calculate the number of particles in a mole (not the animal). In chemistry, a mole is a unit to measure the amount of a substance. The constant is $6.0221413 \times 10^{23}$. Write Avogadro's constant in standard notation.
56. Planck's constant is an important unit of measure in quantum physics describing the relationship between energy and frequency. The constant is written as $6.62606957 \times 10^{-34}$. Write Planck's constant in standard notation.

For the following exercises, perform the operations on a calculator. Write the answers in scientific notation.

57. Estimate the number of atoms in 1 liter of water. Each water molecule contains 3 atoms. The average drop of water contains around $1.32 \times 10^{21}$ molecules, and 1 liter holds about $1.22 \times 10^{4}$ average drops.
58. Most of the stars we can see in the Milky Way are about $8 \times 10^{4}$ light-years away from Earth. If one light year is approximately $5.88 \times 10^{12}$ miles, how many miles are the stars from Earth?
59. The radius of the Earth is $6.371 \times 10^{6}$ meters while the radius of an electron is $4.7 \times 10^{-13}$ meters. How many times greater is the Earth's radius than an electron's radius? (Hint: the solution requires division.)
60. The mass of the sun is about $1.988 \times 10^{30}$ kg. The mass of a particle of dust is about $7.53 \times 10^{-10}$ kg. How many times greater is the Sun's mass than the mass of a dust particle? (Hint: the solution requires division.)
61. The Sun burns about $4.4 \times 10^{6}$ tons of hydrogen per second and there are $3.1536 \times 10^{7}$ seconds in a year. How much hydrogen does the Sun burn in one year?

62. $\dfrac{(9.12 \times 10^{8})}{(7.43 \times 10^{15})}$

63. $\dfrac{(6.5 \times 10^{21})}{(1.25 \times 10^{29})}$

64. $\dfrac{(4.8 \times 10^{14})(6.2 \times 10^{-6})}{(3.6 \times 10^{-19})}$

65. $\dfrac{(3.7 \times 10^{22})(1.4 \times 10^{-10})}{(5.5 \times 10^{-9})}$

66. $\dfrac{8.07 \times 10^{15}}{(2.3 \times 10^{-4})(1.9 \times 10^{30})}$

67. $\dfrac{3.11 \times 10^{16}}{(5.8 \times 10^{-5})(4.72 \times 10^{28})}$

# 2.2 Rational Exponents

## Overview

In Section 2.1, we worked with exponents that are integers. In this section, we will investigate the meaning of rational exponents — exponents that are written as fractions in lowest terms. As you study this section, you will learn how to:

- Interpret rational exponents
- Evaluate expressions with rational exponents
- Simplify expressions that have rational exponents

## A. Rational Exponents with Unit Fractions

To develop a definition of fractional exponents, we begin by examining the expression $b^{1/2}$. This expression obeys the same rules for exponents that we studied in Section 2.1, so we use the product property to simplify the following expression:

$$b^{1/2} \cdot b^{1/2} = b^{1/2+1/2} = b^1 = b$$

This suggests that $b^{1/2}$ is a number that when multiplied times itself ($b^{1/2} \cdot b^{1/2}$) produces $b$. A number multiplied by itself that yields $b$ as a product is known as the square root of $b$. It is reasonable, therefore, to define $b^{1/2}$ as $\sqrt{b}$, the **principal square root** of $b$. So if $b = 9$, then:

$$9^{1/2} \cdot 9^{1/2} = 9^{1/2+1/2} = 9^1 = 9$$

That implies that $9^{1/2} = \sqrt{9} = 3$.

We can generalize the meaning of fractional exponents in order to define $b^{1/n}$ for any natural number $n$. This time we use the power property of exponents to simplify the expression:

$$\left(b^{1/n}\right)^n = b^{n/n} = b^1 = b$$

We use $b^{1/n}$ as a factor $n$ times in the expression $\left(b^{1/n}\right)^n$. It is thus reasonable to define $b^{1/n}$ as $\sqrt[n]{b}$, the **principal $n$th root of $b$**. So if $b = 8$ and $n = 3$, then

$$\left(8^{1/3}\right)^3 = 8^{3/3} = 8^1 = 8$$

That implies that $8^{1/3} = \sqrt[3]{8} = 2$. The number 2 is called the third root, or **cube root** of 8. If $b = -8$, then $(-8)^{1/3} = -2$ because $(-2)^3 = -8$. However, we cannot assign a real-number value to $(-9)^{1/2}$ because no real number squared is equal to $-9$.

## Definition of $b^{1/n}$

For the natural number $n$, where $n \neq 1$,

$$b^{1/n} = \sqrt[n]{b}$$

- If $n$ is odd, then $b^{1/n}$ is the number whose $n$th power is $b$, called the $n$th root of $b$.
- If $n$ is even and $b \geq 0$, then $b^{1/n}$ is the nonnegative number whose $n$th power is $b$, called the principal $n$th root of $b$.
- If $n$ is even and $b < 0$, then $b^{1/n}$ is not a real number.

### Example 1

Rewrite the exponential expressions with radical expressions and evaluate the $n$th root.

1. $49^{1/2}$
2. $27^{1/3}$
3. $(-125)^{1/3}$
4. $10000^{1/4}$
5. $32^{1/5}$
6. $(-16)^{1/2}$

Solutions

1. $49^{1/2} = \sqrt{49} = 7$      because $7^2 = 49$.
2. $27^{1/3} = \sqrt[3]{27} = 3$      because $3^3 = 27$.
3. $(-125)^{1/3} = \sqrt[3]{-125} = -5$      because $(-5)^3 = -125$.
4. $10000^{1/4} = \sqrt[4]{10000} = 10$      because $(10)^4 = 10000$.
5. $32^{1/5} = \sqrt[5]{32} = 2$      because $2^5 = 32$.
6. $(-16)^{1/2} = \sqrt{-16}$, which is not a real number because the square of a real number is not negative.

We can check our solutions using a graphing calculator. See Figure 1 for solutions to the first three problems in Example 1. If the expression entered into the calculator produces a non-real number, such as in problem 6 of the example, the calculator will give you an error message. See Figure 2.

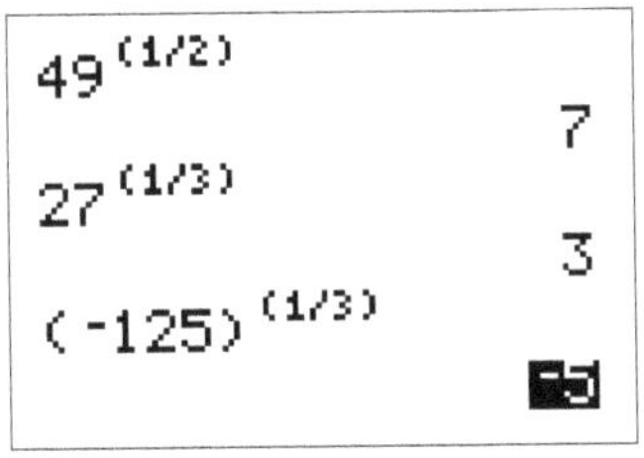

Figure 1. Checks for Problems 1, 2, and 3.

Figure 2. Error message for non-real numbers.

## Practice A

Rewrite the exponential expressions with radical expressions and evaluate the *n*th root. Check your solutions on your calculator and then turn the page to confirm them.

1. $64^{1/2}$
2. $64^{1/3}$
3. $64^{1/6}$
4. $-64^{1/2}$
5. $(-64)^{1/2}$
6. $(-64)^{1/3}$

## B. Definition of Rational Exponents

We can also have fractional exponents with numerators other than 1. The exponent must be a fraction in lowest terms. Because a fraction is a *ratio* between two numbers, we refer to all fractional exponents as **rational exponents**.

The product property of integer exponents from Section 2.1 works for rational exponents. Using the product property and the fact $b^{1/n} = \sqrt[n]{b}$, we can write expressions with rational exponents in two ways. Because the exponent $m/n = m(1/n) = (1/n)m$ we have

$$b^{m/n} = b^{m(1/n)} = (b^m)^{(1/n)} = \sqrt[n]{b^m}$$

and

$$b^{m/n} = b^{(1/n)m} = (b^{(1/n)})^m = \left(\sqrt[n]{b}\right)^m$$

### Definition of Rational Exponents

If $m$ is an integer, $n$ is a natural number, and $b^{1/n}$ is a real number, then

$$b^{m/n} = (b^m)^{(1/n)} = \sqrt[n]{b^m} \text{ and } b^{m/n} = (b^{(1/n)})^m = \left(\sqrt[n]{b}\right)^m$$

A power of the form $b^{m/n}$ is said to have a **rational exponent**.

The numerator of the exponent tells us the power on the base, and the denominator tells us the index of the root of the base. We can perform these operations on the base in either order. It's generally easier to take the *n*th root of the base first and then raise that root to the power. For example, $9^{3/2}$ can be written as $(9^3)^{1/2}$ or $(9^{1/2})^3$. Below, we simplify each expression:

$$(9^3)^{1/2} = 729^{1/2} = \sqrt{729} = 27$$
$$(9^{1/2})^3 = (\sqrt{9})^3 = 3^3 = 27$$

The calculation is easier to do, especially *without* a calculator, when we use the second form of the expression. We can check our work on a calculator if desired. We just have to make sure we use parentheses where needed. Example 2 illustrates how we do this.

## Example 2

Evaluate the following expressions, and then check your work using a calculator.

1. $16^{3/4}$
2. $25^{3/2}$
3. $(-8)^{5/3}$
4. $(-27)^{2/3}$
5. $36^{-1/2}$
6. $81^{-3/4}$

Solutions

1. $16^{3/4} = \left(16^{1/4}\right)^3 = \left(\sqrt[4]{16}\right)^3 = 2^3 = 8$

2. $25^{3/2} = \left(25^{1/2}\right)^3 = \left(\sqrt{25}\right)^3 = 5^3 = 125$

3. $(-8)^{5/3} = \left((-8)^{1/3}\right)^5 = \left(\sqrt[3]{-8}\right)^5 = (-2)^5 = -32$

4. $(-27)^{2/3} = \left((-27)^{1/3}\right)^2 = \left(\sqrt[3]{-27}\right)^2 = (-3)^2 = 9$

5. $36^{-1/2} = \dfrac{1}{36^{1/2}} = \dfrac{1}{\sqrt{36}} = \dfrac{1}{6}$

6. $81^{-3/4} = \dfrac{1}{81^{3/4}} = \dfrac{1}{\left(81^{1/4}\right)^3} = \dfrac{1}{\left(\sqrt[4]{81}\right)^3} = \dfrac{1}{3^3} = \dfrac{1}{27}$

We can again check our solutions using a graphing calculator. The checks for problems 1–5 are show below. For problems 5 and 6, change a decimal answer to the fraction form by pressing the MATH button and selecting 1:Frac.

```
16^(3/4)
            8
25^(3/2)
          125
```

```
(-8)^(5/3)
           -32
(-27)^(2/3)
             9
```

```
36^(-1/2)
     .1666666667
Ans▶Frac
              1/6
```

## Practice B

Evaluate the following expressions, and then turn the page to check your work.

7. $36^{3/2}$
8. $(-32)^{2/5}$
9. $49^{-1/2}$
10. $9^{-3/2}$
11. $216^{2/3}$
12. $81^{-1/4}$

## C. Properties of Rational Exponents

All of the properties of exponents that we learned for integer exponents also hold for rational exponents. Next we will apply those rules to simplify expressions involving rational exponents.

### Example 3

Using the product property, simplify the following expressions. Assume that $a$ and $b$ are positive.

1. $9^{3/8} \cdot 9^{1/8}$
2. $(5a^{1/2}b^{4/3})(7a^{1/2}b^{2/3})$
3. $b^{2/3} \cdot b^{1/4}$

**Solutions**

1. $9^{3/8} \cdot 9^{1/8}$

$$9^{3/8} \cdot 9^{1/8} = 9^{3/8 + 1/8}$$  Use the product property to add the exponents.
$$= 9^{4/8}$$  Add the fractions to simplify the exponent.
$$= 9^{1/2}$$
$$= 3$$  3 is the principal square root of 9.

2. $(5a^{1/2}b^{4/3})(7a^{1/2}b^{2/3})$

$$(5a^{1/2}b^{4/3})(7a^{1/2}b^{2/3}) = 35(a^{1/2}b^{4/3})(a^{1/2}b^{2/3})$$  Multiply the coefficients.
$$= 35a^{1/2+1/2}b^{4/3+2/3}$$  Use the product property to add the exponents.
$$= 35a^{2/2}b^{6/3}$$  Add the fractions to simplify the exponents.
$$= 35ab^2$$

3. $3 \cdot b^{2/3} \cdot b^{1/4}$

$$b^{2/3} \cdot b^{1/4} = b^{2/3 + 1/4}$$  Use the product property to add the exponents.
$$= b^{8/12 + 3/12}$$  Use common denominators to add the fractions.
$$= b^{11/12}$$  Add the fractions to simplify the exponent. This expression cannot be simplified further.

### Practice A — Answers

1. $64^{1/2} = \sqrt{64} = 8$
2. $64^{1/3} = \sqrt[3]{64} = 4$
3. $64^{1/6} = \sqrt[6]{64} = 2$
4. $-64^{1/2} = -\sqrt{64} = -8$
   Note that the negative sign is not in parentheses and is applied after the exponent.
5. $(-64)^{1/2} = \sqrt{-64}$, which is not a real number.
6. $(-64)^{1/3} = \sqrt[3]{-64} = -4$

## Example 4

Simplify the following expressions. Assume that $a$ and $b$ are positive.

1. $(8^{5/3})^{2/5}$
2. $\dfrac{15\,b^{5/4}}{3\,b^{3/4}}$
3. $\left(\dfrac{a^{2/5}}{b^4}\right)^{1/2}$

### Solutions

1. $(8^{5/3})^{2/5}$

$$(8^{5/3})^{2/5} = 8^{10/15} \quad \text{Multiply the exponents (power to a power): } \tfrac{5}{3}\cdot\tfrac{2}{5}=\tfrac{10}{15}$$
$$= 8^{2/3} \quad \text{Simplify the fraction.}$$
$$= (8^{1/3})^2 \quad \text{Definition of rational exponent.}$$
$$= 2^2 \quad \text{Simplify.}$$
$$= 4$$

2. $\dfrac{15\,b^{5/4}}{3\,b^{3/4}}$

$$\dfrac{15\,b^{5/4}}{3\,b^{3/4}} = \dfrac{5\,b^{5/4}}{b^{3/4}} \quad \text{Divide the coefficients.}$$
$$= 5b^{5/4-3/4} \quad \text{Use the quotient property to subtract the exponents.}$$
$$= 5b^{2/4} \quad \text{Simplify the exponent.}$$
$$= 5b^{1/2}$$

3. $\left(\dfrac{a^{2/5}}{b^4}\right)^{1/2}$

$$\left(\dfrac{a^{2/5}}{b^4}\right)^{1/2} = \dfrac{a^{2/5\cdot 1/2}}{b^{4\cdot 1/2}} \quad \text{Raise a quotient to a power by multiplying the exponents.}$$
$$= \dfrac{a^{2/10}}{b^{4/2}} \quad \text{Multiply the fractions.}$$
$$= \dfrac{a^{1/5}}{b^2} \quad \text{Simplify the fractions.}$$

## Practice C

Use the properties of exponents to simplify the following expressions. Assume $a$ and $b$ are positive. Check your work on the next page.

13. $(8a)^{1/3}(14\,a^{4/3})$
14. $\dfrac{2\,b^{3/4}}{16\,b^{1/2}}$
15. $(81\,a^2\,b)^{-1/2}$

### Practice B — Answers

7. $36^{3/2} = 6^3 = 216$
8. $(-32)^{2/5} = (-2)^2 = 4$
9. $49^{-1/2} = \dfrac{1}{49^{1/2}} = \dfrac{1}{7}$
10. $9^{-3/2} = \dfrac{1}{9^{3/2}} = \dfrac{1}{3^3} = \dfrac{1}{27}$
11. $216^{2/3} = 6^2 = 36$
12. $81^{-1/4} = \dfrac{1}{81^{1/4}} = \dfrac{1}{3}$

## Exercises 2.2

Simplify without using a calculator. Then use a calculator to check your work.

1. $4^{1/2}$
2. $125^{1/3}$
3. $81^{1/4}$
4. $49^{3/2}$
5. $8^{4/3}$
6. $16^{3/4}$
7. $32^{2/5}$
8. $27^{4/3}$
9. $27^{-1/3}$
10. $64^{-1/6}$
11. $4^{-3/2}$
12. $36^{-3/2}$
13. $-8^{2/3}$
14. $-32^{3/5}$
15. $(-64)^{1/3}$
16. $(-27)^{2/3}$
17. $3^{2/5} \cdot 3^{4/5}$
18. $2^{1/4} \cdot 2^{5/4}$
19. $(5^{1/2} \cdot 4^{3/2})^2$
20. $(2^{2/3} \cdot 7^{1/3})^3$
21. $\dfrac{2^{7/4}}{2^{3/4}}$
22. $\dfrac{4^{5/2}}{4^{1/2}}$
23. $\dfrac{8^{1/3}}{8^{2/3}}$
24. $\dfrac{27^{2/3}}{27^{5/3}}$

Use the properties of exponents to simplify the following expressions. Assume all variables are positive.

25. $a^{1/5} a^{2/5}$
26. $k^{5/6} k^{7/6}$
27. $b^{1/4} b^{-3/4}$
28. $a^{4/7} a^{-5/7}$
29. $(25 a^6)^{1/2}$
30. $(27 c^{12})^{1/3}$
31. $(10 a^{1/2} b^{3/5})(3 a^{1/2} b^{7/5})$
32. $(7 b^{1/4} c^{2/3})(4 b^{7/4} c^{5/3})$
33. $\dfrac{a^{3/4}}{a^{7/4}}$
34. $\dfrac{2 c^{5/8}}{c^{3/8}}$
35. $\dfrac{14 b^{6/11}}{2 b^{7/11}}$
36. $\dfrac{21 a^{4/7}}{7 a^{5/7}}$
37. $\dfrac{a^{-1/6}}{a^{5/6}}$
38. $\dfrac{b^{-2/3}}{b^{4/3}}$
39. $\dfrac{12 b^{2/5} c^{-3/10}}{18 b^{-1/5} c^{3/10}}$
40. $\dfrac{15 a^{-1/4} b^{3/5}}{20 a^{7/4} b^{-3/5}}$

**Practice C — Answers**

13. $(8 a^{1/3})(14 a^{4/3}) = 28 a^{5/3}$
14. $\dfrac{1}{8} b^{3/4 - 1/2} = \dfrac{b^{1/4}}{8}$
15. $\dfrac{1}{(81 a^2 b)^{1/2}} = \dfrac{1}{9 a b^{1/2}}$

# 2.3 Exponential Functions

## Overview

India is the second most populous country in the world with a population of about 1.25 billion people in 2013. The population is growing at a rate of about 1.2% each year. China is currently the most populous country, but if India's rate continues, India's population will exceed China's by 2031. When populations grow more and more rapidly, we often say that the growth is "exponential." To a mathematician, the term "exponential growth" has a very specific meaning.

In this section, we will take a look at exponential functions, which model this kind of rapid growth. Along the way, you will learn to:

- Identify exponential functions
- Evaluate exponential functions
- Sketch the graph of an exponential function
- Know the graphical significance of $a$ and $b$ for a function of form $f(x) = ab^x$
- Know the base multiplier property
- Distinguish between exponential growth and decay

## A. Definition of an Exponential Function

When we explored linear growth, we observed a constant rate of change, a constant number by which the output increased or decreased for each unit increase in input. For example, in the equation $f(x) = 3x + 4$, the slope, $m = 3$, tells us the output increases by 3 each time the input increases by 1. The scenario in the example of India's population is different because we have a *percent change* per unit of time in the number of people.

Exponential functions are the perfect tool to model this type of change. Examples of exponential functions include:

$$f(x) = 2(3)^x$$
$$g(t) = 22(1.045)^t$$
$$h(x) = 16\left(\frac{3}{5}\right)^x$$

Notice that one defining feature of these equations is that the independent variable is in the exponent position. The base is a constant. Contrast this with the function $q(x) = x^2$, where the base is the variable, $x$, and the exponent, 2, is a constant. Function $q$ is *not* an exponential function.

## Exponential Functions

For any real number $x$, an **exponential function** is a function with the form

$$f(x) = ab^x$$

where the constant $a$ is a non-zero real number, and the constant $b$, the base, is any positive real number such that $b \neq 1$.

In the preceding definition of an exponential function of form $f(x) = ab^x$ we limit the base $b$ to positive values. We do this to ensure the outputs will be real numbers. For example, if $b = -9$, then $(-9)^{1/2} = \sqrt{-9}$, which is not a real number. We also limit the base with the condition $b \neq 1$. If $b = 1$, the equation $f(x) = a(1)^x$ would be a constant function because $1^x = 1$ for all values of $x$.

### Evaluating Exponential Functions

To evaluate an exponential function with the form $f(x) = ab^x$, we replace $x$ with the given value and calculate according to the order of operations. It's a common mistake to multiply the $a$ and $b$ values together before raising to the power. Don't do this! The correct order of operations is to apply the exponent before multiplying.

### Example 1

For $f(x) = 2(3)^x$ and $g(x) = 10(0.25)^x$, find the following.

1. $f(4)$
2. $f(-2)$
3. $g(2)$
4. $g(-1)$

**Solutions**

1. $f(4)$

$$\begin{aligned} f(x) &= 2(3)^x \\ f(4) &= 2(3)^4 \quad \text{Substitute } x = 4. \\ &= 2(81) \quad \text{Evaluate the power.} \\ &= 162 \quad \text{Multiply.} \end{aligned}$$

2. $f(-2)$

$$\begin{aligned} f(-2) &= 2(3)^{-2} \quad \text{Substitute } x = -2. \\ &= 2\left(\tfrac{1}{9}\right) \quad \text{Evaluate the power. Recall that } 3^{-2} = \left(\tfrac{1}{3^2}\right) \\ &= \tfrac{2}{9} \quad \text{Multiply.} \end{aligned}$$

We can check our answers to 1 and 2 using a calculator. We can also use a calculator to evaluate $g(x)$ for the given values.

3. $g(2) = 10(0.25)^2$
4. $g(-1) = 10(.25)^{-1} = 40$

## Practice A

Now it's your turn. When you're done, turn the page to check your work.

1. Let $h(x) = 6(2)^x$. Evaluate without using a calculator, $h(3)$ and $h(-3)$.

2. Let $f(x) = 200(1.08)^x$. Evaluate $f(5)$ and $f(10)$ using a calculator. Round to three decimal places.

## B. Graphing Exponential Functions

Next we explore the graphs of exponential functions. When we graph a new type of function by hand, it's useful to first create a table of input-output pairs. We do this in the following example.

### Example 2

Graph $f(x) = 2^x$ by hand.

**Solution**

First we make a table of values. We choose small integer values for $x$ and use $f$ to calculate the corresponding output values. See Figure 1.

| $x$ | −3 | −2 | −1 | 0 | 1 | 2 | 3 |
|---|---|---|---|---|---|---|---|
| $f(x) = 2^x$ | $\frac{1}{8}$ | $\frac{1}{4}$ | $\frac{1}{2}$ | 1 | 2 | 4 | 8 |

Figure 1.

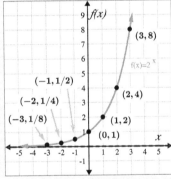

We graph $f$ by plotting the ordered pairs from the table in Figure 1. We connect these points with a smooth curve and add arrows on both ends because the variable $x$ can be any real number — not just the small integers we chose for our table. See Figure 2.

Figure 2. The graph of $f(x) = 2^x$.

Look back at the table for $f(x) = 2^x$ in Example 2 and notice that as $x$ increases by 1 unit, the output value $y$ is multiplied by 2, the base. As $x$ decreases, the output values grow smaller and smaller. Let's look at a few more input-output pairs for this function as $x$ becomes increasingly negative:

$$f(-4) = 2^{-4} = \frac{1}{2^4} = \frac{1}{16}$$

$$f(-5) = 2^{-5} = \frac{1}{2^5} = \frac{1}{32}$$

$$f(-6) = 2^{-6} = \frac{1}{2^6} = \frac{1}{64}$$

It appears that although the output values grow smaller and smaller, they will never equal zero. Therefore, while the domain of this function is all real numbers, the range of the function is only positive real numbers. Using interval notation, we write the domain as $(-\infty, \infty)$ and the range as $(0, \infty)$.

The left side of graph of the exponential function in Example 2 gets close to the x-axis but never touches it. When the graph of a function gets closer and closer to a horizontal line as the input value either increases or decreases, we call that line a **horizontal asymptote**. For the exponential function $f(x) = 2^x$ in Figure 2, the x-axis, which is described by the equation $y = 0$, is the horizontal asymptote.

### Example 3

Graph $g(x) = \left(\frac{1}{2}\right)^x$ by hand.

**Solution**

First we make a table of values. We choose small integer values for $x$ and use $g$ to calculate the corresponding output values. See Figure 3.

| $x$ | $-3$ | $-2$ | $-1$ | $0$ | $1$ | $2$ | $3$ |
|---|---|---|---|---|---|---|---|
| $g(x) = \left(\frac{1}{2}\right)^x$ | 8 | 4 | 2 | 1 | $\frac{1}{2}$ | $\frac{1}{4}$ | $\frac{1}{8}$ |

Figure 3.

We graph $g$ by plotting the ordered pairs from the table in Figure 3. We then connect these points with a smooth curve and add arrows on both ends because the input values can be any real number. The table in Figure 3 shows the y-values of $g$ decreasing by a factor of $\frac{1}{2}$ as the x-values increase by one. As $x$ increases, the output values grow smaller and smaller but will never equal zero. Therefore the range of $g$ is $(0, \infty)$. This exponential function, like $f$ in Example 2, has a horizontal asymptote at $y = 0$, the x-axis. This time, the asymptotic behavior occurs on the right side of the graph rather than on the left side. See Figure 4.

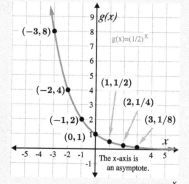

Figure 4. The graph of $g(x) = \left(\frac{1}{2}\right)^x$.

When graphing many functions, it's useful to first determine the y-intercept. Let's examine the y-intercepts of exponential functions with the form $f(x) = ab^x$. When a curve crosses the y-axis, the x-value is zero. So in our general form, if $x = 0$, then

$$f(0) = ab^0 = a(1) = a$$

This result is due to the zero exponent rule. *Any* base $b$ (except 0) to the zero power is equal to 1. This implies that the y-intercept of these functions will be $(0, a)$. Here are a few examples:

For the function $y = 2(5)^x$, the y-intercept is $(0, 2)$.

For the function $y = 7(0.8)^x$, the y-intercept is $(0, 7)$.

For the function $y = 390(1.25)^x$, the y-intercept is $(0, 390)$.

This concept seems simple, but be careful! For the function $y = 3^x$, the y-intercept is *not* $(0, 3)$. If we evaluate the function for $x = 0$, we see that $y = 3^0 = 1$, so the y-intercept is $(0, 1)$. For the function $y = 3^x$, the value of the base is $b = 3$. The value of $a$ isn't seen in the function, but because $3^x = 1(3)^x$, it's understood that $a = 1$.

## Expanded Definition of an Exponential Function

For any real number $x$, an exponential function is a function with the form

$$f(x) = ab^x \text{ where}$$

- $a$ is a non-zero real number.
- $b$ is any positive real number such that $b \neq 1$.
- The $y$-intercept is $(0, a)$.
- the domain of $f$ is all real numbers.
- the range of $f$ is all positive real numbers if $a > 0$.
- the horizontal asymptote is $y = 0$.

Let's take a look at the graphs of a few exponential functions on our calculators. We begin with the function from Example 2, $f(x) = 2^x$. Enter the equation and adjust the viewing window to reflect the fact that the range ($y$-values) will be positive real numbers. We set the values for $x$ to include relatively small positive and negative values. See Figures 5 and 6. Now graph the function and use your trace TRACE button to confirm that the $y$-intercept is $(0, 1)$. See Figure 7.

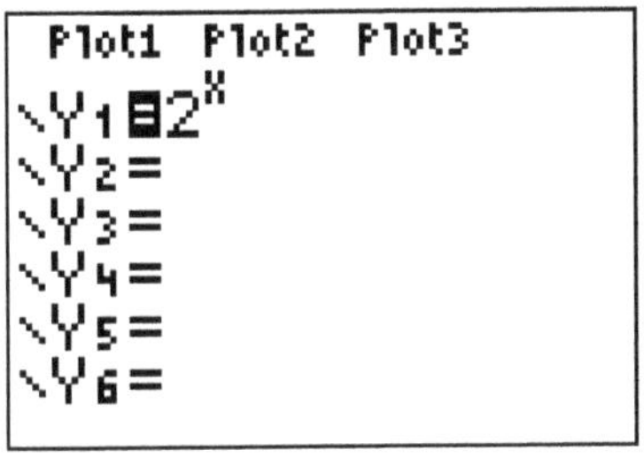

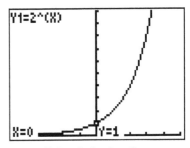

Figure 5. Enter the function.

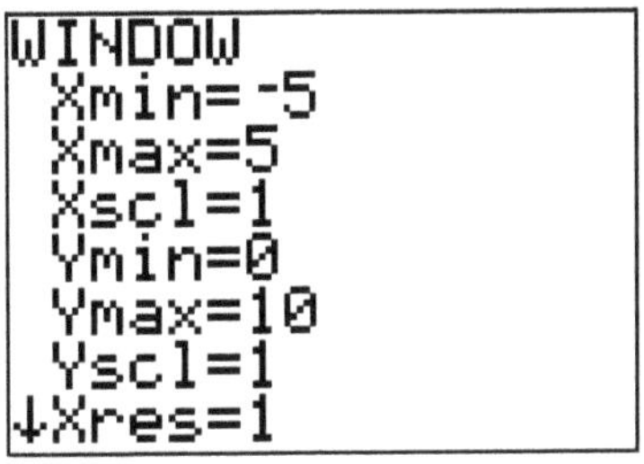

Figure 6. Set up the viewing window.

Figure 7. Graph the function.

Notice in Figure 7 that it looks like the curve touches the $x$-axis on the left. From the discussion on horizontal asymptotes, remember that this curve gets closer and closer to the $x$-axis but never touches it.

Let's do some more graphing on the calculator and compare a few exponential functions with the same $b$-value but different $a$-values.

### Practice A — Answers

1. $h(3) = 6(2)^3 = 6(8) = 48$
   $h(-3) = 6(2)^{-3} = 6\left(\frac{1}{2^3}\right) = 6\left(\frac{1}{8}\right) = \frac{6}{8} = \frac{3}{4}$

2. $f(5) = 200(1.08)^5 \approx 293.866$
   $f(10) = 200(1.08)^{10} \approx 431.785$

### Example 4

Graph the following functions on a calculator. Use a viewing window of $x$: $[-5, 5, 1]$ and $y$: $[0, 10, 1]$.

$$y_1 = 2(1.25)^x \quad y_2 = 4(1.25)^x \quad y_3 = 7(1.25)^x$$

Compare the graphs.

### Solution

The graphs of the three exponential functions look similar. They are all increasing functions. The graphs have different $y$-intercepts, namely $(0, 2)$, $(0, 4)$, and $(0, 7)$. See Figure 8.

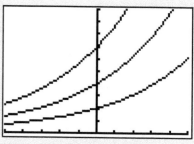

Figure 8. Comparing graphs of exponential functions.

In Example 5, we graph a few exponential functions with the same $a$-value but different $b$-values.

### Example 5

Graph the following functions on a calculator. Use a viewing window of $x$: $[-5, 5, 1]$ and $y$: $[0, 10, 1]$.

$$y_1 = 2(1.3)^x \quad y_2 = 2(1.9)^x \quad y_3 = 2(0.5)^x$$

Compare the graphs.

### Solution

The graphs of the three exponential functions all have the same $y$-intercept, $(0, 2)$. The graphs also have a horizontal asymptote, the $x$-axis.

The first two graphs are increasing functions. The function $y_2$ has a larger $b$-value and increases more quickly than the function $y_1$. The graph of $y_3$, where the $b$-value is less than 1, depicts a decreasing function. See Figure 9.

Figure 9. Comparing graphs of exponential functions.

## Practice B

Now it's your turn. When you're finished, turn the page to check your work.

3. Sketch the graph of $f(x) = 4^x$ by hand. State the domain, range, and asymptote.
4. Sketch a graph of $f(x) = 0.25^x$ by hand. State the domain, range, and asymptote.
5. Use your calculator to compare the graphs of $f(x) = 4(1.1)^x$ and $g(x) = 4(1.8)^x$.
6. Use your calculator to compare the graphs of $f(x) = 3(0.85)^x$ and $f(x) = 6(0.85)^x$.

## C. Base Multiplier Property; Growth vs. Decay

The table of values in Figure 1 of Example 2 shows that as $x$ increases by 1 unit, $y$ values are multiplied by the base $b = 2$. The table of values in Figure 3 of Example 3 shows that as $x$ increases by 1 unit, $y$ values are multiplied by the base $b = \frac{1}{2}$. These observations suggest the **base multiplier property**.

### Base Multiplier Property

For an exponential function of the form $y = ab^x$, if the value of the independent variable increases by 1, the value of the dependent variable is multiplied by the base.

Here are two more examples of the base multiplier property:

1. For the function $y = 2(5)^x$, $b = 5$, so if the value of $x$ increases by 1, the value of $y$ is multiplied by 5.
2. For the function $y = 72(0.8)^x$, $b = 0.8$, so if the value of $x$ increases by 1, the value of $y$ is multiplied by 0.8.

When the base $b$ of an exponential function is larger than 1, the function is increasing. We call this **exponential growth**. When the value of the base is between 0 and 1, the function is decreasing. We call this **exponential decay**. Figure 10 compares the graphs of exponential growth and decay functions.

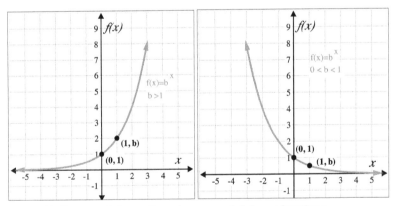

Figure 10. Exponential growth and exponential decay functions.

### Exponential Growth and Decay

Let $f(x) = ab^x$, where $a > 0$, $b > 0$, $b \neq 1$. Then:
 If $b > 1$, the function is increasing and grows exponentially.
 If $0 < b < 1$, the function is decreasing and decays exponentially.

### Example 6

For the following exponential functions, determine whether they represent exponential growth or decay. State the y-intercept and sketch a rough graph of the function. Check the graph on your graphing calculator. $f(x) = 8(0.65)^x$

1. $f(x) = 8(0.65)^x$
2. $g(x) = 5(1.5)^x$
3. $y = \frac{2}{3}(3)^x$
4. $f(x) = 6\left(\frac{4}{5}\right)^x$

**Solution**

1. $b = 0.65 < 1$ indicates exponential decay. $a = 8$ means the y-intercept is $(0, 8)$.

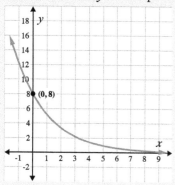

3. $b = 3 > 1$ indicates exponential growth. $a = \frac{2}{3}$ means the y-intercept is $(0, \frac{2}{3})$.

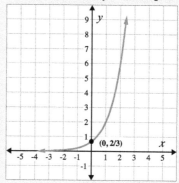

2. $b = 1.5 > 1$ indicates exponential growth. $a = 5$ means the y-intercept is $(0, 5)$.

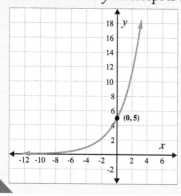

4. $b = \frac{4}{5} < 1$ indicates exponential decay. $a = 6$ means the y-intercept is $(0, 6)$.

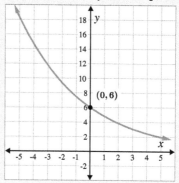

The exponential functions we study in this course all have positive values for $a$. Therefore, whether the function is exponential growth or decay, there will never be an x-intercept because there will never be an exponent that makes the base to the power equal to zero. In other words, the equation $b^x = 0$ has no real solutions. However, the exponential functions we study all have a horizontal asymptote at $y = 0$, the x-axis.

Let's investigate further the properties of the base multiplier $b$ in an exponential function. Consider the growth of an initial investment of $1000 in a savings account at a bank that offers interest at an annual rate of 4% compounded annually. What does this mean?

Because this account pays a 4% annual interest rate, the bank will add 4% of $1000 to the original investment of $1000 at the end of the first year. To find 4% of 1000, we first change from percent to decimal form by moving the decimal two places left (4% = 0.04). Then we multiply 0.04 times 1000. The interest is added to the original investment.

The calculation looks like this:

$$1000 + .04(1000) = 1040$$

This means that the account will be worth $1040 after the first year. Because 1000 = 1(1000), we can simplify this calculation by factoring 1000 from both terms.

$$1(1000) + .04(1000) = 1000(1 + .04)$$

The expression $1 + .04$ can be simplified further because $1 + .04 = 1.04$:

$$1000(1.04) = 1040$$

## Practice B — Answers

3. The domain is $(-\infty, \infty)$; the range is $(0, \infty)$; the horizontal asymptote is $y = 0$.

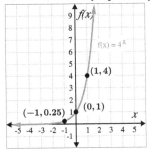

4. The domain is $(-\infty, \infty)$; the range is $(0, \infty)$; the horizontal asymptote is $y = 0$.

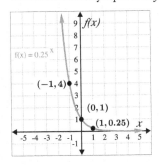

5. The graphs of $f$ and $g$ have the same $y$-intercept at $(0, 4)$. They are both increasing functions, but $g$ increases more quickly.

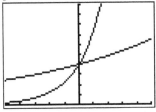

6. The graphs of $f$ and $g$ are both decreasing. They have different $y$-intercepts, namely $(0, 3)$ and $(0, 6)$.

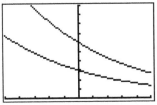

Notice that we get the same result, $1040 after one year. The account earned $40 in interest at the end of the first year. When we multiply any amount by 1.04, the result is 100% of the amount plus 4% of the amount — calculated in one step.

Now, what happens if the money is left in the savings account for a second year? The phrase "earns interest at an annual rate of 4% compounded annually" means the account will earn 4% interest on the amount in the account in the second year as well. Assuming no money (other than the interest) is added or subtracted from the account, the calculation for the second year looks like this:

$$\underbrace{1000(1.04)}_{\text{amt. after 1 year}} (1.04) = 1081.6$$

Notice after year 2, the bank applies the 4% interest rate not just to the original $1000 investment but to the investment *plus* the interest earned in the first year. This is what is meant by the term "compounded interest." The account earns interest on the interest already earned. At the end of the second year, there is $1081.60 in the account.

Here is the calculation for the amount in the account at the end of the third year:

$$\underbrace{1000(1.04)(1.04)}_{\text{amt. after 2 years}} (1.04) \approx 1124.86$$

At the end of year 3, the investment will be worth $1124.86.

Notice that we can use exponents to simplify our last two calculations. The factor 1.04 is applied twice in year 2 and three times in year 3. We'll add the calculation for year 4 as well.

$$1000(1.04)(1.04) = 1000(1.04)^2$$

$$1000(1.04)(1.04)(1.04) = 1000(1.04)^3$$

$$\underbrace{1000(1.04)(1.04)(1.04)}_{\text{amt. after 3 years}}(1.04) = 1000(1.04)^4$$

The exponential expressions on the right side of the previous equations suggest we can write an equation for an exponential function of form $f(x) = ab^x$ to represent the amount of money in the account. The $a$-value is 1000 and the base multiplier, $b$, is 1.04.

$$f(x) = 1000(1.04)^x$$

$f(x)$ represents the amount of money in a savings account after $x$ years, where the initial investment is $1000 and an annual interest rate of 4% is compounded annually.

We continue our study of this function in Example 7.

## Example 7

The exponential function $f(x) = 1000(1.04)^x$ models the amount of money in a savings account after $x$ years. With this in mind, do the following:

a. Find the value of the account after 10 years.

b. Use the calculator table feature to compare the interest earned after the first year to the interest earned at the end of the tenth year.

### Solution

a. To find the value after 10 years, we substitute $x = 10$:

$$f(10) = 1000(1.04)^{10} \approx 1480.24$$

The value will be $1480.24 in 10 years.

b. Enter the equation in the calculator and look at the table. We can scroll down to find values for later years. Arrow right to highlight a value for $y$. This allows us to see more decimal places of the output value at the bottom of the screen.

After the first year, the interest earned is 1040 − 1000 = 40 dollars.

At the end of the tenth year, the interest earned is 1480.24 − 1423.31 = 56.93 dollars.

The amount of interest earned increases each year because the account is earning interest on the interest earned in previous years.

At the beginning of this section, we learned that the population of India was about 1.25 billion in the year 2013, with an annual growth rate of about 1.2%. This situation can be represented with an exponential function in a manner similar to the previous scenario involving compounded interest on an investment account. Just as an investment account will earn interest on the interest earned in previous years, a population that grows exponentially will have more people with which to generate even more people as the years go by. We use this population growth model in Example 8.

## Example 8

The population of India, in billions of people, can be modeled by the exponential growth function $P(t) = 1.25(1.012)^t$, where $t$ is the number of years since 2013. Use this model to predict the population of India in 2031.

### Solution

To estimate the population in 2031, we evaluate the model for $t = 18$ because 2031 is 18 years after 2013.

$$P(18) = 1.25(1.012)^{18} \approx 1.549$$

There will be about 1.549 billion people in India in the year 2031.

## Practice C

For the following exponential functions, determine whether they represent exponential growth or decay. State the $y$-intercept and sketch a rough graph of the function. Check the graph on your graphing calculator. When you're finished, turn the page to check your work.

7. $f(x) = 4(1.2)^x$
8. $g(x) = 12(0.75)^x$
9. The population of China was about 1.39 billion in the year 2013, with an annual growth rate of about 0.6%. This situation can be modeled by the function $P(t) = 1.39(1.006)^t$, where $t$ is the number of years since 2013. Predict the population of China for the year 2031. How does this compare to the population prediction we made for India in example 8?

## Exercises 2.3

1. Given a formula for an exponential function, is it possible to determine whether the function grows or decays exponentially just by looking at the formula? Explain.
2. Explain the base multiplier property.
3. How do you find the $y$-intercept of an exponential function?
4. How can we tell the difference between an equation that represents a linear function and one that represents an exponential function?

Identify whether the equation represents an exponential function. Explain.

5. $y = 1.5(3)^x$
6. $y = 13x + 25$
7. $f(t) = 13(0.875)^t$
8. $g(t) = 275(1.035)^t$
9. $f(x) = 140 - 5x$
10. $h(t) = -4.9t^2 + 18t + 40$

For $f(x) = 24(3)^x$ and $g(x) = 32(0.75)^x$, find the following.

11. $f(4)$
12. $f(10)$
13. $f(-3)$
14. $f(-2)$
15. $g(2)$
16. $g(-1)$
17. $g(-4)$
18. $g(6)$

## Section 2.3: Exponential Functions

For the following exercises, graph the function by hand. Find the domain and range of the function.

19. $g(x) = 20(0.65)^x$    20. $h(x) = 6(1.75)^x$    21. $f(x) = 5(3)^x$    22. $g(x) = 40(0.8)^x$

Determine whether the following equations represent exponential growth or exponential decay. Then determine the y-intercept of the graph of the function.

23. $y = 300(0.08)^x$
24. $y = 22(1.06)^x$
25. $y = 0.5(1.025)^x$
26. $y = 700(0.97)^t$
27. $f(x) = 3\left(\frac{1}{2}\right)^x$
28. $f(x) = 1672\left(\frac{1}{4}\right)^x$
29. $f(x) = 4^x$
30. $f(x) = 10^x$

For the following exercises, match each function with one of the graphs in Figure 11.

31. $f(x) = 2(0.69)^x$
32. $f(x) = 2(1.28)^x$
33. $f(x) = 2(0.81)^x$
34. $f(x) = 4(1.28)^x$
35. $f(x) = 2(1.59)^x$
36. $f(x) = 4(0.69)^x$

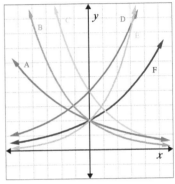

Figure 11.

For the following exercises, use the graphs shown in Figure 12. All have the form $f(x) = ab^x$.

37. Which graph has the largest value for $b$?
38. Which graph has the smallest value for $b$?
39. Which graph has the largest value for $a$?
40. Which graph has the smallest value for $a$?

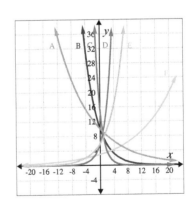

Figure 12.

## Chapter 2: Exponential Functions

For the exercises 41 to 44, consider the following scenario. For each year $t$, the population of a forest of trees is represented by the function $A(t) = 115 (1.025)^t$. In a neighboring forest, the population of the same type of tree is represented by the function $B(t) = 82 (1.029)^t$. Answer the following questions, rounding your answers to the nearest whole number.

41. Which forest's population is growing at a faster rate?

42. Which forest had a greater number of trees initially? By how many?

43. Assuming the population growth models continue to represent the growth of the forests, which forest will have a greater number of trees after 20 years? By how many?

44. Assuming the population growth models continue to represent the growth of the forests, which forest will have a greater number of trees after 100 years? By how many?

45. Let $f(x) = 2700 (1.056)^x$ represent the value of a savings account after $x$ years with an annual interest rate of 5.6% compounded annually.
    a. What is the initial value of the account? Hint: find $f(0)$.
    b. What is the value of the account after 8 years?

46. Let $f(x) = 1800 (1.072)^x$ represent the value of a savings account with an annual interest rate of 7.2% compounded annually.
    a. What is the initial value of the account? Hint: find $f(0)$.
    b. What is the value of the account after 5 years?

47. Let $f(t) = 3970 (0.89)^t$ represent the population of small town $t$ years after 2000.
    a. Is the population of the town growing or shrinking?
    b. What was the population of the town in 2000?
    c. Predict the population in 2025.

48. Let $f(t) = 5205 (0.93)^t$ represent the population of small town $t$ years after 2000.
    a. Is the population of the town growing or shrinking?
    b. What was the population of the town in 2000?
    c. Predict the population in 2030.

### Practice C — Answers

7. $b = 1.2 > 1$ indicates exponential growth. $a = 4$ means the $y$-intercept is $(0, 4)$.

8. $b = 0.75 < 1$ indicates exponential decay. $a = 12$ means the $y$-intercept is $(0, 12)$.

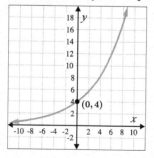
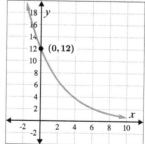

9. To estimate the population in 2031, we evaluate the model for $t = 18$, because 2031 is 18 years after 2013: $P(18) = 1.39(1.006)^{18} \approx 1.548$. The model predicts about 1.548 billion people in China in the year 2031. In that year, India's population will exceed China's by about 0.001 billion or 1 million people.

# 2.4 Finding Equations of Exponential Functions

## Overview

We continue our study of exponential functions by learning to write equations for these functions. Just as you saw with linear models, we have different methods for writing equations. The method to use depends on the information given in the problem. In this section, you will learn to:

- Use the base multiplier property to write an equation for an exponential function
- Solve an exponential equation for the base
- Use two points to write an equation for an exponential function

## A. Using the Base Multiplier to Find Exponential Functions

Recall from section 2.3 that for an exponential function of form $f(x) = ab^x$, if the value of the independent variable increases by 1, then the value of the dependent variable is multiplied by the base $b$. One way to find an equation for an exponential function of the form $y = ab^x$ is to use the base multiplier property.

### Example 1

An exponential function contains the points listed in the table below.

| $x$ | 0 | 1 | 2 | 3 | 4 |
|---|---|---|---|---|---|
| $f(x)$ | 4 | 12 | 36 | 108 | 324 |

Use this table of values to find an equation to represent this function.

### Solution

We recall from Section 2.3 that for the form $f(x) = ab^x$, the y-intercept of the graph is $(0, a)$. From the table above, then, we see that the point $(0, 4)$ is on the curve, so $a = 4$.

As the value of $x$ increases by 1, the value of $y$ is multiplied by 3. So the base multiplier $b = 3$. Putting these pieces together we have an equation for our function: $f(x) = 4(3)^x$

For the next example, recall that for linear functions of the form $y = mx + b$, as $x$ increases by 1, the value of $y$ increases by the slope $m$. We can differentiate between linear and exponential functions by observing the behavior of the dependent variable.

### Example 2

The tables below show points on the graphs of two curves. One represents a linear function, and one represents an exponential function.

| $x$ | 0 | 1 | 2 | 3 | 4 |
|---|---|---|---|---|---|
| $f(x)$ | -7 | -1 | 5 | 11 | 17 |

| $x$ | 0 | 1 | 2 | 3 | 4 |
|---|---|---|---|---|---|
| $g(x)$ | 1280 | 320 | 80 | 20 | 5 |

Find an equation for each. Verify the equations with a calculator table.

#### Solution

For $f$, as the value of $x$ increases by 1, the value of $y$ increases by 6. This indicates that $f$ is a linear function with slope $m = 6$. The $y$-intercept is at $(0, -7)$. We can represent this function with the equation $f(x) = 6x - 7$.

For $g$, as the value of $x$ increases by 1, the value of $y$ is divided by 4, which is equivalent to multiplying by $\frac{1}{4}$. This indicates that $g$ is an exponential function with base multiplier $b = \frac{1}{4}$. The $y$-intercept is $(0, 1280)$, so $a = 1280$. We represent this function with the equation $g(x) = 1280\left(\frac{1}{4}\right)^x$.

Next we can use a graphing calculator and the table feature to check our work. Type each equation in the [Y=] screen to check.

### Practice A

The tables in below show points on the graphs of two functions. Follow the instructions below, and then turn the page to check your work.

1. Find an equation for $f(x)$.

| $x$ | 0 | 1 | 2 | 3 | 4 |
|---|---|---|---|---|---|
| $f(x)$ | 500 | 100 | 20 | 4 | .08 |

2. Find an equation for $g(x)$.

| $x$ | 0 | 1 | 2 | 3 | 4 |
|---|---|---|---|---|---|
| $g(x)$ | 30 | 22 | 14 | 6 | -2 |

## B. Solving Exponential Equations for the Base

Sometimes we're asked to write the equation for an exponential function and don't have a table of values with the *x*-values conveniently increasing by 1. If we're given just two points for an exponential function, however, we can write a formula rule if we know how to solve equations of the form $b^n = k$ for the base $b$. Next we explore how to solve such equations before applying the process to writing equations for exponential functions.

Let's begin with a simple example, such as $b^2 = 16$. There are two solutions to this equation because $4^2 = 16$ and $(-4)^2 = 16$. We write $b = \pm 4$ to show both solutions.

Now let's look at the equation $b^4 = 10{,}000$. This equation also has two solutions because $10^4 = 10{,}000$ and $(-10)^4 = 10{,}000$. We write $b = \pm 10$ to show both solutions.

For equations of the form $b^n = k$ where $n$ is an odd integer, there will only be one solution. For example the equation $b^3 = 8$ has one solution, $b = 2$, because $2^3 = 8$. Notice that if we let $b = -2$, we don't have a solution to the original equation because $(-2)^3 = -8$.

The base $b$ of an exponential equation doesn't have to be an integer. We can use one of the properties of exponents to help us solve exponential equations for the base. We'll do so in the next example.

### Example 3

Find all real-number solutions to the following equations.

1. $b^2 = 144$
2. $b^3 = 200$
3. $12b^3 - 50 = 46$
4. $10b^4 = 6250$
5. $b^4 = -81$
6. $3b^6 = 17.5$

**Solution**

1. $b^2 = 144$

   With this equation, $b = 12$ or $b = -12$ because $12^2 = 144$ and $(-12)^2 = 144$. We can say $b = \pm 12$.

2. $b^3 = 200$

   Because there is no integer whose cube is 200, we use a property of exponents to solve.

   $b^3 = 200$

   $(b^3)^{1/3} = 200^{1/3}$    Raise both sides to the reciprocal power.

   $b = 200^{1/3}$    Simplify using a property of exponents: $(b^3)^{1/3} = b^1 = b$

   $b \approx 5.848$    Use a calculator and round to 3 decimal places.

3. $12b^3 - 50 = 46$

   This equation requires two solving steps before we deal with the exponent.

   $12b^3 - 50 = -146$

   $12b^3 = -96$    Add 50 to both sides.

   $b^3 = -8$    Divide both sides by 12.

   $(b^3)^{1/3} = (-8)^{1/3}$    Raise both sides to the reciprocal power.

   $b = -2$    Simplify.

4. $10b^4 = 6250$

The exponent is even, so we have two solutions, one positive and one negative.

$$10b^4 = 6250$$
$$b^4 = 625 \quad \text{Divide both sides by 10.}$$
$$(b^4)^{1/4} = 625^{1/4} \quad \text{Raise both sides to the reciprocal power.}$$
$$b = \pm 5 \quad \text{Simplify. Include both solutions: } 5^4 = 625 \text{ and } (-5)^4 = 625.$$

5. $b^4 = -81$

The equation $b^4 = -81$ has no real-number solutions because any base raised to an even-numbered exponent always gives a positive result.

6. $3b^6 = 17.5$

The exponent is even and the exponential expression equals a positive value, so we have two solutions:

$$3b^6 = 17.5$$
$$b^6 = 5.8\overline{3} \quad \text{Divide both sides by 3.}$$
$$(b^6)^{1/6} = 5.8\overline{3}^{\,1/6} \quad \text{Raise both sides to the reciprocal power.}$$
$$b \approx \pm 1.342 \quad \text{Simplify. Include both positive and negative solutions.}$$

We generalize the results of the previous examples to specify the type of solution(s) we get when solving equations of the form $b^n = k$ for $b$.

## Solving Exponential Equations for the Base

When solving an equation of the form $b^n = k$ for $b$,
- If $n$ is odd, the real-number solution is $k^{1/n}$.
- If $n$ is even and $k \geq 0$, the real-number solutions are $\pm k^{1/n}$.
- If $n$ is even and $k < 0$, there is no real-number solution.

## Practice B

Find all real-numbered solutions. Round any results to three decimal places. When you're finished, you are welcome to turn the page and check your work.

3. $-2b^3 = -128$  
4. $b^5 - 19 = 13$  
5. $-5b^4 + 11 = -1.2$  
6. $b^6 + 37 = 25$

### Practice A — Answers

1. The table for $f(x)$ indicates an exponential function with base multiplier $b = \frac{1}{5}$ and $y$-intercept $(0, 500)$. $f(x) = 500\left(\frac{1}{5}\right)^x$.

2. The table for $g(x)$ indicates a linear function with slope $m = -8$ and $y$-intercept $(0, 30)$. $g(x) = -8x + 30$.

## C. Using Two Points to Find Equations of Exponential Functions

Now that we know how to solve an exponential equation for the base, we can look at another way to write equations for exponential functions. If we know two points on an exponential curve, and if one of them is the $y$-intercept, $(0, a)$, then we have the $a$-value in the form of $y = ab^x$. After replacing $a$ with the $y$-value of the $y$-intercept, we substitute the second point into the equation $y = ab^x$ and solve for $b$.

### ▶ Example 4

Find an equation of the form $y = ab^x$ that contains the two points given. If necessary, round the value of $b$ to three decimal places

1. $(0, 5)$ and $(4, 405)$
2. $(0, 200)$ and $(8, 86)$

**Solution**

1. Because the $y$-intercept is $(0, 5)$, we know that $a = 5$ and that the equation has the form $y = 5b^x$. Next, we substitute the point $(4, 405)$ in the equation and solve for $b$.

   | | |
   |---|---|
   | $y = 5b^x$ | The equation with $a = 5$. |
   | $405 = 5b^4$ | Substitute 4 for $x$ and 405 for $y$. |
   | $81 = b^4$ | Divide both sides by 5. |
   | $81^{1/4} = (b^4)^{1/4}$ | Raise both sides to the reciprocal power. |
   | $b = \pm 3$ | Simplify. The exponent was even so there are two solutions. |

   The base of an exponential function is positive, so we use the solution $b = 3$ to write the equation $y = 5(3)^x$. This equation models exponential growth.

   We can now use a graphing calculator and the [TRACE] button to verify our work.

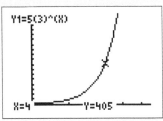

2. Because the $y$-intercept is $(0, 200)$, we know that $a = 200$ and that the equation has the form $y = 200b^x$. We substitute the point $(8, 86)$ in the equation and solve for $b$.

   | | |
   |---|---|
   | $y = 200b^x$ | The equation with $a = 200$. |
   | $86 = 200b^8$ | Substitute 8 for $x$ and 86 for $y$. |
   | $0.43 = b^8$ | Divide both sides by 200. |
   | $0.43^{1/8} = (b^8)^{1/8}$ | Raise both sides to the reciprocal power. |
   | $b \approx \pm 0.900$ | Simplify. The exponent was even so there are two solutions. |

   Again, we use only the positive solution $b = 0.9$ to write the equation $y = 200(0.9)^x$. This equation models exponential decay. You can verify the equation on your graphing calculator.

## Chapter 2: Exponential Functions

We can apply the process of finding an equation for an exponential curve to many real-life situations.

### Example 5

In 2000, 80 deer were introduced into a wildlife refuge. By 2006, the population had grown to 180. The population was growing exponentially. Assuming the trend continued, write an equation of an exponential function that models the growth of the deer population. Let $f(t)$ represent the population of deer $t$ years after 2000. Use your model to predict the number of deer in 2014.

### Solution

The independent variable $t$ is the number of years after 2000, so we can write the information given in the problem as input-output pairs, (0, 80) and (6, 180). The first point represents the $y$-intercept of the graph of the function, so our $a$-value is 80. We can substitute the second point into the equation $f(t) = 80b^t$ and solve to find $b$.

$f(t) = 80 b^t$

$180 = 80 b^6$   Substitute using the point (6, 180).

$2.25 = b^6$   Divide both sides by 80.

$2.25^{1/6} = (b^6)^{1/6}$   Raise both sides to the reciprocal exponent.

$b \approx 1.1447$   Only use the positive solution. Round to 4 decimal places.

The exponential model for the population of deer is
$f(t) = 80(1.1447)^t$

To predict the population in 2014, let $t = 14$:
$f(14) = 80(1.1447)^{14} \approx 530.6$

We predict the population of deer was about 531 in the year 2014. However, this exponential function may only model short-term growth. As the input becomes larger, the output will get increasingly larger — so much so that the model may not be useful in the long term.

We graph our model to observe the population growth of deer in the refuge over time. Notice that the graph in Figure 1 passes through both points given in the problem, (0, 80) and (6, 180).

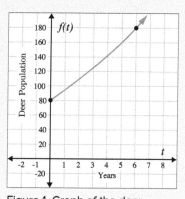

Figure 1. Graph of the deer population $t$ years after 2000.

### Practice B — Answers

3. $b = -4$

4. $b = 2$

5. $b \approx \pm 1.250$

6. There are no real-numbered solutions.

## Practice C

Using the information given, write an exponential function of the form $y = ab^x$. If necessary, round the value of $b$ to three decimal places. To check your solutions, turn the page.

7. Points on the curve $(0, 3)$ and $(6, 192)$
8. Points on the curve $(0, 180)$ and $(5, 21)$
9. The value of new online company is growing exponentially. In 2010, the company was worth 52 million dollars. By 2015, the value of the company is 96 million dollars. Let $y$ be the value of the company in millions of dollars $x$ years after 2010.

## Exercises 2.4

For each table of values, determine if it represents a linear function or an exponential function. Then write an equation for the function. For a linear function use the form $f(x) = mx + b$. For an exponential function, use $f(x) = ab^x$. You should check your solutions using the table feature on your graphing calculator.

1.
| $x$ | $f(x)$ |
|---|---|
| 0 | 8 |
| 1 | 16 |
| 2 | 32 |
| 3 | 64 |
| 4 | 128 |

2.
| $x$ | $f(x)$ |
|---|---|
| 0 | 448 |
| 1 | 224 |
| 2 | 112 |
| 3 | 56 |
| 4 | 28 |

3.
| $x$ | $f(x)$ |
|---|---|
| 0 | 11 |
| 1 | 4 |
| 2 | -3 |
| 3 | -10 |
| 4 | -17 |

4.
| $x$ | $f(x)$ |
|---|---|
| 0 | 2 |
| 1 | 10 |
| 2 | 50 |
| 3 | 250 |
| 4 | 1250 |

5.
| $x$ | $f(x)$ |
|---|---|
| 0 | 243 |
| 1 | 81 |
| 2 | 27 |
| 3 | 9 |
| 4 | 3 |

6.
| $x$ | $f(x)$ |
|---|---|
| 0 | -9 |
| 1 | 1 |
| 2 | 11 |
| 3 | 21 |
| 4 | 31 |

Find all real-number solutions. Round your results to three decimal places.

7. $b^2 = 49$
8. $b^4 = 16$
9. $b^5 = 243$
10. $b^3 = 216$
11. $-1000 b^3 = 125$
12. $3 b^6 = 340$
13. $12.3 b^4 - 65.2 = 557$
14. $-20 b^5 - 14.1 = 27.4$
15. $b^6 - 28.5 = -40.5$
16. $94 b^2 + 61 = 5822$
17. $3.9 b^3 + 9.1 = 11.4$
18. $6 b^4 + 45 = 21$

For the following exercises, find the formula for an exponential function of the form $y = ab^x$ that passes through the given points. Round the value of $b$ to three decimal places. Verify your equation with a graphing calculator.

19. $(0, 6)$ and $(3, 750)$
20. $(0, 2000)$ and $(2, 20)$
21. $(0, 450)$ and $(4, 18)$
22. $(0, 14)$ and $(5, 629)$
23. $(0, 4.5)$ and $(5, 11.2)$
24. $(0, 2.9)$ and $(3, 11.2)$
25. $(0, 11.77)$ and $(8, 1.18)$
26. $(0, 93.67)$ and $(6, 5.24)$

The following exercises give verbal descriptions of exponential models. For each,

a. Use the information to write two points on the curve.
b. Find an equation of the curve in the form $f(t) = ab^t$.
c. Use your exponential model to make the prediction asked for.

27. Research biologists released 300 chicken turtles (*deirochelys reticularia*) into a wetland with hopes of repopulating the habitat. After 7 years, the research team estimated that the population had grown to 550 chicken turtles. Predict the population of chicken turtles in the habitat 12 years after the initial release.

28. Bacteria are growing in a Petri dish in a biology lab such that the original population of 200 bacteria have grown to 7800 bacteria in just 4 hours. Predict the number of bacteria after 6 hours.

29. In 2007, Linda purchased a new Toyota Prius priced at $19,500. In 2016, the Kelly Blue Book value of her car was just $5600. Estimate the value of her car in 2020.

30. In 2010, the value of a share of certain stock on the NASDAQ was $97.50, but the price per share of that stock has been decreasing each year so that the stock was only valued at $73.40 in 2016. Predict the value of the stock in 2025.

### Practice C — Answers

7. $a = 3, b = 2$, and $y = 3(2)^x$
8. $a = 180, b \approx 0.651$, and $y = 180(0.651)^x$
9. We can write two points: $(0, 52)$ and $(5, 96)$. $a = 52, b \approx 1.130$, and $y = 52(1.13)^x$

# 2.5 Using Exponential Functions to Model Data

## Overview

Exponential functions can model many types of data, including investment growth, radioactive decay, and the temperature of a cooling object. In this section, we will take what we've learned in Section 2.4 about writing exponential equations and apply those skills to real world situations. You will learn to:

- Find and use the percent change of an exponential model
- For a model $f(t) = ab^t$, interpret the meaning of $a$ and $b$ in the context of the problem
- Find an equation of an exponential model by using data from real world situations
- Model a half-life situation
- Use exponential regression to find an exponential model for a set of data
- Make estimates and predictions using an exponential model

## A. Using the Base Multiplier to Find a Model

We begin with a look at the meaning of the base multiplier, $b$, in equations of the form $f(t) = ab^t$. We want to understand exactly how the value of $b$ is related to the percent rate of growth or decay over time. So let's take a second look at exponential growth and decay.

**Exponential growth** refers to an increase based on a constant multiplicative rate of change over equal increments of time. It is a constant *percent increase* of the original amount over time.

**Exponential decay** refers to a decrease based on a constant multiplicative rate of change over equal increments of time. It is a constant *percent decrease* of the original amount over time.

**Percent change** refers to a change based on a percent of the original amount. An exponential growth model has the percent change, in decimal form, *added* to 1 to indicate a percent increase of the original amount. An exponential decay model has the percent change, in decimal form, *subtracted* from 1 to indicate a percent decrease of the original amount over time.

> ### Percent Rate of Change of an Exponential Model
> 
> If $f(t) = ab^t$, where $a > 0$, models a quantity at time $t$, then the percent rate of change is constant. Specifically, for a percent in decimal form,
> 
> - If $b > 1$, then the quantity grows exponentially at a rate of $b - 1$ percent per unit of time.
> - If $0 < b < 1$, then the quantity decays exponentially at a rate of $1 - b$ percent per unit of time.

## Example 1

State whether the functions below represent exponential growth or decay of a quantity over time. Then determine the percent change per unit of time.

1. $P(t) = 45(1.07)^t$
2. $f(t) = 6600(0.92)^t$
3. $g(t) = 812(0.75)^t$
4. $A(t) = 3000(1.038)^t$
5. $P(t) = 3(2)^t$

### Solution

1. $b = 1.07$, so $b > 1$, indicating exponential growth. $1.07 - 1 = 0.07$, which indicates a 7% increase per unit of time.

2. $b = 0.92$, so $0 < b < 1$, indicating exponential decay. $1 - 0.92 = 0.08$, which indicates an 8% decrease per unit of time. Although there's an 8% decrease over time, this number is not apparent in the equation. What we see is $b = 0.92$, which indicates that although we *lose* 8% over time, what *remains* is the other 92%.

3. $b = 0.75$, so $0 < b < 1$, indicating exponential decay. $1 - 0.75 = 0.25$, which indicates a 25% decrease per unit of time. 75% remains.

4. $b = 1.038$, so $b > 1$, indicating exponential growth. $1.038 - 1 = 0.038$, which indicates a 3.8% increase per unit of time.

5. $b = 2$, so $b > 1$, indicating exponential growth. $2 - 1 = 1$, which indicates a 100% increase per unit of time.

The last equation from Example 1, $P(t) = 3(2)^t$, shows a 100% increase, which means that 100% is added to the quantity already present. In essence, 100% of a quantity has grown to 200% of that quantity. It is easier to think of this as "doubling." In fact, an exponential model with base $b = 2$ is often referred to as a "doubling function." Likewise, if the base $b = 3$, we say a quantity "triples" over time.

## B. Finding a Model Using Data Described in Words

When $f(t) = ab^t$ models change over a quantity of time $t$, we can find the initial value of the quantity by letting $t = 0$.

$$f(0) = ab^0 = a(1) = a$$

The initial quantity for this model is the $a$-value in the equation. This corresponds to the $y$-intercept of the graph of the function, $(0, a)$. In Section 2.4, for example, we found the exponential model for a population of deer over $t$ years: $f(t) = 80(1.1447)^t$. For this model, when $t = 0$, $f(0) = 80$, which means the initial quantity of deer was 80.

In the equation $f(t) = 80(1.1447)^t$, the value of the base $b = 1.1447 = 1 + 0.1447$. This indicates a 14.47% increase in the population per year.

We can use the initial quantity of an exponential model together with the constant percent rate of change to write and interpret exponential equations.

## Exponential Models

If $f(t) = ab^t$, where $a > 0$, models a quantity at time $t$, then
- $a$ represents the initial quantity at time $t = 0$.
- $b$ represents $1 \pm r$, where $r$ is the percent increase or decrease of the quantity over time.

### Example 2

Write an exponential function of the form $f(t) = ab^t$ to represent the following situations.

1. The initial value is $4545, increasing 6% annually.
2. The original amount is 132 grams, decaying 7.6% every hour.
3. The beginning population is 2.7 million, growing 1.89% annually.
4. The value when purchased is $16,000, depreciating by 13% annually.

**Solutions**

1. $a = 4545$ and $b = 1 + 0.06 = 1.06$, so $f(t) = 4545\,(1.06)^t$, where $t$ is measured in years.
2. $a = 132$ and $b = 1 - 0.076 = 0.924$, so $f(t) = 132\,(0.924)^t$, where $t$ is measured in hours.
3. $a = 2.7$ and $b = 1 + 0.0189 = 1.0189$, so $f(t) = 2.7\,(1.0189)^t$, where $t$ is measured in years.
4. $a = 16000$ and $b = 1 - 0.13 = 0.87$, so $f(t) = 16000\,(0.87)^t$, where $t$ is measured in years.

In Section 2.4, we saw an exponential function that represents the value of an interest-bearing account over time. If an account earns an annual interest rate of $r$ percent compounded annually, we can use $r$ to write a formula for the value of the account after $t$ years. The next two examples involve investments earning interest that is compounded annually.

### Example 3

A person invests $3000 in an account that earns 4.5% interest compounded annually.

a. Let $f(t)$ be the value (in dollars) of the investment after $t$ years. Write an equation for $f$.
b. What will be the value of the investment after 5 years? After 10 years?

**Solution**

a. We use the model for exponential growth $f(t) = ab^t$. The initial value $a = 3000$. The base $b$ represents a 4.5% increase, so $b = 1 + .045 = 1.045$.

$$f(t) = 3000\,(1.045)^t$$

b. To find the value in 5 years and in 10 years, we let $t = 5$ and then $t = 10$, so:

$f(5) = 3000(1.045)^5 \approx 3738.55$
$f(10) = 3000(1.045)^{10} \approx 4658.91$

The value of the investment will be $3738.55 in 5 years and $4658.91 in 10 years.

We can represent the exponential function in Example 3 graphically and in table form. Figure 1 presents a graphing calculator graph of $y = 3000(1.045)^x$. The $y$-intercept $(0, 3000)$ corresponds to the initial investment, and the function increases exponentially.

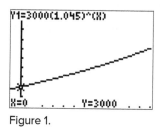

Figure 1.

| $t$ | $f(t) = 3000(1.045)^t$ |
|---|---|
| 0 | $3000(1.045)^0 = 3000$ |
| 1 | $3000(1.045)^1 = 3135$ |
| 2 | $3135(1.045)^2 = 3276.08$ |
| 3 | $3276.08(1.045)^3 = 3423.50$ |
| 4 | $3423.50(1.045)^4 = 3577.56$ |
| 5 | $3577.56(1.045)^5 = 3738.55$ |

Figure 2. Values of a compounded interest account.

Figure 2 shows a table of the first years of the investment from Example 3. Notice that after the first year, the interest earned on the account is $135. We find this by subtracting the initial investment from the value of the investment at the end of year one.

In the second year, the account earns almost $141 in interest. In the third year, the account earns over $147 in interest. The amount of interest earned continues to increase with each year. This is why we say that an exponential growth function "grows more and more rapidly" as the independent variable increases.

## Example 4

A stockbroker encourages his clients to invest in a certain mutual fund by claiming, "You can double your money every 8 years." With that in mind, do the following:

a. Write an equation for an exponential function that models the growth of an initial investment of $5000 in a mutual fund that doubles in value every 8 years.

b. Use your exponential model to determine the average percent increase per year of the investment.

### Solution

a. Let $f(t)$ be the value of the investment $t$ years after the initial amount is invested. The initial amount is $5000, so $a = 5000$ when $t = 0$. Assuming the investment grows approximately exponentially, the equation is of the form

$f(t) = 5000 b^t$

If the value of the investment has doubled after 8 years to $10,000, then when $t = 8, f(t) = 10000$. We substitute the coordinates of the ordered pair $(8, 10000)$ into our equation and solve for $b$.

$$f(t) = 5000 b^t$$
$$10000 = 5000 b^8 \quad \text{Substitute 8 for } t \text{ and 10000 for } f(t).$$
$$2 = b^8 \quad \text{Divide both sides by 5000.}$$
$$2^{1/8} = (b^8)^{1/8} \quad \text{Raise both sides to the reciprocal exponent.}$$
$$1.0905 \approx b \quad \text{Simplify. Use the positive solution only.}$$

Next we substitute 1.0905 for $b$ in the equation.

$$f(t) = 5000(1.0905)^t$$

c. The base of the exponential model is $b = 1.0905$. The percent increase is calculated by subtracting 1 from $b$:

$$1.0905 - 1 = 0.0905$$

The decimal value 0.0905 indicates a percent increase of 9.05% per year.

## Practice B

State whether the functions below represent exponential growth or decay. Then determine the percent change. Turn the page to check your solutions.

1. $g(t) = 400(0.88)^t$
2. $P(t) = 9.9(1.35)^t$
3. $f(t) = 7000(1.022)^t$

A person invests $2500 in an account that earns 6.33% interest compounded annually. Use this situation to answer questions 4 and 5.

4. Let $f(t)$ be the value (in dollars) of the investment after $t$ years. Write an equation for $f$.
5. What will be the value of the investment after 4 years? After 8 years?

## C. Exponential Decay and Half-Life Applications

We've seen that if the base of our exponential function is between 0 and 1, we have a model of exponential decay, a decreasing function.

### Example 5

A car tire that initially has air pressure of 33 psi (pounds per square inch) has begun leaking. The rate at which pressure in the tire is lost is about 4% per minute.

a. Let $f(t)$ represent the air pressure in the tire (in psi) $t$ minutes after it begins leaking. Write an equation for $f$.
b. Predict the air pressure in the tire after 5 minutes and then after a half hour.

### Solution

**a.** The initial amount $a = 33$. The base is found by subtracting the percent decrease from 1: $b = 1 - 0.04 = 0.96$. If 4% of the air pressure is lost each minute, then 96% of the air pressure remains. The equation is: $f(t) = 33(0.96)^t$

**b.** After 5 minutes, $t = 5$, and after a half hour, $t = 30$, so:

$f(5) = 33(0.96)^5 \approx 26.9$. After 5 minutes, the air pressure is 26.9 psi.
$f(30) = 33(0.96)^{30} \approx 9.7$. After 30 minutes, the air pressure is 9.7 psi.

We can also describe how quickly a quantity decays by its **half-life**. If a quantity decays exponentially, the half-life is the amount of time it takes for that quantity to be reduced to half its initial value. The term is often used in physics to describe how quickly unstable atoms undergo radioactive decay. The medical sciences often refer to the biological half-life of drugs and other chemicals in the body.

For example, carbon-14 is a radioactive isotope of carbon that is often used to date organic remains from archaeological sites. The half-life of carbon-14 is about 5730 years, which means that 100 grams of carbon-14 would decay to just 50 grams of carbon-14 over a period of 5730 years.

### Example 6

An archaeologist estimates that the initial amount of carbon-14 in a fossil sample was 100 milligrams. The half-life of carbon-14 is about 5730 years.

**a.** Write an equation for $f(t)$ representing the amount of carbon-14 (in mg) remaining in the fossil after $t$ years.

**b.** Use your model to estimate the amount of carbon-14 remaining in the fossil after 20,000 years.

### Solution

**a.** At time $t = 0$, the initial quantity of carbon-14 is 100 milligrams so the value of $a = 100$. After 5730 years, there will be $100\left(\frac{1}{2}\right) = 50$ milligrams. After a second 5730-year period, 11,460 years, there will be $100\left(\frac{1}{2}\right)\left(\frac{1}{2}\right) = 25$ milligrams. After a third 5730-year period, 17,190 years, there will be $100\left(\frac{1}{2}\right)\left(\frac{1}{2}\right)\left(\frac{1}{2}\right) = 12.5$ milligrams. We simplify these calculations using exponents and organize the results in the Figure 3 table.

In the table, each exponent on the base is equal to the value of $t$ divided by 5730 and gives the number of 5730-year periods. In other words, the exponent represents the number of

| $t$, years | $f(t)$, milligrams |
|---|---|
| 0 | $100\left(\frac{1}{2}\right)^0 = 100$ |
| 5730 | $100\left(\frac{1}{2}\right)^1 = 50$ |
| 11,460 | $100\left(\frac{1}{2}\right)^2 = 25$ |
| 17,190 | $100\left(\frac{1}{2}\right)^3 = 12.5$ |
| $t$ | $100\left(\frac{1}{2}\right)^{t/5730} = f(t)$ |

Figure 3

times the initial quantity of carbon-14 will be reduced by half. An exponential function that models this radioactive decay situation is:

$$f(t) = 100 \left(\frac{1}{2}\right)^{t/5730}$$

c. For $t = 20000$, $f(20000) = 92 \left(\frac{1}{2}\right)^{20000/5730} \approx 8.19$.

After 20,000 years, there will be about 8.19 mg of carbon-14 remaining in the fossil.

The base $b$ in the equation in Example 6 can be written in decimal form, $b = 0.5$, however it is conventional to write the base as a fraction, $b = \frac{1}{2}$, when modeling a half-life application. In either form, the base represents a 50% decrease in the amount of quantity per half-life.

Furthermore, the $t$ variable in the model in Example 6 is divided by 5730. We can rewrite the equation using exponent rules:

$$f(t) = 100 \left(\frac{1}{2}\right)^{t/5730} = 100 \left(\left(\frac{1}{2}\right)^{1/5730}\right)^t \approx 100 (0.9998790)^t$$

We have to write the base $b$ rounded to 7 decimal places to avoid rounding the base to 1. This is one reason that half-life applications often use $b = \frac{1}{2}$ for the base and a rational expression for the exponent.

Using the second form of the model, with $b \approx 0.9998790$, we see each year 99.9879% of the previous year's carbon-14 remains. In other words, the carbon-14 decays by about 0.0121 % per year.

Here are a few more examples of exponential models for half-life applications:

1. Hydrogen-3 has a half-life of 12.3 years. Let $f(t)$ be the quantity remaining if an initial amount of 75 milligrams decays for $t$ years.

$$f(t) = 75 \left(\frac{1}{2}\right)^{t/12.3}$$

2. Cesium-137 has a half-life of 30 years. Let $f(t)$ be the quantity remaining if an initial amount of 2.3 grams decays for $t$ years.

$$f(t) = 2.3 \left(\frac{1}{2}\right)^{t/30}$$

3. Caffeine in the bloodstream has a half-life of about 6 hours. Let $f(t)$ be the quantity remaining $t$ hours after a person consumes 180 milligrams.

$$f(t) = 180 \left(\frac{1}{2}\right)^{t/6}$$

### Practice B — Answers

1. $b = 0.88$, indicating exponential decay.
   $1 - 0.88 = 0.12$, a 12% decrease over time.

2. $b = 1.35$, indicating exponential growth.
   $1.35 - 1 = 0.35$, a 35% increase over time.

3. $b = 1.022$, indicating exponential growth.
   $1.022 - 1 = 0.022$, a 2.2% increase over time.

4. $f(t) = 2500 (1.0633)^t$

5. $f(4) = 2500 (1.0633)^4 \approx 3195.68$ and
   $f(8) = 2500 (1.0633)^8 \approx 4084.95$
   The value of the investment is thus $3195.68 after 4 years and $4084.95 after 8 years.

## Practice C

Now it's your turn to work with real-world models. When you are finished, turn the page to check your solutions.

6. A new computer that cost $2200 depreciates in value by about 13% per year. Write an exponential equation that models the value of the computer (in dollars) after $t$ years. Use the model to estimate the value after 6 years.

7. The half-life of caffeine in a person's bloodstream is about 6 hours. Write an exponential functions that models the amount of caffeine (in mg) remaining in a person's bloodstream after $t$ hours if they began with 100 milligrams of caffeine. Use the model to predict the amount of caffeine remaining after 8 hours.

## D. Finding a Model Using a Data Table and Exponential Regression

In Section 2.4, we found equations for exponential functions using two points when one of those points was the y-intercept (0, a). In this book, if we have several data points, or if we don't know the initial value, we will use **exponential regression** on a graphing calculator to find a model for our data.

To do that, we enter the data into Lists 1 and 2 and use the command ExpReg on the calculator to fit an exponential function to a set of data points. This returns an equation of the form $y = ab^x$.

From this equation, we know the following:

- The initial value of the model is $y = a$.
- If $b > 1$, the function models exponential growth.
- If $0 < b < 1$, the function models exponential decay.

These are the steps you'll need to follow to perform exponential regression using a graphing calculator:

1. Use the [STAT] button and then the Edit menu to enter given data. Clear any existing data from the lists. List the input values in the L1 column. List the output values in the L2 column.

2. Graph and observe a scatter plot of the data using the [STAT PLOT] feature. Use [ZOOM] [9] to adjust axes to fit the data. Then verify that the data follows an exponential pattern.

3. Find the equation that models the data. Select ExpReg from the [STAT] then [CALC] menu. Use the values returned for $a$ and $b$ to record the model, $y = ab^x$.

4. Graph the model in the same window as the scatter plot to verify that it is a good fit for the data.

## Example 7

In 2007, Indiana University published a study investigating the crash risk of alcohol-impaired driving. The study used data from 2,871 crashes to measure the association of a person's blood alcohol level (BAC) with the risk of being in an accident. Figure 4 presents some results from the study. The relative risk is a measure of how many times more likely a person is to crash. So, for example, a person with a BAC of 0.09 is 3.54 times as likely to crash as a person who has not been drinking alcohol.

| BAC | 0 | 0.01 | 0.03 | 0.05 | 0.07 | 0.09 | 0.11 | 0.13 | 0.15 | 0.17 | 0.19 | 0.21 |
|---|---|---|---|---|---|---|---|---|---|---|---|---|
| Relative Risk of Crashing | 1 | 1.03 | 1.06 | 1.38 | 2.09 | 3.54 | 6.41 | 12.6 | 22.1 | 39.05 | 65.32 | 99.78 |

Figure 4.

Use the data from Figure 4 to do the following:

a. Let $x$ represent the BAC level, and let $y$ represent the corresponding relative risk. Use exponential regression to fit a model to these data.

b. After 6 drinks, a person weighing 160 pounds will have a BAC of about 0.16. How many times more likely is this person to crash if they drive after having a 6-pack of beer?

## Solution

a. *Step 1:* Using the [STAT] button and then the Edit menu on a graphing calculator, list the BAC values in L1 and the relative risk values in L2.

*Step 2:* Use the [STAT PLOT] feature to verify that the scatter plot follows an exponential pattern.

*Step 3:* Use the ExpReg command from the [STAT] and then the [CALC] menus to obtain the exponential model,
$y = 0.58304829(2.20720213\text{E}10)^x$.

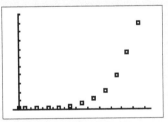

*Step 4:* Converting the base from scientific notation, we have $y = 0.58304829(22{,}072{,}021{,}300)^x$. We can then graph the model in the same window as the scatter plot to verify it is a good fit.

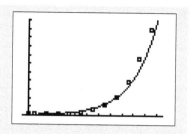

c. To estimate the risk associated with a BAC of 0.16, we substitute 0.16 for $x$ in the model and calculate $y$.

$$y = 0.58304829(22{,}072{,}021{,}300)^x$$
$$= 0.58304829(22{,}072{,}021{,}300)^{0.16}$$
$$\approx 26.35$$

If a 160-pound person drives after having 6 drinks, he or she is about 26.35 times more likely to crash than if driving while sober.

## Example 8

The data in Figure 5 shows the world population in billions over the last century. For this problem, let $f(x)$ be the world population in billions $x$ years after 1900, so that $x = 30$ corresponds to the year 1930, $x = 100$ corresponds to the year 2000, and so on.

a. Use exponential regression to fit a model to this data.
b. Interpret the meaning of the base $b$ in the context of the problem.
c. Interpret the meaning of the $y$-intercept in the context of the problem.
d. Evaluate and interpret $f(125)$.

| Year | Population (billions) |
|---|---|
| 1930 | 2.070 |
| 1940 | 2.295 |
| 1950 | 2.500 |
| 1960 | 3.050 |
| 1970 | 3.700 |
| 1980 | 4.454 |
| 1990 | 5.279 |
| 2000 | 6.080 |
| 2010 | 6.866 |

Figure 5. World population figures from the U.S. Census Bureau.

### Solution

a. Rounding the values of $a$ and $b$ to three significant digits, the regression model is $y = f(x) = 1.210(1.0161)^x$.

### Practice C — Answers

6. $a = 2200$, $b = 1 - 0.13 = 0.87$, and $f(t) = 2200(0.87)^t$
$f(6) = 2200(0.87)^6 \approx 953.98$, so the computer is worth about $954 after 6 years.

7. $a = 100$, $b = \frac{1}{2}$, and $f(t) = 100\left(\frac{1}{2}\right)^{t/6}$
$f(8) = 100\left(\frac{1}{2}\right)^{8/6} \approx 39.7$, so after 8 hours, there are 39.7 mg of caffeine in the bloodstream.

b. The model represents exponential growth because $b > 1$. More specifically, $b = 1.0161 = 1 + 0.0161$, which tells us the world population is increasing by 1.61% per year.

c. The $y$-intercept is $(0, 1.210)$ and indicates that the world population in 1900, the initial year in this problem, was about 1.210 billion.

d. $y = f(125) = 1.210(1.0161)^{125} \approx 8.909$

Our model predicts the world population will be about 8.909 billion in 2025. Notice that if you use the regression equation without rounding $a$ and $b$, the prediction is 8.926 billion people. This discrepancy gives you an idea about how sensitive a model can be when it comes to rounding.

## Practice D

The table shows a recent graduate's credit card balance each month after graduation. Use this data to solve the following problems. When you are finished, turn the page to check your solutions.

| Month | 1 | 2 | 3 | 4 | 5 | 6 | 7 | 8 |
|---|---|---|---|---|---|---|---|---|
| Debt ($) | 620.00 | 761.88 | 899.80 | 1039.93 | 1270.63 | 1589.04 | 1851.31 | 2154.92 |

8. Use exponential regression to fit a model to these data.
9. Determine the percent increase of the model.
10. If spending continues at this rate, what will the graduate's credit card debt be one year after graduating?

## Exercises 2.5

State whether the functions below represent exponential growth or decay of a quantity over time. Then determine the percent change per unit of time.

1. $f(t) = 225(1.19)^t$
2. $P(t) = 4.1(1.64)^t$
3. $g(t) = 2000(0.98)^t$
4. $h(t) = 7.48(0.87)^t$
5. $h(t) = 9.13(1.0285)^t$
6. $f(t) = 290(1.0854)^t$
7. $A(t) = 300\left(\frac{1}{2}\right)^{t/12}$
8. $Q(t) = 12\left(\frac{1}{2}\right)^{t/5}$

Write an exponential function of the form $f(t) = ab^t$ to represent the situation described. Assume that $t$ is measured in years.

9. The beginning population is 42,000, growing 2.54% annually.
10. The beginning population is 1.6 million, decreasing 4.1% annually.
11. The initial quantity is 250 milligrams, with a half-life of 28 years.
12. The initial value is $7200, increasing 5% annually
13. A stock investment of $1500 doubles every 12 years.
14. The original amount is 16 grams, decaying 7.6% every hour.

15. The initial value is $2340, depreciating by 9.5% per year.
16. An investment of $450 triples every 22 years.
17. A person invests $2700 in an account that earns 3.25% interest compounded annually.
    a. Let $f(t)$ be the value in dollars of the account after $t$ years. Write an equation for $f$.
    b. What will be the value of the account after 5 years? After 10 years?
18. A person invests $3500 in an account that earns 7.1% interest compounded annually.
    a. Let $f(t)$ be the value in dollars of the account after $t$ years. Write an equation for $f$.
    b. What will be the value of the investment after 6 years? After 12 years?
19. Someone invests $1000 in stocks today, and the stocks' value doubles every 7 years.
    a. Let $f(t)$ be the value in dollars of the investment after $t$ years. Write an equation for $f$.
    b. What will be the value of the investment 25 years from now?
20. Someone invests $9000 in stocks today and the stocks' value doubles every 12 years.
    a. Let $f(t)$ be the value in dollars of the investment after $t$ years. Write an equation for $f$.
    b. What will be the value of the investment after 5 years? After 10 years?
21. A small coastal fishing town in Oregon has been losing job opportunities and losing population since 1990. In 1990, the population was 9740 but it has been decreasing by about 7.2% per year.
    a. Write an exponential function $P$ that models the population $t$ years since 1990.
    b. If this trend continues, predict the population in 2025.
22. A small logging town in Oregon has been losing job opportunities and losing population since 2000. In 2000, the population was 13,400 but it has been decreasing by about 8.1% per year.
    a. Write an exponential function that models the population $t$ years since 2000.
    b. If this trend continues, predict the population in 2020.
23. A sport utility vehicle (SUV) that cost $32,000 new depreciates in value by about 24% per year after it is purchased. Write an equation to model the value $V$ of the SUV $t$ years after it is purchased and use the model to predict the value of the car after 4 years.
24. The value of a new printing press is $375,000. This value is expected to depreciate by about 22% per year for the first 7 years of operation. Write an equation to model the value $V$ of the press after $t$ years of operation and use the model to predict the value of the press after 7 years.
25. The half-life of radioactive radium (Ra-226) is 1620 years. Write an equation to model the amount of Ra-226 left after $t$ years in a sample that originally contained 40 mg of the isotope. Use the model to estimate the amount of Ra-226 remaining after 2000 years.

## Practice D — Answers

8. Rounding to three decimal places, the exponential regression model that fits these data is $y = 522.886(1.196)^x$.

9. $1.196 - 1 = 0.196$, so the percent increase is about 19.6% per month.

10. If spending continues at this rate, the graduate's credit card debt will be about $4,479 after one year ($x = 12$). Yikes.

26. An archaeologist estimates that the initial amount of carbon-14 in a fossil sample was 26 milligrams. The half-life of carbon-14 is about 5730 years. Write an equation to model the amount of carbon-14 (in mg) remaining in the fossil after $t$ years. Use the model to estimate the amount of carbon-14 remaining after 12,000 years.

27. The half-life of aspirin in a person's bloodstream is about 20 minutes. If a person's bloodstream contains 200 milligrams of aspirin, how much of that aspirin will remain after 2 hours?

28. The half-life of caffeine in a person's bloodstream is about 6 hours. A person who consumes an energy drink has about 242 milligrams of caffeine in their bloodstream. How much caffeine remains after 5 hours?

Use a graphing calculator for the remaining problems.

29. The table below shows U.S. population data from the U.S. Census Bureau for selected years over the last century. Use this data to do the following.

| Year | 1920 | 1940 | 1960 | 1980 | 2000 | 2014 |
|---|---|---|---|---|---|---|
| US Population (millions) | 106.5 | 132.1 | 180.7 | 226.5 | 282.2 | 318.9 |

   a. Let $y$ be the U.S. population in millions $x$ years since 1900. Create a scatter plot of the data and calculate an exponential regression model. Round values to three decimal places.
   b. Graph the regression equation in the same window as the scattergram.
   c. Determine the approximate percent change of the population over time.
   d. Predict the population of the U.S. in 2025.

30. The table below shows the U.S. gross domestic product (GDP) in trillions of U.S. dollars (USD) for selected years.

| Year | 1960 | 1970 | 1980 | 1990 | 2000 | 2010 | 2015 |
|---|---|---|---|---|---|---|---|
| U.S. GDP (Trillion $) | 0.543 | 1.076 | 2.862 | 5.980 | 10.280 | 14.960 | 17.914 |

   a. Let $y$ be the U.S. GDP in trillions of USD $x$ years since 1900. Create a scatter plot of the data and calculate an exponential regression model. Round values to three decimal places.
   b. Graph the regression equation in the same window as the scattergram.
   c. Determine the approximate percent change of the GDP over time.
   d. Predict the U.S. GDP in 2030.

31. The table shows the revenue in billions of dollars collected by the U.S. Internal Revenue Service (IRS) for selected years.

| Year | 1960 | 1965 | 1970 | 1975 | 1980 | 1985 | 1990 | 1995 | 2000 |
|---|---|---|---|---|---|---|---|---|---|
| Revenue (Billion $) | 91.8 | 114.4 | 293.8 | 293.8 | 519.4 | 742.9 | 1056.4 | 1375.7 | 2096.9 |

a. Let $y$ be revenue collected by the IRS (in billions of dollars) $x$ years since 1900. Create a scatter plot of the data and calculate an exponential regression model. Round values to three decimal places.
b. Graph the regression equation in the same window as the scattergram.
c. Interpret the meaning of the values of $a$ and $b$ in the context of the problem.
d. Estimate the revenue collected by the IRS in 2018.

32. The table below hows the number of movie theater screens (in thousands) in the U.S. for selected years.

| Year | 1975 | 1980 | 1985 | 1990 | 1995 | 2000 |
|---|---|---|---|---|---|---|
| Screens (thousands) | 11 | 14 | 18 | 23 | 27 | 37 |

a. Let $y$ be number of movie screens (in thousands) $x$ years since 1975. Create a scatter plot of the data and calculate an exponential regression model. Round values to three decimal places.
b. Graph the regression equation in the same window as the scattergram.
c. Interpret the meaning of the values of $a$ and $b$ in the context of the problem.
d. Predict the number of movie screens in 2018.

33. The table to the right presents the number of computers connected to the Internet for selected years.

a. Let $y$ be the number of computers connected to the Internet $x$ years since 1983. Create a scatter plot of the data and calculate an exponential regression model. Round values to three decimal places.
b. Graph the regression equation in the same window as the scattergram.
c. Estimate the number of computers on the Internet in 2000.

| Year | Approximate Number of Computers on the Internet |
|---|---|
| 1983 | 562 |
| 1984 | 1,024 |
| 1985 | 1,961 |
| 1986 | 2,308 |
| 1987 | 5,089 |
| 1988 | 28,174 |
| 1989 | 80,000 |
| 1990 | 290,000 |
| 1991 | 500,000 |
| 1992 | 727,000 |
| 1993 | 1,200,000 |
| 1994 | 2,217,000 |

CHAPTER 3
# Logarithmic Functions

In Chapter 2, we studied exponential functions. In this chapter, we study *logarithmic* functions, which are related to exponential functions. Logarithmic functions are useful in helping us solve exponential equations for the variable, and that will be our main focus in Sections 3.2 and 3.3. We will also study some properties of logarithms and discuss how they relate to the properties of exponents.

In addition to being a sort of solving tool for exponential equations, logarithmic functions have many real-world applications of their own. One of the most familiar uses of logarithms is to measure the relative magnitude of earthquakes using the Richter scale. The Richter scale is a base-ten logarithmic scale. It describes how an earthquake of magnitude 8 is not twice as great as an earthquake of magnitude 4. It's *10,000 times* as great because $10^{8-4} = 10^4 = 10,000$. We will investigate the nature of the Richter scale and the base-ten logarithm upon which it is defined.

In this chapter, we'll cover the following topics:

3.1 Introduction to Logarithmic Functions ................................................................. page 142

3.2 Properties of Logarithms ................................................................................... page 152

3.3 Natural Logarithms............................................................................................ page 170

# 3.1 Introduction to Logarithmic Functions

## Overview

In this section, we define a new function called a *logarithm*. This new function generates exponents on a base to help us answer questions such as, "10 to what exponent equals 500?" This is equivalent to solving the equation $10^x = 500$ for $x$.

In this section, you will learn to:

- Understand the meaning of logarithm and logarithmic function
- Find logarithms
- Evaluate logarithmic functions
- Understand the basic properties of logarithms

## A. Definition of Logarithm

Consider the function $y = f(x) = 2^x$. The table in Figure 1 shows some input-output pairs for $f$. For this function, the input is the exponent on base 2, and the output is the power of 2. So if the input is $x = 3$, then 3 is the exponent on base 2, and the output is $y = 2^3 = 8$.

A **logarithm base $b$**, written $\log_b(x)$, is a function that inverts or exchanges the input-output pairs of an exponential function with the same base. For a logarithm, the input is the power of the base and the output is the exponent. Using our example, if $x = 8$ is the input to a logarithm with base 2, then the output is 3, because $2^3 = 8$. For $\log_2(x) = y$, then:

| $x$ | $y = 2^x$ |
|---|---|
| 0 | $1 = 2^0$ |
| 1 | $2 = 2^1$ |
| 2 | $4 = 2^2$ |
| 3 | $8 = 2^3$ |
| 4 | $16 = 2^4$ |

Figure 1. Some input-output pairs for $y = 2^x$.

$\log_2(1) = 0$  because $2^0 = 1$
$\log_2(2) = 1$  because $2^1 = 2$
$\log_2(4) = 2$  because $2^2 = 4$
$\log_2(8) = 3$  because $2^3 = 8$

A logarithm base 2 gives back the exponent when the input is a power of 2. In fact, a logarithm *always* equals an exponent on the base.

### Logarithm Base $b$

For $b > 0$, $b \neq 1$, and $x > 0$, the logarithm $\log_b(x)$ is the number $y$ such that $b^y = x$.
In words, we say "log base $b$ of $x$" is the exponent we raise $b$ to in order to get $x$.

## Evaluating Logarithms

Knowing the squares, cubes, and roots of numbers allows us to evaluate many logarithms mentally. Given a logarithm of the form $\log_b(x)$, we can evaluate it mentally by following these steps:

1. Rewrite the input $x$ as a power of $b$, $b^y = x$.
2. Use previous knowledge of powers of $b$ to identify $y$ by asking, "To what exponent should $b$ be raised in order to get $x$?"

For example, consider $\log_7(49)$. We ask, "To what exponent must 7 be raised in order to get 49?" We know $7^2 = 49$. Therefore, we know that $\log_7(49) = 2$.

### ▸ Example 1

Find the logarithms mentally.

1. $\log_3(81)$
2. $\log_8(64)$
3. $\log_5(125)$
4. $\log_2(32)$
5. $\log_{10}(1{,}000{,}000)$
6. $\log_9 1$

#### Solutions

1. $\log_3(81) = 4$      because $3^4 = 81$
2. $\log_8(64) = 2$      because $8^2 = 64$
3. $\log_5(125) = 3$      because $5^3 = 125$
4. $\log_2(32) = 5$      because $2^5 = 32$
5. $\log_{10}(1{,}000{,}000) = 6$      because $10^6 = 1{,}000{,}000$
6. $\log_9(1) = 0$      because $9^0 = 1$

In order to tackle some trickier logarithms, let's recall the meaning of negative exponents. The general rule, $b^{-n} = \frac{1}{b^n}$, reminds us that if the power of the base is in the denominator of a fraction, it can also be written with a negative exponent. So, for example:

$$\log_9\left(\tfrac{1}{81}\right) = -2 \qquad \text{because } 9^{-2} = \tfrac{1}{9^2} = \tfrac{1}{81}$$

### ▸ Example 2

Find the logarithms.

1. $\log_2\left(\tfrac{1}{16}\right)$
2. $\log_4\left(\tfrac{1}{64}\right)$

#### Solutions

1. $\log_2\left(\tfrac{1}{16}\right) = -4$      because $2^{-4} = \tfrac{1}{2^4} = \tfrac{1}{16}$
2. $\log_4\left(\tfrac{1}{64}\right) = -3$      because $4^{-3} = \tfrac{1}{4^3} = \tfrac{1}{64}$

Now recall the relationship between fractional exponents and roots:

$$b^{1/2} = \sqrt{b} \quad \text{and} \quad b^{1/3} = \sqrt[3]{b}$$

To find $\log_{25}(5)$, for example, we ask, "25 to what exponent equals 5?" We see $\log_{25}(5) = \frac{1}{2}$ because $25^{1/2} = \sqrt{25} = 5$.

### Example 3

Find the logarithms.

1. $\log_8(2)$
2. $\log_{49}(7)$
3. $\log_{25}\left(\frac{1}{5}\right)$

**Solution**

1. $\log_8(2) = \frac{1}{3}$ because $8^{1/3} = \sqrt[3]{8} = 2$
2. $\log_{49}(7) = \frac{1}{2}$ because $49^{1/2} = \sqrt{49} = 7$
3. $\log_{25}\left(\frac{1}{5}\right) = -\frac{1}{2}$ because $25^{-1/2} = \frac{1}{25^{1/2}} = \frac{1}{\sqrt{25}} = \frac{1}{5}$

The equations in problem 3 of the above example incorporate both the rule for negative exponents and the definition of fractional exponents.

### Practice A

Find the logarithm. When you are finished, turn the page to check your work.

1. $\log_3(27)$
2. $\log_6(36)$
3. $\log_2(32)$
4. $\log_4\left(\frac{1}{16}\right)$
5. $\log_5\left(\frac{1}{125}\right)$
6. $\log_7(\sqrt{7})$

## B. Common Logarithms

Sometimes you will see a logarithm written without a base. In this case, assume that the base is 10. In other words, the expression $\log(x)$ means $\log_{10}(x)$. The base-10 logarithm is also known as the **common logarithm**. The reason for the name "common" is that the numeration system we all use in daily life is structured on the base of 10. We read $\log(x)$ as "the logarithm with base 10 of $x$" or just "log base 10 of $x$."

Common logarithms are used to define the Richter Scale, a system used to measure the relative strength of earthquakes. Scales for measuring the brightness of stars and the pH of acids and bases also use common logarithms.

## The Common Logarithm

A **common logarithm** is a logarithm with base 10. We write $\log_{10}(x)$ simply as $\log(x)$. The common logarithm of a positive number $x$ satisfies the following definition.

If $x > 0$, $y = \log(x)$ is equivalent to $10^y = x$.

The logarithm $y$ is the exponent to which 10 must be raised to get $x$.

Given a common logarithm of the form $y = \log(x)$, we can evaluate it mentally by following these steps:

1. Rewrite the input $x$ as a power of 10: $10^y = x$.
2. Use previous knowledge of powers of 10 to identify $y$ by asking, "To what exponent must 10 be raised in order to get $x$?"

Let's make the process a little bit easier by looking at Figure 2, a table of some of the powers of 10.

| Exponent | Power of 10 |
| --- | --- |
| -3 | $10^{-3} = 1/10^3 = 1/1000 = 0.001$ |
| -2 | $10^{-2} = 1/10^2 = 1/100 = 0.01$ |
| -1 | $10^{-1} = 1/10^1 = 1/10 = 0.1$ |
| 0 | $10^0 = 1$ |
| 1 | $10^1 = 10$ |
| 2 | $10^2 = 100$ |
| 3 | $10^3 = 1000$ |
| 4 | $10^4 = 10{,}000$ |
| 5 | $10^5 = 100{,}000$ |
| 6 | $10^6 = 1{,}000{,}000$ |

Figure 2. Some powers of 10.

### Example 4

Find the value of the common logarithm mentally.

1. $\log(100{,}000)$
2. $\log(1{,}000{,}000{,}000)$
3. $\log\left(\frac{1}{10000}\right)$
4. $\log(.01)$
5. $\log(1)$

**Solution**

1. $\log(100{,}000) = 5$    because $10^5 = 100{,}000$
2. $\log(1{,}000{,}000{,}000) = 9$    because $10^9 = 1{,}000{,}000{,}000$
3. $\log\left(\frac{1}{10000}\right) = -4$    because $10^{-4} = \frac{1}{10^4} = \frac{1}{10000}$
4. $\log(.01) = -2$    because $10^{-2} = \frac{1}{100} = .01$
5. $\log(1) = 0$    because $10^0 = 1$

In order to compare the magnitudes of two different earthquakes, we need to be able to convert between logarithmic and exponential form. For example, suppose the amount of energy released from one earthquake is 500 times greater than the amount of energy released from another. We want to calculate the difference in magnitude or the difference in the Richter numbers of the two quakes. Because a Richter number is a base-10 logarithm, the equation that represents this problem is $10^x = 500$, where $x$ represents the difference in magnitudes on the Richter scale.

How would we solve for $x$?

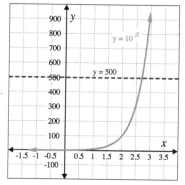

We haven't learned a method yet for solving exponential equations. None of the algebraic tools discussed so far can solve $10^x = 500$. But we do know that $10^2 = 100$ and $10^3 = 1000$, so it's clear that $x$ must be some value between 2 and 3 because $y = 10^x$ is an increasing function, specifically an exponential growth function.

We can examine a graph, as in Figure 3, to better estimate the solution. We see that the value of $x$ that satisfies $10^x = 500$ is between 2.5 and 3. Estimating from a graph, however, is still imprecise. To find an algebraic solution, we must use a common logarithm so that the input is the power on base 10, namely 500, and the output is the exponent.

Figure 3. Graph of $y = 10^x$.

We have $\log_{10}(500) = \log(500) = x$. By definition, this is equivalent to the equation $10^x = 500$. Now we can use a calculator to evaluate a common logarithm with the form $y = \log(x)$ by following these steps:

1. Press [LOG].
2. Enter the value given for $x$, followed by [)].
3. Press [ENTER].

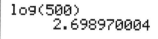

We enter log(500) and see that the value is about 2.6990, as in Figure 4. This means that $x \approx 2.6990$ is a solution to the equation $10^x = 500$.

If the amount of energy released from one earthquake is 500 times greater than the amount of energy released from another, then according to the Richter scale, the difference in magnitudes is about 2.699.

Figure 4. Calculating a common logarithm.

### Practice A — Answers

1. $\log_3(27) = 3$
2. $\log_6(36) = 2$
3. $\log_2(32) = 5$
4. $\log_4\left(\frac{1}{16}\right) = -2$
5. $\log_5\left(\frac{1}{125}\right) = -3$
6. $\log_7(\sqrt{7}) = \frac{1}{2}$

## Example 5

One of the strongest earthquakes ever recorded occurred in the Guerrero region of Mexico in March of 1787. This earthquake released approximately 25 times more energy than a 2014 earthquake in the same region. With this in mind, answer these questions:

a. What is the difference in magnitude on the Richter scale between the two earthquakes?

b. If the more recent earthquake had magnitude 7.2, what was the magnitude of the 1787 quake?

### Solution

a. Because the Richter scale is a base-10 logarithm, the equation represented by this problem is $10^x = 25$. By definition, this exponential equation is equivalent to the logarithmic equation $\log(25) = x$. We can use our calculators to find the value of the common logarithm:

b. $\log(25) \approx 1.4$

The difference in magnitude between the two earthquakes is about 1.4. To find the magnitude of the 1787 quake, we add $1.4 + 7.2 = 8.6$. The 1787 earthquake in Guerrero, Mexico had a magnitude of 8.6 on the Richter scale.

With the next example, you may find it useful to refer to the table in Figure 2 in this section.

## Example 6

1. Consider $\log(3215)$. Name the two integers that are above and below the value of the logarithm. Then find $\log(3215)$ to four decimal places using a calculator.

2. Consider $\log(7)$. Name the two integers that are above and below the value of the logarithm. Then find $\log(7)$ to four decimal places using a calculator.

### Solutions

1. Note that $10^3 = 1000$ and $10^4 = 10{,}000$. 3215 is between 1000 and 10,000, so we know that $\log(3215)$ is between $\log(1000)$ and $\log(10000)$. In other words, $\log(3215)$ is between 3 and 4.

$$1000 < 3215 < 10{,}000$$
$$3 < \log(3215) < 4$$

Using a calculator, we find that $\log(3215) \approx 3.5072$.

2. Note that $10^0 = 1$ and $10^1 = 10$. Because 7 is between 1 and 10, we know that $\log(7)$ must be between $\log(1)$ and $\log(10)$. That is $\log(7)$ is between 0 and 1.

$$1 < 7 < 10$$
$$0 < \log(7) < 1$$

Using a calculator we find $\log(7) \approx 0.8451$.

## Practice B

Now it's your turn to work with a common logarithm. When you're finished, turn the page to check your work.

7. Without using a calculator, evaluate $y = \log(1{,}000{,}000)$.
8. Using a calculator, evaluate $y = \log(123)$ to four decimal places.
9. The amount of energy released from one earthquake was 8,500 times greater than the amount of energy released from another. The equation $10^x = 8500$ represents this situation, where $x$ is the difference in magnitudes on the Richter scale. To the nearest thousandth, what was the difference in magnitudes?

## C. Basic Properties of Logarithms

Let's look at some basic properties of logarithms. Assume that $b > 0$ and $b \neq 1$. Now consider $\log_5(5)$ and $\log_{12}(12)$. The value of both is 1 because $5^1 = 5$ and $12^1 = 12$. In general, we know that $b^1 = b$, so $\log_b(b) = 1$.

Next, consider $\log_5(1)$ and $\log_{12}(1)$. The value of both logarithms is 0 because $5^0 = 1$ and $12^0 = 1$. In general, we know that $b^0 = 1$, so $\log_b(1) = 0$.

### Properties of Logarithms

For $b > 0$ and $b \neq 1$, $\log_b(b) = 1$ and $\log_b(1) = 0$

We've seen that a logarithm is actually a function or rule that returns an exponent. For example, $\log_2(x)$ is a function that pairs the input $x$ with an output that is the exponent on base 2 that equals $x$. We write the general definition of a logarithmic function as you see in the definition below.

### Logarithmic Function

For $x > 0$, $b > 0$, $b \neq 1$, $y = \log_b(x)$ if and only if $x = b^y$.
The function given by $f(x) = \log_b(x)$ is called the **logarithmic function with base $b$**.

The base of an exponential function is always positive, so no power of that base can ever be negative. Therefore, the log of a negative number is not a real number. In addition, we cannot take the logarithm of zero, so in general, the domain of a logarithmic function $\log_b(x)$ is the set of all positive real numbers or $(0, \infty)$.

## Example 7

Let $f(x) = \log_3(x)$.

1. Find $f(9)$.
2. Find $f(81)$.
3. Find $f(3)$.
4. Find $f(1)$.

**Solutions**

1. $f(9) = \log_3(9) = 2$ because $3^2 = 9$
2. $f(81) = \log_3(81) = 4$ because $3^4 = 81$
3. $f(3) = \log_3(3) = 1$ because $\log_b(b) = 1$
4. $f(1) = \log_3(1) = 0$ because $\log_b(1) = 0$

Notice that we justified the last two solutions in the previous example with the two basic properties of logarithms introduce in this section.

## Exercises 3.1

Find the logarithm.

1. $\log_5(125)$
2. $\log_4(16)$
3. $\log_3(243)$
4. $\log_6(216)$
5. $\log_8(64)$
6. $\log_2(128)$
7. $\log_4(256)$
8. $\log_5(625)$
9. $\log(10,000,000)$
10. $\log(100)$
11. $\log(1)$
12. $\log(10)$
13. $\log_3\left(\frac{1}{27}\right)$
14. $\log_2\left(\frac{1}{32}\right)$
15. $\log_6\left(\frac{1}{36}\right)$
16. $\log_8\left(\frac{1}{8}\right)$
17. $\log\left(\frac{1}{100,000}\right)$
18. $\log\left(\frac{1}{100}\right)$
19. $\log_{13}(1)$
20. $\log_6(1)$
21. $\log_7(7)$
22. $\log_3(3)$
23. $\log_{16}(4)$
24. $\log_{49}(7)$
25. $\log_{125}(5)$
26. $\log_{64}(4)$
27. $\log_6(\sqrt{6})$
28. $\log_3(\sqrt{3})$
29. $\log_8(\sqrt[4]{8})$
30. $\log_{16}(\sqrt[5]{16})$
31. $\log_{11}(11^5)$
32. $\log_9(9^7)$

For the following exercises, evaluate each expression using a calculator. Round to the nearest thousandth.

33. $\log(6,775)$
34. $\log(89,900)$
35. $\log(29)$
36. $\log(333)$
37. $\log(0.249)$
38. $\log(0.058)$

For the following exercises, let $g(x) = \log_4(x)$.

39. Find $g(256)$.
40. Find $g(64)$.
41. Find $g(\frac{1}{16})$.
42. Find $g(2)$.

## Chapter 3: Logarithmic Functions

For the following exercises, let $f(x) = \log_5(x)$.

43. Find $f(125)$.
44. Find $f(3125)$.
45. Find $f(\sqrt{5})$.
46. Find $f\left(\frac{1}{625}\right)$.

For the following exercises, refer to Example 5.

47. The famous San Francisco earthquake in April 1906 released 316 times more energy than the more recent Borrego Springs earthquake which had a magnitude of 5.4 on the Richter scale. Find the difference in magnitudes between the earthquakes. What was the magnitude of the 1906 San Francisco quake?

48. In 1993, there were two earthquakes in Oregon. The Scotts Mills quake occurred in March and was 2.5 times less powerful than the Klamath Falls quake, which occurred in September of that year. If the Scotts Mills earthquake had a magnitude of 5.6 on the Richter scale, what was the magnitude of the Klamath Falls quake?

49. In 2010, there were two earthquakes in Midwestern states that registered a magnitude of 3.8 on the Richter scale. In September 2016, Oklahoma had an earthquake that released 100 times the energy of the 2010 earthquakes. What was the difference in magnitudes between these quakes? What was the magnitude of the Oklahoma quake?

50. One of the most deadly earthquakes of the last century occurred in Tohoku, Japan in 2011. This earthquake released 158.5 times more energy than an earthquake that occurred in Fukushima five years later. The Fukushima quake had a magnitude of 6.9 on the Richter scale. What was the magnitude of the deadly quake of 2011?

### Practice B — Answers

7. $\log(1{,}000{,}000) = 6$
8. $\log(123) \approx 2.0899$
9. The equation $10^x = 8500$ is equivalent to $\log(8500) = x$. We find $\log(8500) \approx 3.929$. The difference in magnitudes was about 3.929 on the Richter scale.

# 3.2 Properties of Logarithms

## Overview

In this section we will study some more properties of logarithms. We'll focus on how to use these properties to solve exponential and logarithmic equations. As you study this section, you will learn how to:

- Convert between exponential and logarithmic form
- Use the power rule for logarithms
- Use properties of logarithms to solve logarithmic equations
- Use properties of logarithms to solve exponential equations
- Use the change-of-base formula for logarithms

## A. Converting Between Exponential and Logarithmic Forms

In Section 3.1, we learned how to find logarithms, and we often justified our solutions using the exponential form of the equation. For example, to find $\log_3(81)$, we said:

$$\log_3(81) = 4 \quad \text{because} \quad 3^4 = 81$$

The equation $\log_3(81) = 4$ is in **logarithmic form**, and the equation $3^4 = 81$ is in **exponential form**. These two forms are equivalent, and either one can replace the other when solving a problem.

### Logarithmic and Exponential Forms Property

For $a > 0$, $b > 0$, $b \neq 1$, the equations $\log_b(a) = c$ and $b^c = a$ are equivalent.

The table below offers more examples of equivalent equations in both logarithmic and exponential form.

| Logarithmic Form | Exponential Form |
|---|---|
| $\log_b(a) = c$ | $b^c = a$ |
| $\log_4(16) = 2$ | $4^2 = 16$ |
| $\log_5(125) = 3$ | $5^3 = 125$ |
| $\log_2(64) = 6$ | $2^6 = 64$ |
| $\log(100{,}000) = 5$ | $10^5 = 100{,}000$ |

## Example 1

Write the following logarithmic equations in exponential form.

1. $\log_6(216) = 3$
2. $\log_3(243) = 5$
3. $\log\left(\frac{1}{100}\right) = -2$
4. $\log_q(k) = m$

**Solutions**

1. $\log_6(216) = 3$ becomes $6^3 = 216$
2. $\log_3(243) = 5$ becomes $3^5 = 243$
3. $\log\left(\frac{1}{100}\right) = -2$ becomes $10^{-2} = \frac{1}{100}$ Remember: common log has base 10.
4. $\log_q(k) = m$ becomes $q^m = k$

## Example 2

Write the following exponential equations in logarithmic form.

1. $4^6 = 4096$
2. $2^{3.808} \approx 14$
3. $10^4 = 10{,}000$
4. $81^{1/2} = 9$

**Solutions**

1. $4^6 = 4096$ becomes $\log_4(4096) = 6$
2. $2^{3.808} \approx 14$ becomes $\log_2(14) \approx 3.808$
3. $10^4 = 10{,}000$ becomes $\log(10{,}000) = 4$
4. $81^{1/2} = 9$ becomes $\log_{81}(9) = \frac{1}{2}$

## Practice A

Write the following logarithmic equations in exponential form.

1. $\log_{10}(1{,}000{,}000) = 6$
2. $\log_5(25) = 2$

Write the following exponential equations in logarithmic form.

3. $3^2 = 9$
4. $5^3 = 125$
5. $2^{-1} = \frac{1}{2}$

When you're done, turn the page and check your solutions.

## B. Solving Equations in Logarithmic Form

Converting between equations in exponential and logarithmic forms is an essential skill for solving equations involving exponents or logarithms. The conversion often turns a seemingly impossible problem into a very simple problem to solve.

### Example 3

Solve the following equations.

1. $\log_6(x) = 4$
2. $\log_2(5x - 4) = 8$

**Solutions**

1. We write $\log_6(x) = 4$ in exponential form and simplify.

$$6^4 = x \quad \text{Write in exponential form.}$$
$$1296 = x \quad \text{Simplify.}$$

2. We write $\log_2(5x - 4) = 8$ in exponential form and solve for $x$.

$$2^8 = 5x - 4 \quad \text{Write in exponential form.}$$
$$256 = 5x - 4 \quad \text{Simplify.}$$
$$260 = 5x \quad \text{Add 4 to both sides.}$$
$$52 = x \quad \text{Divide both sides by 5.}$$

Many of the solving steps for logarithmic equations are familiar to us from previous algebra courses. It's important to execute the solving steps in the correct order.

In the second equation for Example 3, for example, we didn't add 4 to both sides of the equation and divide both sides by 5 until *after* we rewrote the equation in exponential form. That's is because the expression $5x - 4$ is in parentheses.

In the next example, we need to apply one or two solving steps *before* we rewrite the equation in exponential form. Pay attention to the order of the steps as we work the next two examples.

### Practice A — Answers

1. $10^6 = 1{,}000{,}000$
2. $5^2 = 25$
3. $\log_3(9) = 2$
4. $\log_5(125) = 3$
5. $\log_2(\tfrac{1}{2}) = -1$

## Example 4

Solve for $x$.

1. $\log_3(2x - 5) + 7 = 11$
2. $4\log(x - 7) + 1 = 9$

**Solutions**

1. We get $\log_3(2x - 5)$ alone on the left side of the equation before we write it in exponential form and solve for $x$:

   | | | |
   |---|---|---|
   | $\log_3(2x - 5) + 7 = 11$ | | Original equation. |
   | $\log_3(2x - 5) = 4$ | | Subtract 7 from both sides. |
   | $3^4 = 2x - 5$ | | Write in exponential form. |
   | $81 = 2x - 5$ | | Simplify. |
   | $86 = 2x$ | | Add 5 to both sides. |
   | $43 = x$ | | Divide both sides by 2. |

2. We get $\log(x - 7)$ alone on the left side of the equation before we write it in exponential form and solve for $x$:

   | | | |
   |---|---|---|
   | $4\log(x - 7) + 1 = 9$ | | Original equation. |
   | $4\log(x - 7) = 8$ | | Subtract 1 from both sides. |
   | $\log(x - 7) = 2$ | | Divide both sides by 4. |
   | $10^2 = x - 7$ | | Write in exponential form. For a common logarithm, the base is 10. |
   | $100 = x - 7$ | | Simplify. |
   | $107 = x$ | | Add 7 to both sides. |

In the next example, we'll solve a logarithmic equation for the base. Again, we write the equation in exponential form as part of the solving process. Then we use a reciprocal exponent to solve for $b$.

## Example 5

Solve for $b$.

1. $\log_b(243) = 5$
2. $\log_b(59) = 4$

**Solution**

1. We write $\log_b(243) = 5$ in exponential form and solve for $b$.

   | | | |
   |---|---|---|
   | $b^5 = 243$ | | Write in exponential form. |
   | $(b^5)^{1/5} = 243^{1/5}$ | | Raise both sides to the reciprocal exponent. |
   | $b = 3$ | | Simplify. |

2. We write $\log_b(59) = 4$ in exponential form and solve for $b$.

$$b^4 = 59 \qquad \text{Write in exponential form.}$$
$$(b^4)^{1/4} = 59^{1/4} \qquad \text{Raise both sides to the reciprocal exponent.}$$
$$b \approx \pm 2.7715 \qquad \text{Simplify. The exponent was even, so there are two solutions.}$$
$$b \approx 2.7715 \qquad \text{The base of a logarithm is positive.}$$

## Practice Set B

Solve the following equations for $x$.

6. $\log_2(x) = 9$
7. $\log_3(2x + 1) = 5$
8. $7\log(x) - 24 = 18$

Solve the following equations for $b$.

9. $\log_b(729) = 3$
10. $\log_b(311) = 6$

When you're done, turn the page and check your solutions.

## C. Using the Power Rule for Logarithms to Solve Exponential Equations

Our next objective is to be able to solve exponential equations with the variable in the exponent. Here are some examples of exponential equations:

$$5^x = 94$$
$$2(1.048)^x = 6.6$$
$$3^{2x+1} = 774$$

The **power rule for logarithms** states that a logarithm of a power is equal to the exponent times the logarithm. This property is useful for solving exponential equations. We'll first investigate the power rule of logarithms. Then we'll use it to help us solve exponential equations for the variable. It's easier to understand in formula form, as you see in the highlighted box, and it's a simple to apply it.

### The Power Rule for Logarithms

For $a > 0$, $b > 0$, and $b \neq 1$, $\log_b(a^p) = p\log_b(a)$

### Example 6

Apply the power rule of logarithms to rewrite the following expressions.

1. $\log_2(x^5)$
2. $\log_6(3.4^t)$
3. $\log(3^{x-1})$

**Solutions**

1. $\log_2(x^5) = 5\log_2(x)$
2. $\log_6(3.4^t) = t\log_6(3.4)$
3. $\log(3^{x-1}) = (x-1)\log(3)$

Some math textbooks include a proof of this important property. We will simply point out that the power rule for logarithms follows the rule for exponents regarding raising a power to a power. This property for exponents states that "when raising an exponent to a power, multiply the exponents":

$$(b^m)^n = b^{mn}$$

When raising a logarithm to an exponent, we also rewrite the expression as a multiplication. After all, a logarithm is an exponent.

Next, let's look at the **logarithm property of equality**, which is also useful in solving exponential equations. This property states that if we have an equation, such as $a = c$, we can take the log base $b$ of both sides of the equation. This property is true for *any* base $b$. In practice, however, we will use base 10, the common logarithm, because our calculators are programmed to find the logarithm of any positive number in base 10. When solving exponential equations, we refer to this step simply as "taking the log of both sides."

## Logarithm Property of Equality

For positive real numbers, $a$, $b$, and $c$, where $b \neq 1$, the following equations are equivalent:
$$a = c \quad \text{and} \quad \log_b(a) = \log_b(c)$$

Now let's use these two new properties of logarithms to solve some equations.

## Example 7

Solve the exponential equation $5^x = 402$.

**Solution**

| | | |
|---|---|---|
| $5^x = 402$ | | Original equation. |
| $\log(5^x) = \log(402)$ | | Take the log of both sides. |
| $x\log(5) = \log(402)$ | | Apply the power rule for logarithms. |
| $x = \dfrac{\log(402)}{\log(5)}$ | | Divide both sides by $\log(5)$. Make sure to use both sets of parentheses on the calculator! |
| $x \approx 3.7258$ | | Compute. |

We can check that 3.7258 is the approximate solution to the equation $5^x = 402$:

$$5^{3.7258} \approx 401.9967 \approx 402$$

When solving an exponential equation of the form $ab^x = k$, it's tempting to multiply the constant $a$ times the base $b$. Don't do it! The constant $a$ is not raised to the exponent. When solving an equation in this form, first divide both sides of the equation by $a$ to isolate the exponential expression. Then take the common logarithm of both sides. Example 8 illustrates this.

## Example 8

Solve the equation $2(3)^x = 52$.

### Solution

| | | |
|---|---|---|
| $2(3)^x = 52$ | | Original equation. |
| $(3)^x = 26$ | | Divide both sides by 2. |
| $\log(3^x) = \log(26)$ | | Take the log of both sides. |
| $x \log(3) = \log(26)$ | | Apply the power rule for logarithms. |
| $x = \dfrac{\log(26)}{\log(3)}$ | | Divide both sides by $\log(3)$. |
| $x \approx 2.9656$ | | Compute. |

We can check that 2.9656 is the approximate solution to the equation $2(3)^x = 52$:

$$2(3)^{2.9656} \approx 51.9973 \approx 52$$

Sometimes the exponent on a base is itself an algebraic expression, as in the next example. In this case, we have to continue solving for the variable after applying the properties of exponents.

## Example 9

Solve the equation $2^{4x-1} = 294$.

### Solution

| | | |
|---|---|---|
| $2^{4x-1} = 294$ | | Original equation. |
| $\log(2^{4x-1}) = \log(294)$ | | Take the log of both sides. |
| $(4x - 1)\log(2) = \log(294)$ | | Apply the power rule for logarithms. The expression on the left is not correct without the parentheses. |
| $4x - 1 = \dfrac{\log(294)}{\log(2)}$ | | Divide both sides by $\log(2)$. |
| $4x = \dfrac{\log(294)}{\log(2)} + 1$ | | Add 1 to both sides. |
| $x = \dfrac{\dfrac{\log(294)}{\log(2)} + 1}{4}$ | | Divide both sides by 4. |
| $x \approx 2.2999$ | | Compute. |

We can check that 2.9656 is the approximate solution to the equation $2^{4x-1} = 294$:

$$2^{4(2.2999)-1} \approx 293.9853 \approx 294$$

After the fourth line in the solving process above, it gets kind of messy to continue solving without calculating the division of the logarithms. We could round the expression $\frac{\log(294)}{\log(2)}$ at this point in the process, but we may lose some accuracy in our final answer.

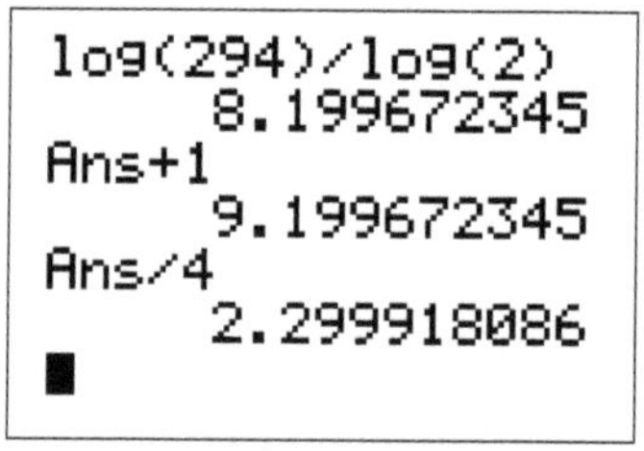

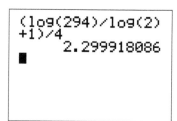

Figure 1. Computing one step at a time.

Figure 2. Computing at the last step.

We can enter the calculations on our calculator one step at a time, as you see in Figure 1, or we can enter it all at once at the last step, as you see in Figure 2. Always be careful to use parentheses correctly.

### Example 10

Solve the equation $11(0.25)^x - 8 = 47$.

**Solution**

| | | |
|---|---|---|
| $11(1.25)^x - 8 = 47$ | | Original equation. |
| $11(1.25)^x = 55$ | | Add 8 to both sides. |
| $(1.25)^x = 5$ | | Divide both sides by 11. |
| $\log(1.25)^x = \log(5)$ | | Take the log of both sides. |
| $x \log(1.25) = \log(5)$ | | Apply the power rule for logarithms. |
| $x = \frac{\log(5)}{\log(1.25)}$ | | Divide both sides by $\log(1.25)$. |
| $x \approx 7.2126$ | | Compute. |

We can check that 7.2126 is the approximate solution to $11(1.25)^x - 8 = 47$:

$$11(1.25)^{7.2126} - 8 \approx 47.0004 \approx 47$$

### Practice B — Answers

6. $2^9 = x \rightarrow x = 512$
7. $3^5 = 2x + 1 \rightarrow x = 121$
8. $\log(x) = 6 \rightarrow 10^6 = x \rightarrow x = 1{,}000{,}000$
9. $b^3 = 729 \rightarrow b = 9$
10. $b^6 = 311 \rightarrow b \approx 2.6029$

## Practice C

Solve for $x$. Round answers to the fourth decimal place. When you are done, turn the page and check your solutions.

11. $4^x = 555$    12. $3(9)^x = 6810$    13. $5^{3x-2} = 147$    14. $822 = 7(2.5)^x - 4$

## D. Solving Equations by Using Graphs

We can also solve exponential equations by using graphs. In Example 10, we solved the equation $11(1.25)^x - 8 = 47$ algebraically. The solution was $x \approx 7.2126$.

Now let's solve this equation graphically. Use the system of equations below, where $y_1$ equals the left side of the equation and $y_2$ equals the right side of the equation. Adjust the viewing window to include the line $y = 47$. Check the graph to make sure that the viewing window includes the intersection of the two curves. See Figure 3.

$$y_1 = 11(1.25)^x - 8$$
$$y_2 = 47$$

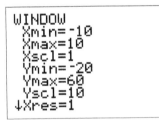
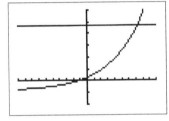

Figure 3.

Next we access the intersect feature of the calculator by selecting [2nd] and [TRACE] and number 5:intersect. The calculator screen asks us to identify which two curves we are interested in. We confirm these by pressing [ENTER] for each. To make a guess, use the left or right arrow keys to move the cursor close to the intersection point before pressing [ENTER]. See Figure 6.

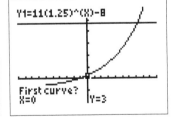
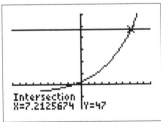

Figure 4.

Using graphing, we find the same approximate solution, $x \approx 7.2126$. Some equations containing exponential terms are impossible to solve algebraically but can be solved graphically. The equation in the next example is like that.

## Example 11

Solve $2^{x-4} + 3 = 4x - 1$ using a graph.

### Solution

The equation has an exponential term and a linear term. It is impossible to solve algebraically. To solve graphically, enter the system of equations below and use the intersect feature to solve. See Figure 3.

$$y_1 = 2^{x-4} + 3$$
$$y_2 = 4x - 1$$

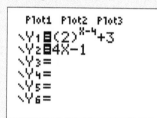 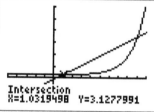 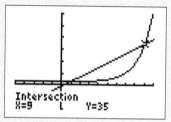

Figure 5. Solve the equation by graphing.

There are two solutions because the curves intersect at two points. The points have (approximate) coordinates (1.0319, 3.1278) and (9, 35). These $x$-$y$ pairs are solutions to the system of equations. The solutions to the original equation are $x \approx 1.0319$ and $x = 9$.

# E. Change-of-Base

In order to evaluate logarithms with a base other than 10, we can use the **change-of-base formula**. This formula allows us to rewrite the logarithm as the quotient of logarithms of any other base. When we're using a calculator to solve an exponential equation, we change the base to common log because our calculators are programmed to evaluate the log base 10 of any positive number.

## Change-of-Base Property

For positive real numbers $a$, $b$, and $c$, where $b \neq 1$ and $c \neq 1$,

$$\log_b(a) = \frac{\log_c(a)}{\log_c(b)}$$

Here are some examples of the change of base formula:

$$\log_2(14) = \frac{\log_3(14)}{\log_3(2)} \qquad \log_4(x) = \frac{\log_9(x)}{\log_9(4)} \qquad \log_5(36) = \frac{\log(36)}{\log(5)}$$

We'll take a closer look at the last of these three examples because the original logarithm was changed to a logarithm base 10, which we can compute on a calculator.

## Example 12

Evaluate $\log_5 36$ using a calculator.

**Solution**

$\log_5 36 = \dfrac{\log(36)}{\log(5)}$    Apply the change of base formula using base 10.

$\phantom{\log_5 36} \approx 2.2266$    Use a calculator to evaluate to 4 decimal places.

Checking our approximate solution with the exponential form we find that $5^{2.2266} \approx 36$.

## Example 13

Evaluate $\log_{1.75} 86.38$ using a calculator.

**Solution**

$\log_{1.75} 86.38 = \dfrac{\log(86.38)}{\log(1.75)}$    Apply the change of base formula using base 10.

$\phantom{\log_{1.75} 86.38} \approx 7.9675$    Use a calculator to evaluate to 4 decimal places.

Checking our approximate solution with the exponential form we find that $1.75^{7.9675} \approx 86.38$.

To see how we can use the change-of-base formula to solve exponential equations, let's revisit a problem that we worked earlier in this section, Example 7. We will compare the original solving method with the method that uses the change-of-base formula.

## Example 14

Solve the equation $5^x = 402$.

**Solution**

*Method 1:* Take the log and both sides and then use the power rule for logarithms.

$5^x = 402$    Original equation.

$\log(5^x) = \log(402)$    Take the log of both sides.

$x \log(5) = \log(402)$    Apply the power rule for logarithms.

$x = \dfrac{\log(402)}{\log(5)}$    Divide both sides by $\log(5)$.

$x \approx 3.7258$    Compute.

*Method 2:* Convert to logarithmic form and then use the change-of-base formula.

| | |
|---|---|
| $5^x = 402$ | Original equation. |
| $\log_5(402) = x$ | Write the equation in logarithmic form. |
| $\dfrac{\log(402)}{\log(5)} = x$ | Change-of-base formula: convert to base 10. |
| $3.7258 \approx x$ | Compute. |

We arrive at the same solution with either method.

### Example 15

Solve for $x$: $4(7)^x = 58$

**Solution**

| | |
|---|---|
| $4(7)^x = 58$ | Original equation. |
| $(7)^x = 14.5$ | Divide both sides by 4. |
| $\log_7(14.5) = x$ | Write the equation in logarithmic form. |
| $\dfrac{\log(14.5)}{\log(7)} = x$ | Apply the change-of-base formula. |
| $1.2345 \approx x$ | Compute. |

## F. Exponential Models

In 1859, an Australian landowner named Thomas Austin released 24 rabbits into the wild for hunting. Because Australia had few predators and ample food, the rabbit population exploded. In fewer than ten years, the rabbit population numbered in the millions.

Uncontrolled population growth, as in the wild rabbits in Australia, can be modeled with exponential functions. Equations resulting from those exponential functions can be solved to analyze and make predictions about exponential growth. To wrap up this section, we'll now take these techniques for solving exponential equations and apply them to real-world situations.

### Practice C — Answers

11. $x = \dfrac{\log(555)}{\log(4)} \approx 4.5582$

12. $x = \dfrac{\log(2270)}{\log(9)} \approx 3.5170$

13. $x = \dfrac{\dfrac{\log(147)}{\log(5)} + 2}{3} \approx 1.7002$

14. $x = \dfrac{\log(118)}{\log(2.5)} \approx 5.2065$

Recall from Section 2.5 that for models of the form $f(t) = ab^t$, the initial value is $a$ and $b = 1 \pm r$ where $r$ is the percent change in decimal form. This model has many real-world applications including problems involving investments with compounded interest.

### Example 16

Let $B = f(t)$ be the balance of a bank account after $t$ years with an initial investment of $5000 and an annual interest rate of 2.4%, compounded annually.

a. Write an exponential equation to model this situation.

b. Find and interpret $f(7)$.

c. To the nearest tenth of a year, predict when the balance of the account will be $8,000.

**Solution**

a. The initial value is $a = 5000$. The base is $b = 1 + 0.024 = 1.024$, so:

$$B = f(t) = 5000(1.024)^t$$

b. $f(7) = 5000(1.024)^7 \approx 5902.96$. After 7 years, the account balance will be $5902.96.

c. Replace $f(t)$ with $8000 and solve for $x$.

| | |
|---|---|
| $5000(1.024)^t = 8000$ | $f(t) = 8000$ |
| $1.024^t = 1.6$ | Divide both sides by 5000. |
| $\log(1.024^t) = \log(1.6)$ | Take the log of both sides. |
| $t \log(1.024) = \log(1.6)$ | Apply the power rule for logarithms. |
| $t = \dfrac{\log(1.6)}{\log(1.024)}$ | Divide both sides by $\log(1.024)$. |
| $t \approx 19.8$ | Compute. |

It will take approximately 19.8 years for the balance of the account to grow to $8000.

In Section 2.5, Example 7, we found an exponential regression equation to model the world population in billions of people. We use this model again for the example below.

### Example 17

Let $f(t)$ represent the world's population in billions $t$ years after 1900:

$$f(t) = 1.21(1.0161)^t$$

Estimates vary, but many experts think the carrying capacity of the planet is about 10 billion people — unless we stop eating meat. Predict when the world population will grow to 10 billion.

## Solution

Replace $f(t)$ with 10 and solve for $t$.

| | |
|---|---|
| $1.21(1.0161)^t = 10$ | $f(t) = 10$ |
| $1.0161^t = 10/1.21$ | Divide both sides by 1.21. |
| $\log(1.0161^t) = \log\left(\frac{10}{1.21}\right)$ | Take the log of both sides. |
| $t\log(1.0161) = \log\left(\frac{10}{1.21}\right)$ | Apply the power rule for logarithms. |
| $t = \dfrac{\log\left(\frac{10}{1.21}\right)}{\log(1.0161)}$ | Divide both sides by $\log(1.0161)$. |
| $t \approx 132.23$ | Compute. |

Approximately 132 years after 1900, the world population will grow to 10 billion.

$1900 + 132 = 2032$

We estimate the planet will reach its carrying capacity in the year 2032.

### Example 18

A 2016 Tesla Model X costs approximately $95,000 when new. It's expected that this unique car will depreciate less quickly than the average automobile. However, the Model X still depreciates in value by about 8.5% per year.

a. Write an equation that models the value of a Tesla Model X $t$ years after it is purchased.
b. What is the value of the car after 3 years?
c. Predict to the nearest tenth of a year when the Model X will be worth $30,000.

### Solution

a. Let $f(t)$ represent the value of the Tesla Model X $t$ years after it is purchased. The initial value is $a = 95{,}000$. The value of the base is $b = 1 - 0.085 = 0.915$:

$$f(t) = 95{,}000(0.915)^t$$

b. $f(3) = 95{,}000(0.915)^3 \approx 72{,}776$

After 3 years, the Tesla Model X will be worth about $72,776.

c. Replace $f(t)$ with $30,000 and solve for $t$.

$$95{,}000(0.915)^t = 30{,}000 \qquad f(t) = 30{,}000$$

$$0.915^t = \left(\tfrac{6}{19}\right) \qquad \text{Divide both sides by 95,000.}$$

$$\log(0.915^t) = \log\left(\tfrac{6}{19}\right) \qquad \text{Take the log of both sides.}$$

$$t\log(0.915) = \log\left(\tfrac{6}{19}\right) \qquad \text{Apply the power rule for logarithms.}$$

$$t = \frac{\log\left(\tfrac{6}{19}\right)}{\log(0.915)} \qquad \text{Divide both sides by } \log(0.915).$$

$$t \approx 12.976 \qquad \text{Compute.}$$

In approximately 13 years, the value of the Tesla Model X will depreciate to $30,000.
If you round the value of the fraction on the second step, hold at least 3 decimal places for accuracy in the final answer.

### Example 19

When twins Julio and Rodrigo turned 21, they received an inheritance of $10,000 each from their grandfather. Julio spent most of his inheritance right away, but before it was all gone, he set aside $500 to invest in a mutual fund. Rodrigo invested all of his inheritance in the same mutual fund which had a 6.75% annual interest rate compounded annually.

a. Write an exponential equation to model Julio's investment.

b. Determine the time it takes for Julio's initial investment to double.

c. Compare the answer to (b) to the amount of time it takes Rodrigo's investment to double.

### Solution

a. Let $f(t)$ represent the value of the investment after $t$ years. Julio's initial investment is $500. The value of the base $b = 1 + 0.0675 = 1.0675$:

$$f(t) = 500(1.0675)^t$$

b. When $500 is doubled it equals $1000. Replace $f(t)$ with 1000 and solve for $t$.

$$500(1.0675)^t = 1000 \qquad f(t) = 1000$$

$$1.0675^t = 2 \qquad \text{Divide both sides by 500.}$$

$$\log(1.0675^t) = \log(2) \quad \text{Take the log of both sides.}$$
$$t \log(1.0675) = \log(2) \quad \text{Apply the power rule for logarithms.}$$
$$t = \frac{\log(2)}{\log(1.0675)} \quad \text{Divide both sides by } \log(1.0675).$$
$$t \approx 10.6 \quad \text{Compute.}$$

It will take about 10.6 years for Julio's investment of $500 to double.

c. Rodrigo's investment can be modeled by $f(t) = 10{,}000 \, (1.0675)^t$.

Because the brothers invested in the same fund, the only part of the equations modeling the growth of their investments that has changed is the initial value. To find the doubling time for Rodrigo's $10,000 investment, replace $f(t)$ with 20,000 and solve for $t$.

$$10{,}000 \, (1.0675)^t = 20{,}000 \quad f(t) = 20{,}000$$
$$1.0675^t = 2 \quad \text{Divide both sides by } 10{,}000.$$
$$\log(1.0675^t) = \log(2) \quad \text{Take the log of both sides.}$$
$$t \log(1.0675) = \log(2) \quad \text{Apply the power rule for logarithms.}$$
$$t = \frac{\log(2)}{\log(1.0675)} \quad \text{Divide both sides by } \log(1.0675).$$
$$t \approx 10.6 \quad \text{Compute.}$$

After the first step, the calculation is exactly the same as it was in Julio's case! It takes about 10.6 years for Rodrigo's $10,000 investment to double.

What we see in this last example is something interesting about the behavior of exponential growth functions. The doubling time is the same no matter where you begin in the growth process.

# Exercises 3.2

Write the following logarithmic equations in exponential form. Assume all constants are positive and not equal to 1.

1. $\log_4(64) = 3$
2. $\log_{11}(121) = 2$
3. $\log_5(625) = 4$
4. $\log_3(243) = 5$
5. $\log\left(\frac{1}{8}\right) = -3$
6. $\log_9\left(\frac{1}{81}\right) = -2$
7. $\log(1) = 0$
8. $\log(1000) = 3$
9. $\log_w(m) = r$
10. $\log_z(y) = k$

Write the following exponential equations in logarithmic form.

11. $2^8 = 256$
12. $5^6 = 15625$
13. $49^{1/2} = 7$
14. $27^{1/3} = 3$
15. $10^4 = 10{,}000$
16. $10^{-1} = 0.1$
17. $4^x = 97$
18. $12^x = 44$

Solve for $x$.

19. $\log_{12}(x) = 2$
20. $\log_7(x) = 3$
21. $\log_8(x) = 0$
22. $\log_{11}(x) = 0$
23. $\log(x) = -3$
24. $\log(x) = -2$
25. $\log_2(7x + 4) = 5$
26. $\log_3(10x - 9) = 4$
27. $6\log_5(x) - 15 = -3$
28. $5\log_4(x) + 11 = 26$
29. $24\log_4(x) - 17 = -5$
30. $3\log_8(x) + 17 = 18$

Solve for $b$. If needed, round solutions to the fourth decimal place.

31. $\log_b(2401) = 4$
32. $\log_b(216) = 3$
33. $\log_b(9) = 1/2$
34. $\log_b(1000) = 1/3$
35. $\log_b(188) = 7$
36. $\log_b(75) = 6$

Solve for $x$. If needed, round solutions to the fourth decimal place. Check your solutions in the original equation.

37. $3^x = 101$
38. $6^x = 340$
39. $4(5)^x = 708$
40. $56 = 2(9)^x$
41. $17.16 = 6.24(1.15)^x$
42. $9.25(1.065)^x = 31.82$
43. $7(4)^x - 55 = 241.8$
44. $98 = 3(5)^x + 40$
45. $3^{2x+4} = 695$
46. $4^{3x-7} = 879$
47. $4000 = 25(1.2)^{3x}$
48. $38(1.04)^{x+1} = 133$
49. $50(2)^{x-3} + 66 = 786$
50. $7.7(3)^{2x} - 48 = 337$

Use the intersect feature on a graphing calculator to solve the following equations. If needed, round solutions to the fourth decimal place.

51. $4^x + 3 = \frac{1}{5}x + 8$
52. $3^x - 5 = -2x + 7$
53. $1.5x - 2 = 2.4^{x-3}$
54. $-\frac{1}{2}x + 4 = 0.8^{2x-7}$
55. $0.49^x = -3x - 8$
56. $1.4^x = x - 6$

Apply what you have learned in this chapter to solve the following real-world applications.

57. An investment of $2200 is made in a stock that increases in value by about 4.7% per year.
    a. Write an equation to model the value of the investment over $t$ years.
    b. Use the model to predict the value of the investment in 12 years.
    c. How long it will take for the initial investment to be worth $5000?

58. An investment of $750 is made in an account that earns a 3.9% interest rate compounded yearly.
    a. Write an equation to model the value of the investment over $t$ years.
    b. Use the model to predict the value of the investment in 7 years.
    c. How long it will take for the initial investment to be worth $1200?

59. A person invests $14,000 in an account that earns 2.25% interest compounded annually. When will the investment be worth $25,000?

60. A person invests $8000 in an account that earns 7.45% interest compounded annually. When will the investment be worth $12,000?

61. The population of a species of whales is estimated to be decreasing at a rate of 4% per year. The current population is approximately 15,000.
    a. Write an equation to model the population of whales over $t$ years.
    b. Use the model to predict the population after 10 years.
    c. When will the whale population be 8000?

62. The population of white owls in a wilderness area is estimated to be decreasing at a rate of 7% per year. The current population is approximately 98.
    a. Write an equation to model the population of owls over $t$ years.
    b. Use the model to predict the population after 5 years.
    c. When will the owl population be below 50?

63. The value of a certain car model is $22,500 when new, but it depreciates at a rate of 16% per year. When will the value of the car depreciate to $10,000?

64. The value of a construction crane is $375,000 when new, but depreciates at a rate of 13% per year. When will the value of the crane depreciate to $150,000?

65. The following equation models the amount of caffeine in a person's bloodstream in milligrams, $t$ hours after 7:00 a.m.: $f(t) = 180\left(\frac{1}{2}\right)^{t/6}$ How much caffeine is in the person's bloodstream at 7:00 a.m.?
    a. Find and interpret $f(12)$.
    b. When will the caffeine level be less than 10 milligrams?

66. The following equation models the amount of aspirin in a person's blood in milligrams, $t$ hours after ingesting a dose: $f(t) = 300\left(\frac{1}{2}\right)^{t/20}$
   a. How much aspirin is in the dose?
   b. Find and interpret $f(60)$.
   c. When will the aspirin level be less than 10 milligrams?

67. The population of a small coastal town is modeled by the following equation where $f(t)$ represents the population in thousands $t$ years after 2000: $f(t) = 7.8\,(1.013)^t$
   a. What does the value of the base mean in the context of the problem?
   b. Find and interpret $f(8)$.
   c. Predict when the population of the town will be 10 thousand.

68. A population of suburban town is model by the following equation where $f(t)$ represents the population in thousands $t$ years after 2010: $f(t) = 33\,(1.021)^t$
   a. What does the value of the base mean in the context of the problem?
   b. Find and interpret $f(5)$.
   c. Predict when the population of the town will be 100 thousand.

69. The number of passengers traveling by airplane each year has increased according to the model, $f(t) = 466\,(1.035)^t$, where $f(t)$ represents the number of airline passengers in millions $t$ years after 1990. *(US Census Bureau)*
   a. How many passengers travel by air in 2001?
   b. In what year is it predicted that 2 billion (2000 million) passengers will travel by airplane?

70. The average price of a ticket to a major league baseball game has increased according to the model, $f(t) = 1.22\,(1.051)^t$, where $f(t)$ represents the price of a ticket in dollars $t$ years after 1950.
   a. What was the price of a ticket in 1960? In 2017?
   b. In what year is it predicted that the price of a ticket will be $50?

71. The number of bankruptcies (in millions) filed per year since 1994 can be modeled by the equation $f(t) = 0.798\,(1.164)^t$. *(Admin Office of the U.S. Courts)* In what year is it predicted that 50 million bankruptcies will be filed?

72. The amount of money (in billions of dollars) spent on health care in the U.S. per year since 1970 can be modeled by the equation $f(t) = 78.16\,(1.11)^t$. In what year is it estimated that $40 trillion will be spent on health care expenditures?

73. A venture capitalist invests $1.2 million into a start-up company whose sales pitch promises an 8% per year growth on the investment. How long will it take for the initial investment to double?

74. A start-up company is valued at $24.3 million and increases in value by 6% per year. How long will it take for the value of the company to double? A venture capitalist invests $1.2 million into a start-up company whose sales pitch promises an 8% per year growth on the investment. How long will it take for the initial investment to double?

# 3.3 Natural Logarithms

## Overview

In many of the sciences, the most frequently used base for logarithms is a special number, $e$. Base $e$ logarithms, known as natural logarithms, and the related exponential expressions with base $e$ are studied in this section. The base $e$ logarithm, $\log_e(x)$, has its own notation, $\ln(x)$.

Equations using base $e$ are uniquely suited to describing phenomenon in the natural world. Quantities like animal populations and radioactive substances change *continuously*, not just at the end of a specified period of time. Exponential models using base $e$ describe continuous growth or decay.

As you study this section, you will learn to:

- Understand the meaning of a natural logarithm
- Evaluate natural logarithms
- Use exponential models with base $e$ to make estimates and predictions

## A. Definition of Natural Logarithm

A **natural logarithm** is a logarithm with a specific base called $e$, where $e$ is an irrational number. Rounded to twenty decimal places:

$$e \approx 2.71828182845904523635 \ldots \approx 2.72$$

Notice the value of $e$ is between the whole numbers 2 and 3. The letter $e$ in this case does *not* represent a variable; rather, it represents a constant – the same number all the time. In this way it is similar to another irrational number that you may be familiar with, the number $\pi$ (pi). The number $\pi$ is represented by a Greek letter but is also a constant and *not* a variable.

We write $\log_e(x)$ simply as $\ln(x)$. We read $\ln(x)$ as "the logarithm with **base $e$** of $x$" or most commonly, "the natural log of $x$."

The working definition is that a natural logarithm, $y = \ln(x)$ is the exponent to which $e$ must be raised to get $x$. The equation $y = \ln(x)$ is in logarithmic form and the equation $e^y = x$ is in exponential form.

For example, we can rewrite the logarithmic equation $y = \ln(99)$ in exponential form as $e^y = 99$.

> **Natural Logarithm**
>
> A **natural logarithm** is a logarithm with base $e$. We write $\ln(x)$ to represent $\log_e(x)$.
> For $x > 0$, $y = \ln(x)$ is equivalent to $e^y = x$.

## Example 1

Convert the following logarithmic equations to exponential form.

1. $\ln(2981) \approx 8$
2. $\ln(115) \approx 4.75$
3. $\ln(33) \approx 3.5$

### Solutions

We use the definition of a natural logarithm to write equivalent equations in base $e$.

1. $e^8 \approx 2981$
2. $e^{4.75} \approx 115$
3. $e^{3.5} \approx 33$

We can also convert an equation in exponential form with base $e$ to logarithmic form using a natural logarithm. Remember, we do not need to write the base if we use "ln" rather than "log."

## Example 2

Convert the following exponential equations to logarithmic form.

1. $e^2 \approx 7.39$
2. $e^{4.2} \approx 66.69$
3. $e^{-3} \approx 0.05$

### Solutions

We use natural logarithms because in each equation the base is $e$.

4. $\ln(7.39) \approx 2$
5. $\ln(66.69) \approx 4.2$
6. $\ln(0.05) \approx -3$

Most values of $\ln(x)$ can only be found by using a calculator. One exception is $\ln(1) = 0$ because the logarithm of 1 is always 0 in any base. For other natural logarithms, we can use the [LN] key on a graphing calculator.

Given a natural logarithm with the form $y = \ln(x)$, evaluate it using a calculator by following these steps:

1. Press [LN].
2. Enter the value given for $x$, followed by [)].
3. Press [ENTER].

To calculate a power of $e$, we use the [2nd] button to access the command $e^x$. Notice that this command already sets up an exponent entry. Follow these steps:

1. Press [2nd], and then press [LN].
2. Enter the value given for $x$, followed by [)].
3. Press [ENTER].

## Example 3

Use a calculator to find ln(72).

**Solution**

We find that the calculator shows $\ln(72) \approx 4.2767$.

We use the $e^x$ command to verify our solution. In this case,

$\ln(72) \approx 4.2767$ because $e^{4.2767} \approx 72$

Figure 1 shows the calculation of ln(72) and its verification in exponential form. Notice that if we use the answer command, [2nd] [(-)], the exponential form returns exactly 72. If we round the value of the natural logarithm, the answer will be approximately 72.

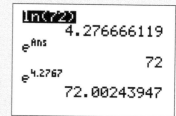

Figure 1. Verifying ln(72).

## Practice A

Convert the following logarithmic equations to exponential form.

1. $\ln(5) \approx 1.609$
2. $\ln(0.135) \approx -2$
3. $\ln(478) \approx 6.17$

Convert the following exponential equations to logarithmic form.

4. $e^4 \approx 54.6$
5. $e^0 = 1$
6. $e^{7.11} \approx 1224$

Use a calculator to find the logarithms to four decimal places. Write the solution in exponential form to verify your work.

7. $\ln(400)$
8. $\ln(0.77)$
9. $\ln(-16)$

When you're finished, turn the page to check your work.

## B. Solving Logarithmic and Exponential Equations in Base e

In Section 3.2 we solved equations in logarithmic form by rewriting the equation in exponential form. Remember, sometimes it is necessary to simplify the equation first.

### Example 4

Solve the following equations for $x$.

1. $\ln(x) = 6$
2. $5 \ln(x + 1) = 22$
3. $\ln(2x) + 7 = 10.3$

**Solutions**

1. We rewrite the expression $\ln(x) = 6$ in exponential form and compute using a calculator:

$x = e^6 \approx 403.4288$.

2. $5\ln(x+1) = 22$

| | |
|---|---|
| $5\ln(x+1) = 22$ | Original equation. |
| $\ln(x+1) = 4.4$ | Divide both sides by 5. |
| $e^{4.4} = x+1$ | Write in exponential form. |
| $e^{4.4} - 1 = x$ | Subtract 1 from both sides. |
| $x \approx 80.4509$ | Compute. |

We check that 80.4509 is an approximate solution to the original equation:

$5\ln(80.4509+1) = 5\ln(81.4509) \approx 22.0000019 \approx 22$

3. $\ln(2x) + 7 = 10.3$

| | |
|---|---|
| $\ln(2x) + 7 = 10.3$ | Original equation. |
| $\ln(2x) = 3.3$ | Subtract 7 from both sides. |
| $e^{3.3} = 2x$ | Write in exponential form. |
| $\dfrac{e^{3.3}}{2} = x$ | Divide both sides by 2. |
| $x \approx 13.5563$ | Compute. |

We check that 13.5563 is an approximate solution to the original equation:

$\ln(2(13.5563)) + 7 \approx 10.299999 \approx 10.3$

To solve equations involving exponential terms with base $e$, simplify the equation (if necessary), and then change to logarithmic form. After that, evaluate the natural logarithm with a calculator.

### Example 5

Solve the following equations for $x$.

1. $2e^x = 68$
2. $1.5e^{3x} + 10 = 1393$

Solutions

1. $2e^x = 68$

| | |
|---|---|
| $2e^x = 68$ | Original equation. |
| $e^x = 34$ | Divide both sides by 2. |
| $\ln(34) = x$ | Write in logarithmic form. |
| $x \approx 3.5264$ | Compute. |

We check that 3.5264 is an approximate solution to the original equation:

$2e^{3.5264} \approx 68.0027 \approx 68$

2. $1.5e^{3x} + 10 = 1393$

$$\begin{align*}
1.5e^{3x} + 10 &= 1393 & &\text{Original equation} \\
1.5e^{3x} &= 1383 & &\text{Subtract 10 from both sides.} \\
e^{3x} &= 922 & &\text{Divide both sides by 1.5.} \\
\ln(922) &= 3x & &\text{Write in logarithmic form.} \\
\frac{\ln(922)}{3} &= x & &\text{Divide both sides by 3.} \\
x &\approx 2.2755 & &\text{Compute.}
\end{align*}$$

We check that 2.2755 is an approximate solution to the original equation:

$1.5e^{3(2.2755)} + 10 \approx 1392.9375 \approx 1393$

Figure 2 shows the verification step on the calculator.

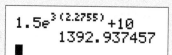

Figure 2. Verify your work.

## Practice Set B

Solve for $x$. Round your answers to four decimal places. Turn the page to check your solutions.

10. $\ln(x) = 9.2$  11. $3\ln(x - 2) = 18$  12. $e^x + 4 = 25$  13. $8e^{2x} = 436$

### Practice A — Answers

1. $e^{1.609} \approx 5$
2. $e^{-2} \approx 0.135$
3. $e^{6.17} \approx 478$
4. $\ln(54.6) \approx 4$
5. $\ln(1) = 0$
6. $\ln(1224) \approx 7.11$
7. $\ln(400) \approx 5.9915$ because $e^{5.9915} \approx 400$
8. $\ln(0.77) \approx -0.2615$ because $e^{-0.2615} \approx 0.77$
9. It's not possible to take the logarithm of a negative number in the set of real numbers.

## C. Exponential Models with Base e

The constant $e$ was named by the Swiss mathematician Leonhard Euler (1707–1783), who first investigated and discovered many of its properties. Scientists and mathematicians use the number $e$ as a base for many real-world exponential models. Exponential models with base $e$ represent *continuous* growth or decay, meaning that the quantity measured changes continuously as time passes.

### Example 6

$P$, the population in thousands of Bellevue, Washington, $t$ years after 2000 is given by:

$$P = f(t) = 110\, e^{0.0198t}$$

a. Estimate the population of Bellevue in 2016.

b. When will the population of Bellevue reach 180 thousand?

**Solution**

a. For the year 2016, $t = 16$.

$$P = f(16) = 110\, e^{0.0198(16)} \approx 151$$

We estimate the population of Bellevue in 2016 is 151 thousand.

b. Replace $f(t)$ with 180 (thousand) and solve for $t$.

| | |
|---|---|
| $110\, e^{0.0198t} = 180$ | $f(t) = 180$ |
| $e^{0.0198t} = \dfrac{18}{11}$ | Divide both sides by 110. |
| $\ln\left(\dfrac{18}{11}\right) = 0.0198t$ | Write in logarithmic form. |
| $\dfrac{\ln\left(\dfrac{18}{11}\right)}{0.0198} = t$ | Divide both sides by 0.0198. |
| $24.9 \approx t$ | Compute. |

The population of Bellevue, Washington, will reach 180 thousand about 24.9 years after the year 2000 — near the end of 2024.

---

**Practice B — Answers**

10. $e^{9.2} \approx 9897.1291$  11. $e^6 + 2 \approx 405.4288$  12. $\ln(21) \approx 3.0445$  13. $\dfrac{\ln(54.5)}{2} \approx 1.9991$

## Example 7

Radon-222 decays at a *continuous* rate of 17.3% per day. This is modeled by the following equation where $A$ represents the amount of Ra-222 remaining from a 100 mg sample after $t$ days.

$$A = f(t) = 100\, e^{-0.0173t}$$

How long does it take for there to be 50 mg remaining in the sample?

### Solution

Set the equation equal to 50 and solve for $t$.

| | |
|---|---|
| $100\, e^{-0.0173t} = 50$ | Substitute $f(t) = 50$. |
| $e^{-0.0173t} = 0.5$ | Divide both sides by 100. |
| $\ln(0.5) = -0.0173t$ | Rewrite in logarithmic form. |
| $\dfrac{\ln(0.5)}{-0.0173} = t$ | Divide both sides by $-0.0173$ |
| $t \approx 40.0663$ | Compute. |

In about 40 days, 50 mg of radon-222 will remain.

Because we began with 100 mg of Ra-222 and ended with 50 mg, this can be interpreted to mean that the *half-life* of radon-222 is about 40 days.

## Exercises 3.3

Use a calculator to find the natural logarithm. Round your result to the fourth decimal place.

1. $\ln(10.83)$
2. $\ln(72{,}110)$
3. $\ln(4600)$
4. $\ln(0.25)$

Convert the following logarithmic equations to exponential form.

5. $\ln(1096.6) \approx 7$
6. $\ln(5000) \approx 8.5$
7. $\ln(0.74) \approx -0.3$
8. $\ln(9) \approx 2.2$
9. $\ln(493) \approx 6.2$
10. $\ln(272) \approx 5.6$

Convert the following exponential equations to logarithmic form.

11. $e^{4.836} \approx 126$
12. $e^{5.916} \approx 371$
13. $e^6 \approx 403.43$
14. $e^9 \approx 8103.1$
15. $e^{-5} \approx 0.0067$
16. $e^{-0.025} \approx 0.9753$

Solve for $x$. Round your solutions to the fourth decimal place.

17. $\ln(x) = 4$
18. $\ln(x) = 7$
19. $9 \ln(x - 4) = 18$
20. $6 \ln(x + 3) = 24$
21. $\ln(6x) + 2 = 7$
22. $\ln(5x) - 8 = -2$
23. $4 e^x = 530$
24. $8 e^x = 936$
25. $2.75\, e^{x+3} = 17{,}050$
26. $8.4\, e^{x-2} = 2801.4$
27. $e^{2x-9} = 795.7$
28. $e^{4x+1} = 6213.5$
29. $9 e^x - 240 = 6 e^x + 270$
30. $8 e^x - 733 = 6 e^x + 422$

31. A culture of bacteria grows according to the continuous growth model $B = f(t) = 500\,e^{0.062t}$ where $B$ is the number of bacteria and $t$ is in hours.
    a. Find and interpret $f(0)$.
    b. Find the number of bacteria after 6 hours.
    c. To the nearest tenth of an hour, determine how long it will take for the population to grow to 1200 bacteria.

32. A population of rabbits in a nature preserve grows according to the continuous growth model $R = f(t) = 8.8\,e^{0.052t}$ where $R$ is the number of rabbits in thousands and $t$ is in years.
    a. Find and interpret $f(0)$.
    b. Find the number of rabbits after 7 years.
    c. To the nearest tenth of a year, determine how long it will take for the population to grow to 50 thousand rabbits.

33. A person invests in an account with interest compounded continuously according to the formula $A = f(t) = 15\,e^{0.044t}$ where $A$ is in thousands of dollars and $t$ is in years.
    a. Find and interpret $f(0)$.
    b. Find the value of the account after 10 years.
    c. To the nearest tenth of a year, determine how long it will take for the investment to grow to 40 thousand dollars.

34. A person invests in account with interest compounded continuously according to the formula $A = f(t) = 12\,e^{0.0375t}$ where $A$ is in thousands of dollars and $t$ is in years.
    a. Find and interpret $f(0)$.
    b. Find the value of the account after 8 years.
    c. To the nearest tenth of a year, how long it will take for the investment to grow to 25 thousand dollars?

35. A person buys a cup of coffee. The coffee's temperature, in °F, is given by $f(t) = 205\,e^{-0.05t}$ where $t$ is in minutes. If the person waits until the coffee is 170°F before they begin to drink it, how long will they have to wait?

36. A person makes a cup of tea. The tea's temperature, in °F, is given by $f(t) = 200\,e^{-0.04t}$ where $t$ is in minutes. If the person waits until the tea is 175°F before they sip it, how long will they have to wait?

37. A medical lab has a 120 mg sample of Cobalt-60 which is a radioactive isotope of cobalt. The amount of Cobalt-60 in milligrams remaining in the sample after $t$ years is given by $C = f(t) = 120\,e^{-0.1216t}$. How long will it take for the sample to have only 25 milligrams of Cobalt-60 remaining?

38. The fossil of the vertebrae of a whale was discovered by paleontologists and analyzed. They determined that the amount of Carbon-14 in the vertebrae at the time the whale was alive was approximately 16.5 milligrams. So the paleontologists knew that $C$, the amount of C-14 in mg in the fossil, would decay according to the equation $C = f(t) = 16.5\,e^{-0.00012t}$ where $t$ is in years. How long ago did the whale live if there is only 0.04 mg of C-14 remaining?

39. $P$, the population of a state, can be modeled by the equation $P = f(t) = P_0\,e^{0.0145t}$ where $P_0$ is the population of the state now and $t$ represents years from now. How long will it take the population to double?

40. $P$, the population $P$ of a country, can be modeled by the equation $P = f(t) = P_0\,e^{0.019t}$ where $P_0$ is the population of the country now and $t$ represents years from now. How long will it take the population to triple?

CHAPTER 4
# Quadratic Functions

A home run hit during a baseball game, a stone tossed off a cliff, a small rocket launched from the backyard — these events are examples of projectile motion, which can be modeled with a quadratic function. In this chapter, we will study quadratic functions, their characteristics, and their applications to real-world situations.

In the first section of this chapter, we'll review some basic algebra skills, focusing on expanding and factoring polynomials. Then we'll define quadratic functions and represent these functions using tables, graphs, and equations. Finally, we will study how to calculate outputs, how to model data, and how to solve problems involving quadratic functions.

4.1  Expanding and Factoring Polynomials ................................................................. page 180

4.2 Quadratic Functions in Standard Form ............................................................. page 205

4.3 The Square Root Property ................................................................................. page 221

4.4 The Quadratic Formula ...................................................................................... page 237

4.5 Modeling with Quadratic Functions .................................................................. page 251

# 4.1 Expanding and Factoring Polynomials

## Overview

This section examines algebraic expressions called polynomials. We've already used simple polynomials, such as $3x - 7$, when we worked with linear functions. Now we'll practice multiplying and factoring polynomials and solving equations involving polynomials.

It's likely that you've studied the topics in this section in a previous algebra class, but refreshing your knowledge of some basic algebra skills will facilitate the study of quadratic functions, which is the focus of the rest of Chapter 4.

In this section, you will learn how to :

- Multiply polynomials
- Use the FOIL method to multiply binomials
- Write the product of two binomial conjugates as the difference of squares
- Factor polynomials with a greatest common factor
- Factor trinomials
- Apply the zero product property

## A. Multiplication of Monomials and Polynomials

We'll start with a review of the vocabulary associated with polynomials. Recall that a **term** is a number, variable, or product of a number and one or more variables that are raised to powers. The following are examples of single terms:

$$2x \quad \text{and} \quad -7x^2y \quad \text{and} \quad a^3b^5 \quad \text{and} \quad 8$$

A **polynomial** or **polynomial expression** is the sum or difference of terms with the condition that all the exponents on the variables must be nonnegative integers. The following are all examples of polynomials:

$$5x^4 - 6x^2 + 9x + 2$$

$$9x^2y - xy + y^2$$

$$4z^2 - 1$$

The first polynomial above contains four terms. The term in a polynomial that has the largest combined powers on its exponents is called the **leading term**. In all three examples above, the leading term has been written first.

The numerical factor of a term — for example, the 5 in $5x^4$ — is called a **coefficient**. The coefficient on the term $y^2$ is 1 because $y^2$ can be written as $1 \cdot y^2$. Coefficients can be positive, negative, or even zero, and they can be whole numbers, decimals, or fractions. The coefficient of the leading term is called the **leading coefficient**. If a term in a polynomial doesn't contain a variable, it's called a **constant**.

A polynomial containing only one term, such as $5x^4$, is called a **monomial**. A polynomial containing two terms, such as $2x - 9$, is called a **binomial**. A polynomial containing three terms, such as $-3x^2 + 8x + 7$, is called a **trinomial**.

Now let's practice multiplying one polynomial by another. When we studied the product property for exponents, we looked at the product of two monomials. Recall that the product property for exponents states that $x^m x^n = x^{m+n}$.

### Example 1

Multiply these monomials and then simplify.

1. $(9x^4 y^2)(-4x^5 y^3)$
2. $(-2a^3 bc^6)(-8ab^7 c^2)$

**Solutions**

1. $(9x^4 y^2)(-4x^5 y^3)$

$$\begin{aligned}(9x^4 y^2)(-4x^5 y^3) &= (9)(-4)(x^4 x^5)(y^2 y^3) && \text{Regroup and reorder factors.} \\ &= -36 x^{4+5} y^{2+3} && \text{Apply the product property for exponents.} \\ &= -36 x^9 y^5 && \text{Simplify.}\end{aligned}$$

2. $(-2a^3 bc^6)(-8ab^7 c^2)$

$$\begin{aligned}(-2a^3 bc^6)(-8ab^7 c^2) &= (-2)(-8)(a^3 a)(bb^7)(c^6 c^2) && \text{Regroup and reorder factors.} \\ &= 16 a^{3+1} b^{1+7} c^{6+2} && \text{Apply the product property for exponents.} \\ &= 16 a^4 b^8 c^8 && \text{Simplify.}\end{aligned}$$

Multiplying a monomial by a polynomial requires us to use the distributive property of multiplication over addition and subtraction. In this case, we multiply each term in the polynomial by the monomial. To simplify each product, we multiply the coefficients and apply the product property for exponents.

### Example 2

Multiply the following monomials by polynomials.

1. $3x^2(7x^5 - 4x^3 + 6)$
2. $(2x^2 + 11xy - 3y^2)(5xy^2)$

**Solutions**

1. $3x^2(7x^5 - 4x^3 + 6)$

$$3x^2(7x^5 - 4x^3 + 6) = 3x^2(7x^5) - 3x^2(4x^3) + 3x^2(6)$$ — Apply the distributive property.
$$= 21(x^2 x^5) - 12(x^2 x^3) + 18x^2$$ — Regroup and reorder factors. Multiply coefficients.
$$= 21x^7 - 12x^5 + 18x^2$$ — Simplify each term, using the product property for exponents.

2. $(2x^2 + 11xy - 3y^2)(5xy^2)$

$$(2x^2 + 11xy - 3y^2)(5xy^2) = 2x^2(5xy^2) + 11xy(5xy^2) - 3y^2(5xy^2)$$ — Apply the distributive property.
$$= 10(x^2 x)(y^2) + 55(xx)(yy^2) - 15x(y^2 y^2)$$ — Regroup and reorder factors. Multiply coefficients.
$$= 10x^3 y^2 + 55x^2 y^3 - 15xy^4$$ — Simplify using the product property for exponents.

Notice in Example 2 that it doesn't matter if the monomial term is to the left or the right of the other polynomial when multiplying. The distributive property works the same way in either case.

### Practice A

Now it's your turn to multiply with monomial factors. When you are done, turn the page and check your solutions.

1. $(-7x^5 y^3)(6x^2 y)$
2. $(11a^9 b^2 c^3)(2ab^4 c)$
3. $(-3x^2)(3x^2 + 8x - 4)$
4. $(4x^3 y - 9x^2 y^2 + y)(5y)$
5. $(8xy^2)(2x^3 y^3 + 3xy - 1)$

## B. Multiplying Binomials

We next use the distributive property to help us find the product of two binomials. Consider the example $(2x + 5)(x + 6)$. To apply the distributive property, we multiply both terms in the second binomial by the first binomial. We continue to apply the distributive property to the resulting factors.

$$
\begin{aligned}
(2x + 5)(x + 6) &= (2x + 5)x + (2x + 5)6 && \text{Apply the distributive property.} \\
&= 2x(x) + 5(x) + 2x(6) + 5(6) && \text{Apply the distributive property.} \\
&= 2x^2 + 5x + 12x + 30 && \text{Simplify.} \\
&= 2x^2 + 17x + 30 && \text{Combine like terms: } 5x + 12x = 17x.
\end{aligned}
$$

Notice in the last step that we combined the **like terms** to further simplify the result. Recall like terms have identical variable factors, including identical exponents. We can add or subtract like terms when simplifying polynomials.

Look back at the four products in the second step of the example above. These four products are found by multiplying each term in the first binomial by each term in the second binomial. This is an expansion of the distributive property and the general rule for multiplying two binomials.

### Example 3

Find the product of these binomials: $(2x^2 + 9)(5x - 3)$

**Solution**

We multiply each term in the first binomial by each term in the second binomial.

$$
\begin{aligned}
(2x^2 + 9)(5x - 3) &= 2x^2(5x) - 2x^2(3) + 9(5x) - 9(3) && \text{Multiply pairs of terms.} \\
&= 10x^3 - 6x^2 + 45x - 27 && \text{Simplify.}
\end{aligned}
$$

When multiplying two binomials, we sometimes use a process called the FOIL method to find the product. It's called the "FOIL" method because we multiply the First terms, the Outer terms, the Inner terms, and then the Last terms of each binomial.

The FOIL method arises out of the distributive property. It's an easy way to remember the multiplication steps used in Example 3. We simply multiply each term of the first binomial by each term of the second binomial and then combine like terms.

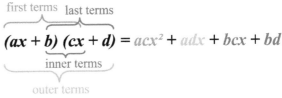

Figure 1. The FOIL method.

Given two binomials, then, we can use the FOIL method to simply the expression by following these six steps:

1. Multiply the first terms of each binomial.
2. Multiply the outer terms of the binomials.
3. Multiply the inner terms of the binomials.
4. Multiply the last terms of each binomial.
5. Add the products obtained in steps 1–4.
6. Combine like terms and simplify.

## Example 4

Use the FOIL method to find the product: $(2x - 7)(3x + 4)$

**Solution**

*Step 1:* Find the product of the **first terms**.

*Step 2:* Find the product of the **outer terms**.

*Step 3:* Find the product of the **inner terms**.

*Step 4:* Find the product of the **last terms**.

*Step 5:* Add the products.

*Step 6:* Combine like terms.

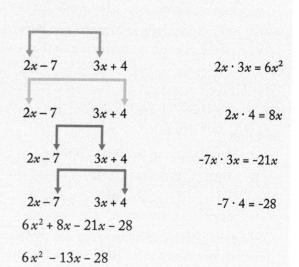

Let's look at a few more examples of multiplying binomials pairs. We'll use binomials of form, $x \pm a$, and give attention to the pattern in the products. Specifically, watch how we find the middle term and the last term of the resulting trinomial.

### Practice A — Answers

1. $-42x^7y^4$
2. $22a^{10}b^6c^4$
3. $-9x^4 - 24x^3 - 12x^2$
4. $20x^3y^2 - 45x^2y^3 + 5y^2$
5. $16x^4y^5 + 24x^2y^3 - 8xy^2$

## Example 5

Use the FOIL method to multiply each pair of binomials.

1. $(x + 9)(x + 2)$
2. $(x + 7)(x - 3)$
3. $(x - 6)(x + 1)$

**Solutions**

1. $(x + 9)(x + 2)$

   $(x + 9)(x + 2) = x^2 + 2x + 9x + 18$    Use the FOIL method.
   
   $\phantom{(x + 9)(x + 2)} = x^2 + 11x + 18$    Combine like terms.

2. $(x + 7)(x - 3)$

   $(x + 7)(x - 3) = x^2 - 3x + 7x - 21$    Use the FOIL method.
   
   $\phantom{(x + 7)(x - 3)} = x^2 + 4x - 21$    Combine like terms.

4. $(x - 6)(x + 1)$

   $(x - 6)(x + 1) = x^2 + x - 6x - 6$    Use the FOIL method.
   
   $\phantom{(x - 6)(x + 1)} = x^2 - 5x - 6$    Combine like terms.

Let's look back at the products in Example 5. With regard to the constants in the original binomials, the coefficient of the middle term of the trinomial is the sum of the constants. The last term of the trinomial is the product of the original constants.

1. $(x + 9)(x + 2) = x^2 + 11x + 18$
   The constants are 9 and 2, the sum is 11, and the product is 18.

2. $(x + 7)(x - 3) = x^2 + 4x - 21$
   The constants are 7 and −3, the sum is 4, and the product is −21.

3. $(x - 6)(x + 1) = x^2 - 5x - 6$
   The constants are −6 and 1, the sum is −5, and the product is −6.

We can use the pattern observed in these examples as a shortcut to finding products of binomials of the form $x \pm a$. For example, let's expand $(x - 8)(x - 5)$ using this shortcut. The constants are −8 and −5. The sum is $-8 + -5 = -13$, the coefficient of the middle term. The product is $(-8)(-5) = 40$, the constant of the trinomial. That gives us:

$$(x - 8)(x - 5) = x^2 - 13x + 40$$

Some students find this shortcut helpful when multiplying binomials.

Be careful, though! If the binomials are not of form $x \pm a$, the mental math used to find the middle term is more involved:

$(2x + 5)(3x - 4) = 6x^2 + 7x - 20$    Expand and mentally combine the outer and inner products: $-8x + 15x = 7x$.

If a product of two binomials looks challenging, you can always write down all four products using the FOIL method before combining like terms. The FOIL method is a dependable way to keep track of the process.

## Practice B

Now it's your turn to multiply binomials. When you are done, turn the page and check your solutions.

6. $(x + 2)(x + 7)$
7. $(x + 3)(x - 9)$
8. $(x - 8)(x + 1)$
9. $(a - 10b)(a - 6b)$
10. $(5x + 4)(2x + 1)$
11. $(3x^2 - x)(4x + 7)$

## C. More Polynomial Products

### Expanding a Binomial Square

We can use the distributive property or the FOIL method to find the square of a binomial by writing the expression as a product. This is often referred to as "expanding a binomial square." The definition of exponents tells us that if we raise an expression to the second power, then the expression is used as a factor twice in the product.

### Example 6

Expand the binomial squares.

1. $(3x - 8)^2$
2. $(x + 5)^2$

**Solution**

1. $(3x - 8)^2$

$$\begin{aligned}(3x - 8)^2 &= (3x - 8)(3x - 8) && \text{Definition of exponents.} \\ &= 9x^2 - 24x - 24x + 64 && \text{Multiply the binomials.} \\ &= 9x^2 - 48x + 64 && \text{Combine like terms.}\end{aligned}$$

2. $(x + 5)^2$

$$\begin{aligned}(x + 5)^2 &= (x + 5)(x + 5) && \text{Definition of exponents.} \\ &= x^2 + 5x + 5x + 25 && \text{Multiply the binomials.} \\ &= x^2 + 10x + 25 && \text{Combine like terms.}\end{aligned}$$

## ☠ Warning! Incorrect Approach! ☠

It's a common error to omit the middle term when squaring a binomial. Students often believe they can "distribute" the exponent across addition or subtraction, but sadly, there's no rule or property that allows this.

$(x + 5)^2 \neq x^2 + 25$     These expressions are *not equal*.

$(x + 5)^2 = x^2 + 10x + 25$     This result is correct.

## Difference of Squares

Sometimes we need to multiply a binomial by another binomial with the same terms but the opposite sign. Pairs of this type are referred to as **binomial conjugates** or simply as "conjugate pairs." For example, $x + 4$ and $x - 4$ are binomial conjugates. Let's expand $(x + 4)(x - 4)$:

$$(x + 4)(x - 4) = x^2 + 4x - 4x - 16$$
$$= x^2 - 16$$

The two middle terms show the sum of opposites, $-4x$ and $4x$. The sum is $0x = 0$. In effect, the middle term drops out, resulting in a **difference of squares**.

We can better understand this description of the result by writing the solution as

$$x^2 - 16 = (x)^2 - (4)^2$$

The second form emphasizes that we're subtracting or finding the difference of two expressions that have been squared. In general, the product of binomial conjugates will look like this:

$$(a + b)(a - b) = a^2 + ab - ab - b^2$$
$$= a^2 - b^2$$

Because the sign is opposite in the second binomial, the outer and inner terms are opposites and add to zero. We're left with the square of the first term minus the square of the last term, $a^2 - b^2$, a difference of squares.

### Difference of Squares

When multiplying binomial conjugates, the result is the square of the first term minus the square of the last term:
$$(a + b)(a - b) = a^2 - b^2$$

We can calculate the product of binomial conjugates using the formula in the definition box. Square the first term of the binomials, square the last term of the binomials, and then subtract the square of the last term from the square of the first term.

### Example 7

Multiply.

1. $(x + 7)(x - 7)$
2. $(9x + 4)(9x - 4)$

**Solutions**

1. Square the first term: $(x)^2 = x^2$, and the last: $7^2 = 49$. Write the difference: $x^2 - 49$.

   $(x + 7)(x - 7) = x^2 - 49$

2. Square the first term, $(9x)^2 = 81x^2$, and the last: $4^2 = 16$. Write the difference: $81x^2 - 16$.

   $(9x + 4)(9x - 4) = 81x^2 - 16$

### Multiplication of Polynomials

We can find the product of two polynomials with any number of terms by applying the distributive property. The key is to multiply each term in the first polynomial by each term in the second polynomial. As always, combine like terms whenever possible.

### Example 8

Multiply the polynomials: $(3x + 4)(x^2 - 6x + 7)$

**Solution**

To begin, we multiply each term in the binomial by each term in the trinomial.

$$\begin{aligned}(3x + 4)(x^2 - 6x + 7) &= 3x(x^2) - 3x(6x) + 3x(7) + 4(x^2) - 4(6x) + 4(7) &&\text{Multiply pairs of terms.}\\ &= 3x^3 - 18x^2 + 21x + 4x^2 - 24x + 28 &&\text{Simplify.}\\ &= 3x^3 - 14x^2 + 3x + 28 &&\text{Combine like terms.}\end{aligned}$$

## Practice C

Now it's your turn. Expand the following. When you are done, turn the page to check your solutions.

12. $(x + 3)^2$
13. $(x - 8)^2$
14. $(2x + 11)^2$

Multiply the following.

15. $(x + 1)(x - 1)$
16. $(x + 12)(x - 12)$
17. $(2x + 7)(2x - 7)$
18. $(x - 5)(4x^2 + 8x - 3)$

### Practice B — Answers

6. $x^2 + 9x + 14$
7. $x^2 - 6x - 27$
8. $x^2 - 7x - 8$
9. $a^2 - 16ab + 60b^2$
10. $10x^2 + 13x + 4$
11. $12x^3 + 17x^2 - 7x$

# D. Factoring Polynomials and the Greatest Common Factor

Many polynomial expressions can be written as a product by factoring. Factoring involves finding polynomial expressions that can be multiplied together to give the original polynomial. We have a variety of methods for factoring polynomial expressions. One way to think of factoring is to see it as "undoing" or reversing multiplication. Below, we use whole numbers to compare multiplying to factoring:

$3 \cdot 5 = 15$     Multiplying: Calculate.
$15 = 3 \cdot 5$     Factoring: Write a product of two numbers.

Here's an example comparing multiplying to factoring with polynomials:

$6x(x + 5) = 6x^2 + 30x$     Multiplying: Apply the distributive property.
$6x^2 + 30x = 6x(x + 5)$     Factoring: Write a product of two expressions.

Both processes above are actually applications of the distributive property. Beginning with $6x^2 + 30x$, we note the expression $6x$ is a common factor of $6x^2$ and $30x$.

$$6x^2 + 30x = 6x(x) + 6x(5)$$

We use the distributive property to *factor out* the common factor $6x$:

$$6x^2 + 6x = 6x(x) + 6x(5) = 6x(x + 5)$$

The expression $6x(x + 5)$ is the original expression written in factored form. Notice that the two terms in parentheses are the result of dividing each term in the original polynomial by the common factor:

$$\frac{6x^2}{6x} = x \quad \text{and} \quad \frac{30x}{6x} = 5$$

When we factor a common factor from a polynomial, we're essentially dividing. The common factor is written first, followed by parentheses that contain the quotients that result when each term is divided by the common factor.

When we write a polynomial in factored form, the first step is to find a common factor of its terms if one exists. We look for the largest number that divides each coefficient. We also identify the largest factor of the variable parts. Multiplied together these determine the **greatest common factor** (GCF) for a polynomial.

## Greatest Common Factor

The **greatest common factor** (GCF) for a polynomial is the monomial with the largest coefficient and the highest exponent(s) that divides (is a factor of) each term of the polynomial.

In the example above, we used the common factor of $6x$ to factor the polynomial $6x^2 + 30x$. Notice that the expression $3x$ is also a common factor of $6x^2$ and $30x$. However, $6x$ is the *greatest* common factor.

In the process of factoring a polynomial expression, we usually don't stop and divide each term by the GCF. Instead, we may find ourselves framing the problem this way: "What do I need to multiply the GCF by so that the results are the original terms in the polynomial?" This process is demonstrated in the next example.

### Example 9

Write each polynomial in factored form using the greatest common factor.

1. $14x^2 + 21x$
2. $6x^4 - 9x^3 + 12x^2$

**Solution**

1. $14x^2 + 21x$

   The GCF of the coefficients is 7 because 7 is the largest integer that divides 14 and 21 evenly.
   The GCF of the variables is $x$ because $x$ is the largest power of the variable that divides $x^2$ and $x$.
   Therefore, the GFC of the polynomial is $7x$.
   Next we ask, "What is $7x$ multiplied by to obtain the terms of the original expression?"

   $$7x(?) + 7x(?) = 14x^2 + 21x$$

   We write each term of the original polynomial as a product of $7x$ and another monomial. Then we apply the distributive property to factor $7x$ from each term.

   $$14x^2 + 21x = 7x(2x) + 7x(3)$$
   $$= 7x(2x + 3)$$

2. $6x^4 - 9x^3 + 12x^2$

   The GCF of the coefficients 6, 9, and 12 is 3.
   The GCF of the variable expressions $x^4$, $x^3$, and $x^2$ is $x^2$.
   So the GCF of the polynomial is $3x^2$.
   We write each term of the original polynomial as a product of $3x^2$ and another monomial. Then we apply the distributive property to write the polynomial in factored form.

   $$6x^4 - 9x^3 + 12x^2 = 3x^2(2x^2) - 3x^2(3x) + 3x^2(4)$$
   $$= 3x^2(2x^2 - 3x + 4)$$

After you factor a polynomial, verify your work by reapplying the distributive property to find the product. It should be the same as the original polynomial.

### Practice C — Answers

12. $x^2 + 6x + 9$
13. $x^2 - 16x - 64$
14. $4x^2 + 44x + 121$
15. $x^2 - 1$
16. $x^2 - 144$
17. $4x^2 - 49$
18. $4x^3 - 12x^2 - 43x + 15$

Next we factor a polynomial with two variables. The process is the same. We just need to make sure we find the monomial that is greatest common factor of the coefficients and both variables.

### Example 10

Factor: $8x^4y^3 + 44x^3y^2 + 20x^2y$

**Solution**

The GCF of the coefficients 8, 44, and 20 is 4.
The GCF of $x^4$, $x^3$, and $x^2$ is $x^2$.
And the GCF of $y^3$, $y^2$, and $y$ is $y$.
So the GCF of the polynomial is $4x^2y$.

$$8x^4y^3 + 44x^3y^2 + 20xy^2 = 4x^2y(2x^2y^2) + 4x^2y(11xy) + 4x^2y(5)$$
$$= 4x^2y(2x^2y^2 + 11xy + 5)$$

### Practice D

Factor each polynomial using the greatest common factor of the terms. When you are done, turn the page and check your solutions.

19. $24x^2 + 80x$

20. $5x^3 - 35x^2 + 10x$

21. $8a^6 - 10a^4 + 6a^3$

22. $12x^3y^2 + 3x^2y^3 + 18xy^4$

## E. Factoring Trinomials

### Factoring a Trinomial with Leading Coefficient 1

When attempting to factor a polynomial, we should always begin by looking for the greatest common factor of its terms. However, factoring out the GCF is not the only way that polynomial expressions can be factored. Some trinomials can be written in factored form as the product of two binomials.

Earlier in this section, we multiplied the following binomials:

$$(x + 9)(x + 2) = x^2 + 11x + 18$$
$$(x + 7)(x - 3) = x^2 + 4x - 21$$
$$(x - 6)(x + 1) = x^2 - 5x - 6$$

In each case, the product of two binomials is a trinomial. With some trinomials, we can reverse this process and factor to write the trinomial as the product of two binomials. To see how to do this, let's recall the shortcut steps for finding the product $(x + 9)(x + 2)$.

Earlier in this section we observed that, if 9 and 2 are the constants of each binomial, then in the resulting trinomial, the coefficient of the middle term is the sum of 9 and 2:

$$(x + 9)(x + 2) = x^2 + 11x + 18$$

The **constants** are 9 and 2, the **sum** is 11, and the **product** is 18.

Now let's find the general product of two binomials whose constants are $p$ and $q$.

$$\begin{aligned}(x + p)(x + q) &= x^2 + qx + px + pq & \text{Multiply pairs of terms.} \\ &= x^2 + px + qx + pq & \text{Reorder the two middle terms.} \\ &= x^2 + (p + q)x + pq & \text{Factor } x \text{ from middle terms.}\end{aligned}$$

Writing this as a factoring problem we have:

$$x^2 + (p + q)x + pq = (x + p)(x + q)$$

To factor a trinomial with a leading coefficient of 1, we find the two integers $p$ and $q$ whose sum is the coefficient of the middle term and whose product is the last term. We then write the binomial factors using $p$ and $q$.

### Factoring a Trinomial with Leading Coefficient 1

For a trinomial of the form $x^2 + bx + c$, if integers $p$ and $q$ exist such that

$$p + q = b \quad \text{and} \quad pq = c$$

In factored form, the trinomial is

$$(x + p)(x + q)$$

### Example 11

Factor the trinomial: $x^2 + 2x - 15$.

**Solution**

We have a trinomial with leading coefficient 1, $b = 2$, and $c = -15$. We need to find two numbers with a product of $-15$ and a sum of 2. In the table below, we list factors until we find a pair with the desired sum, $b = 2$.

| Factors of $-15$ | Sum of Factors |
|---|---|
| 1, $-15$ | $-14$ |
| $-1$, 15 | 14 |
| 3, $-5$ | $-2$ |
| $-3$, 5 | 2 |

Once we have identified $p$ and $q$ as $-3$ and 5, we write the factored form:

$$x^2 + 2x - 15 = (x - 3)(x + 5)$$

We can check our work by multiplying. We expand to confirm that $(x - 3)(x + 5)$ equals $x^2 + 2x - 15$.

The order in which we write the binomial factors doesn't matter, by the way, because multiplication is commutative: $a \cdot b = b \cdot a$.

## Example 12

Write each polynomial in factored form.

1. $x^2 - 4x - 21$
2. $x^2 + 9x + 8$
3. $x^2 - 13x + 30$

### Solution

1. $x^2 - 4x - 21$

   We look for a pair of numbers whose product is $-21$ and whose sum is $-4$. We find $(-7)(3) = -21$ and $-7 + 3 = -4$. So the factors of $x^2 - 4x - 21$ are $(x - 7)$ and $(x + 3)$.
   Now we check our work: $(x - 7)(x + 3) = x^2 - 4x - 21$.

2. $x^2 + 9x + 8$

   We look for a pair of numbers whose product is 8 and whose sum is 9. We find $(8)(1) = 8$ and $8 + 1 = 9$. So the factors of $x^2 + 9x + 8$ are $(x + 8)$ and $(x + 1)$.
   We check our work: $(x + 8)(x + 1) = x^2 + 9x + 8$.

3. $x^2 - 13x + 30$

   We look for a pair of numbers whose product is 30 and whose sum is $-13$. We find $(-10)(-3) = 30$ and $-10 + -3 = -13$. So the factors of $x^2 - 13x + 30$ are $(x - 10)$ and $(x - 3)$.
   Check: $(x - 10)(x - 3) = x^2 - 13x + 30$.

Recall a prime number, such as 7 or 13, has no positive factors other than 1 and itself. Similarly, a polynomial that cannot be factored is called a **prime polynomial**.

Consider the polynomial $x^2 + 5x + 2$. To factor this polynomial, we need to find two numbers whose product is 2 and whose sum is 5. The only factors of 2 are 2 and 1, or $-2$ and $-1$. Since neither of these pairs of factors adds up to 5, this polynomial is not factorable. We conclude that $x^2 + 5x + 2$ is a prime polynomial.

## Example 13

Factor: $x^2 + 9x - 28$

### Solution

We look for a pair of numbers whose product is $-28$ and whose sum is 9. In the table below, we list all the possible factors of $-28$ along with the corresponding sums.

### Practice D — Answers

19. $8x(3x + 10)$
20. $5x(x^2 - 7x + 2)$
21. $2a^3(4a^3 - 5a + 3)$
22. $3xy^2(4x^2 + xy + 6y^2)$

| Factors of –28 | Sum of Factors |
|---|---|
| 1, –28 | –27 |
| –1, 28 | 27 |
| 4, –7 | –3 |
| –4, 7 | 3 |

We see none of the sums equals 9, so we conclude that the trinomial $x^2 + 9x - 28$ is a prime polynomial.

Some trinomials have more than one variable. If a trinomial in two variables is not a prime polynomial, then we apply the factoring strategy we just studied to write such trinomials in factored form.

**Example 14**

Write $x^2 + 10xy + 24y^2$ in factored form.

**Solution**

We can write the polynomial in the form $x^2 + (10y)x + 24y^2$ and look for a two monomials whose product is $24y^2$ and whose sum is $10y$. We find that $6y$ and $4y$ satisfy this condition. Now we write the polynomial in factored form:

$$x^2 + 10xy + 24y^2 = (x + 6y)(x + 4y)$$

As you saw earlier in this section, a difference of squares is a perfect square subtracted from a perfect square. For example, the expression $x^2 - 16$ is a difference of squares because $x^2 = (x)^2$ and $16 = 4^2$. A difference of squares is a binomial expression, but it can be written as a trinomial with a coefficient of 0 on the middle term.

$$x^2 - 16 = x^2 + 0x - 16$$

Let's factor this trinomial using the method of looking for two integers whose product is –16 and whose sum is 0. These integers are 4 and –4.

$$x^2 - 16 = x^2 + 0x - 16 = (x + 4)(x - 4)$$

We see that the factored form of this difference of squares, $(x + 4)(x - 4)$, is the product of binomial conjugates, binomials that differ only by the sign between their terms. When multiplying binomial conjugates, we obtain a difference of squares.

$$(a + b)(a - b) = a^2 - b^2$$

When we reverse the process and factor a difference of squares, we obtain a conjugate pair of binomials.

$$a^2 - b^2 = (a + b)(a - b)$$

We can use this last equation to factor any difference of squares.

## Example 15

Factor the following expressions.

1. $x^2 - 36$.
2. $9x^2 - 25$

### Solution

1. $x^2 - 36$

   Notice that $x^2$ and 36 are perfect squares because $x^2 = (x)^2$ and $36 = 6^2$. So the factored form of the difference of squares $x^2 - 36$ is $(x + 6)(x - 6)$.

2. $9x^2 - 25$

   Notice that $9x^2$ and 25 are perfect squares because $9x^2 = (3x)^2$ and $25 = 5^2$. So the factored form of the difference of squares $9x^2 - 25$ is $(3x + 5)(3x - 5)$.

The second polynomial in Example 15 had a leading coefficient of 9 rather than 1, but we were still able to apply the formula for a difference of squares to write it in factored form. Now we turn our attention to more polynomials with leading coefficients that are not 1.

## Factoring a Trinomial when the Leading Coefficient is not 1

Some trinomials have a greatest common factor that can be factored out before we try to find two binomial factors. If the GCF of a polynomial is *not* 1, factor out the GCF before applying other factoring techniques. Once a polynomial cannot be factored further, it's said to be *completely factored*.

## Example 16

Factor: $4x^3 + 12x^2 - 160x$

### Solution

The GCF of the coefficients is 4, and the GCF of the variables is $x$, so the GCF of the polynomial is $4x$. Factoring out $4x$ we have:

$$4x^3 + 12x^2 - 160x = 4x(x^2 + 3x - 40)$$

In order to factor the resulting trinomial, we look for a pair of numbers whose product is $-40$ and whose sum is 3. We find that 8 and $-5$ satisfy this condition. So this is the complete factored form of the polynomial:

$$4x(x^2 + 3x - 40) = 4x(x + 8)(x - 5)$$

So far the polynomials we have factored have all had positive leading coefficients. When the leading coefficient of a polynomial is negative, we factor out the *opposite* of the greatest common factor. Then, we check to see if we can factor further.

## Example 17

Write the polynomial in complete factored form:

$$-3x^3 + 27x^2 - 42x$$

### Solution

For $-3x^3 + 33x^2 - 42x$, the leading coefficient is $-3$, and the GCF is $3x$. We factor out the opposite of the GCF, which is $-3x$.

$$-3x^3 + 27x^2 - 42x = -3x(x^2 - 9x + 14)$$

Next, to see if the resulting trinomial factors further, we look for two integers whose product is 14 and whose sum is $-9$. These integers are $-2$ and $-7$. Here is the complete factored form:

$$-3x(x^2 - 9x + 14) = -3x(x - 2)(x - 7)$$

We turn our attention to trinomials with leading coefficients other than 1 and with no greatest common factor other than 1. These are slightly more complicated to factor.

Suppose we want to factor $2x^2 - 7x - 15$. If it is factorable into two binomials, the product of the first terms will be $2x^2$, and the product of the last terms will be $-15$. We use trial and error and list some of the possible factors along with their products in the table below.

| Possible Factors | First Term | Middle Term | Last Term |
|---|---|---|---|
| $(2x - 1)(x + 15)$ | $2x^2$ | $30x - x = 29x$ | $-15$ |
| $(2x - 15)(x + 1)$ | $2x^2$ | $2x - 15x = -13x$ | $-15$ |
| $(2x - 3)(x + 5)$ | $2x^2$ | $10x - 3x = 7x$ | $-15$ |
| $(2x + 3)(x - 5)$ | $2x^2$ | $-10x + 3x = -7x$ | $-15$ |

We haven't listed all of the possible factors, but we find that the fourth trial gives the correct middle term. The factors of $2x^2 - 7x - 15$ are $(2x + 3)$ and $(x - 5)$. That means that $2x^2 - 7x - 15 = (2x + 3)(x - 5)$.

When using trial and error to solve a problem, it helps to pay attention to the *incorrect* results during the process. Observing whether the middle term is the wrong sign — positive vs. negative— or too large or too small can help guide the next trial. In this way, we can shorten the process of finding the correct factorization.

## Example 18

Factor $5x^2 + 13x - 6$.

### Solution

If this trinomial factors into two binomials, the product of the first terms will be $5x^2$ and the product of the last terms will be $-6$. In table below, we list possible factors along with their products until we find the correct pair.

| Possible Factors | First Term | Middle Term | Last Term | Observation |
|---|---|---|---|---|
| $(5x + 1)(x - 6)$ | $5x^2$ | $-30x + x = -29x$ | $-6$ | The middle term should be positive. |
| $(5x + 6)(x - 1)$ | $5x^2$ | $-5x + 6x = x$ | $-6$ | The middle term is positive but too small. |
| $(5x + 3)(x - 2)$ | $5x^2$ | $-10x + 3x = -7x$ | $-6$ | The middle term should be positive. |
| $(5x - 2)(x + 3)$ | $5x^2$ | $15x - 2x = 13x$ | $-6$ | This gives the correct middle term. |

We can check our work by multiplying. Expand to confirm that $(5x - 2)(x + 3) = 5x^2 + 13x - 6$.

As an alternative to the trial and error method of factoring trinomials, we can use a technique called factoring by grouping. This new method requires less trial and error but requires us to rewrite the original trinomial with four terms that can be grouped in pairs of two with a common factor. We outline the steps of this process before we provide an example.

### Factoring $ax^2 + bx + c$ by Grouping

If a trinomial of the form $ax^2 + bx + c$ is factorable, then to find the factored form, follow these steps:

1. Form the product $ac$.
2. Find a pair of numbers whose product is $ac$ and whose sum is $b$. Call this pair $p$ and $q$.
3. Rewrite the original trinomial so that the middle term $bx$ is written as the sum $px + qx$.
4. Factor $ax^2 + px + qx + c$ by grouping the four terms in pairs of two and factoring out the GCF from each pair.

## Example 19

Factor the trinomial by grouping: $4x^2 + 23x + 15$

### Solution

The trinomial has the form $ax^2 + bx + c$ where $a = 4$, $b = 23$, and $c = 15$.

*Step 1:* The product $ac$ is $4(15) = 60$.

*Step 2:* We look for two numbers whose product is 60 and whose sum is 23.

| Product | Sum |
|---|---|
| $1(60) = 60$ | $1 + 60 = 61$ |
| $2(30) = 60$ | $2 + 30 = 32$ |
| $3(20) = 60$ | $3 + 20 = 23$ |
| $4(15) = 60$ | $4 + 15 = 19$ |

We have not found all the possible pairs of numbers whose product is 60, but we *have* found a pair with a sum of 23. The numbers $p$ and $q$ are 3 and 20.

*Step 3:* We rewrite the original trinomial so the middle term is written as the sum of $3x$ and $20x$.

$$4x^2 + 23x + 15 = 4x^2 + 3x + 20x + 15$$

*Step 4:* To factor by grouping, use the distributive property to factor out the common factor of the first two terms. Repeat this process with the last two terms.

$$\underbrace{4x^2 + 3x}_{\text{GCF}\,=\,x} + \underbrace{20x + 15}_{\text{GCF}\,=\,5} = x(4x + 3) + 5(4x + 3)$$

Finally, we factor out the common factor $4x + 3$ from the remaining terms.

$$x(4x + 3) + 5(4x + 3) = (4x + 3)(x + 5)$$

The factored form of $4x^2 + 23x + 15$ is $(4x + 3)(x + 5)$.

▶ We can multiply the binomials to confirm we obtain the original trinomial.

## Example 20

Factor the trinomial by grouping: $9x^2 - 6x - 8$

**Solution**

Here $a = 9$, $b = -6$, and $c = -8$.

*Step 1:* The product $ac$ is $9(-8) = -72$.

*Step 2:* We look for two numbers whose product is $-72$ and whose sum is $-6$.

| Product | Sum |
|---|---|
| $1(-72) = -72$ | $1 + -72 = -71$ |
| $2(-36) = -72$ | $2 + -36 = -34$ |
| $3(-24) = -72$ | $3 + -24 = -21$ |
| $4(-18) = -72$ | $4 + -18 = -14$ |
| $6(-12) = -72$ | $6 + -12 = -6$ |

The numbers $p$ and $q$ are 6 and $-12$.

*Step 3:* We rewrite the original trinomial so that the middle term, $-6x$, is written as the sum of $6x$ and $-12x$.

$$9x^2 - 6x - 8 = 9x^2 + 6x - 12x - 8$$

*Step 4:* Factor by grouping:

$$\underbrace{9x^2 + 6x}_{\text{GCF}\,=\,3x}\;\; \underbrace{-12x - 8}_{\substack{\text{Opposite of}\\ \text{GCF}\,=\,-4}} = 3x(3x + 2) - 4(3x + 2)$$

$$= (3x + 2)(3x - 4)$$

## Example 21

Factor the trinomial by grouping: $25x^2 + 20x + 4$

**Solution**
Here $a = 25$, $b = 20$, and $c = 4$.

*Step 1:* The product $ac$ is $25(4) = 100$.

*Step 2:* We look for two numbers whose product is 100 and whose sum is 20. The numbers $p$ and $q$ are 10 and 10.

*Step 3:* We write $25x^2 + 20x + 4 = 25x^2 + 10x + 10x + 4$

*Step 4:* Factor by grouping:
$$25x^2 + 10x + 10x + 4 = 5x(5x + 2) + 2(5x + 2)$$
$$= (5x + 2)(5x + 2)$$

The result is the product of two identical binomials. This product can be written in simplest form with an exponent on the common factor.

$$25x^2 + 20x + 4 = (5x + 2)^2$$

### Practice E

Now it's time for you to do some factoring. When you are done, turn the page and check your solutions.

23. $x^2 + 7x + 12$
24. $x^2 - 6x - 16$
25. $2x^2 + 9x + 9$
26. $6x^2 + x - 1$
27. $49x^2 - 14x + 1$.
28. $x^2 - 4$
29. $81x^2 - 100$

## F. Zero Product Property

A **quadratic equation** is an equation in one variable that contains a polynomial having 2 as the highest power on the variable. Quadratic equations are used in many ways in engineering, architecture, finance, biological science, and, of course, mathematics. Here are some examples of quadratic equations:

$$2x^2 + 3x + 1 = 0$$
$$-16t^2 + 135t + 8 = 0$$
$$x^2 - 4 = 0$$

Sometimes the easiest method for solving a quadratic equation involves factoring. Solving a quadratic equation by factoring depends on the **zero-product property**, which states that if the product of two numbers or two expressions equals zero, then at least one of the numbers or one of the expressions *must* equal zero.

## The Zero-Product Property

If $a \cdot b = 0$, where $a$ and $b$ are real numbers or algebraic expressions, then

$a = 0$     or     $b = 0$     or     both $= 0$

Given a quadratic equation that has a factorable expression when set equal to 0, we can solve it by factoring and applying the zero-product property. If $ax^2 + bx + c$ is factorable, then to solve $ax^2 + bx + c = 0$, follow these steps:

1. Factor the polynomial expression.
2. Use the zero-product property to set each factor equal to zero.
3. Solve the resulting linear equations.

### Example 22

Solve the quadratic equation by factoring: $x^2 + x - 6 = 0$

**Solution**

*Step 1*

$x^2 + x - 6 = 0$     Original equation
$(x - 2)(x + 3) = 0$     Factor the trinomial.

*Step 2:* By the zero-product property, one or both of these factors must be 0. We set each factor equal to zero.

$x - 2 = 0$ or $x + 3 = 0$

*Step 3:* Solve.

$x = 2$ or $x = -3$

We can check our solutions in the original equation as follows:

Check that $x = 2$: $(2)^2 + 2 - 6 = 4 + 2 - 6 = 0$
Check that $x = -3$: $(-3)^2 + -3 - 6 = 9 - 3 - 6 = 0$

In both cases we get a true statement, which means that both 2 and −3 are solutions to the original equation.

## Example 23

Solve the following quadratic equations by factoring. Then check each solution.

1. $2x^2 + 11x - 6 = 0$
2. $x^2 - 8x + 16 = 0$
3. $x^2 - 144 = 0$

**Solutions**

1. $2x^2 + 11x - 6 = 0$

   $(x - 2)(x + 3) = 0$     Factor the trinomial.

   $2x - 1 = 0$ or $x + 6 = 0$     Set each factor equal to zero.

   $2x = 1$ or $x = -6$     Solve both equations.

   $x = \frac{1}{2}$

   We can use a calculator to check our solutions in the original equation:

   Check that $x = \frac{1}{2}$:    $2\left(\frac{1}{2}\right)^2 + 11\left(\frac{1}{2}\right) - 6 = 0$

   Check that $x = -6$:    $2(-6)^2 + 11(-6) - 6 = 0$

   So $\frac{1}{2}$ and $-6$ are solutions to the original equation.

2. $x^2 - 8x + 16 = 0$

   $(x - 4)(x - 4) = 0$     Factor the trinomial.

   $(x - 4) = 0$     Apply the zero product property. Both equations are the same.

   $x = 4$     There is only one solution to the equation.

   Remember to check the solution on your calculator.

3. $x^2 - 144 = 0$

   $(x + 12)(x - 12) = 0$     Factor the difference of squares.

   $x + 12 = 0$ or $x - 12 = 0$     Set each factor equal to zero.

   $x = 12$ or $x = -12$     Solve.

   Check both solutions on your calculator.

Some single variable polynomial equations have a greatest common factor (GCF) and we can use the distributive property to rewrite the polynomial as a product having the GCF as a factor. If the GCF is a monomial containing the variable, then when we apply the zero-product property, we find one of the solutions is always 0.

**Practice E — Answers**

23. $(x + 3)(x + 4)$
24. $(x - 8)(x + 2)$
25. $(2x + 3)(x + 3)$
26. $(3x - 1)(2x + 1)$
27. $(7x - 1)^2$
28. $(x + 4)(x - 4)$
29. $(9x + 10)(9x - 10)$

## Example 24

Use the zero-product property to solve the following polynomial equations with a GCF.

1. $4x^2 - 28x = 0$
2. $x^3 + 14x^2 + 45x = 0$

**Solutions**

1. $4x^2 - 28x = 0$

    $4x(x-7) = 0$ \qquad Factor out the GCF.

    $4x = 0$ or $x - 7 = 0$ \qquad Apply the zero-product property.

    $x = 0$ or $x = 7$ \qquad Solve.

2. $x^3 + 14x^2 + 45x = 0$

    $x(x^2 + 14x + 45) = 0$ \qquad Factor out the GCF.

    $x(x+5)(x+9) = 0$ \qquad Factor the trinomial.

    Expanding the zero product property to a product of three factors we know:

    $x = 0$ or $x + 5 = 0$ or $x + 9 = 0$

    $x = 0$ or $x = -5$ or $x = -9$

    You can check all three solutions in the original equation.

## Practice F

Now it's your turn. Solve the following equations by factoring. When you are done, turn the page and check your solutions.

30. $x^2 - 5x - 6 = 0$
31. $x^2 - 4x - 21 = 0$
32. $x^2 - 25 = 0$
33. $x^3 + 11x^2 + 10x = 0$

# Exercises 4.1

For the following exercises, find the product.

1. $(6x^2y^7)(8x^5y)$
2. $(-3x^8y^2)(9xy^3)$
3. $(-7a^4b)(2ab^6)$
4. $(-2ab^3)(-a^2b)$
5. $(4x)(x^2+9x-5)$
6. $(8x)(2x^2-3x+1)$
7. $(3x^2)(-7x^2+11x-2)$
8. $(10x^2)(-6x^2+3x-8)$
9. $(6a^2b-5ab+4b)(-3ab)$
10. $(-5ab)(3a^2b-8ab+2b)$
11. $(2x^2y)(x^2+9xy-y^2)$
12. $(x^2+2xy+y^2)(7xy^2)$

Multiply the binomials.

13. $(x+2)(x-4)$
14. $(x+4)(x-3)$
15. $(x+11)(x+8)$
16. $(x+12)(x+2)$
17. $(x-9)(x-6)$
18. $(x-7)(x-5)$
19. $(x^2-4)(x+8)$
20. $(x^2-6)(x+7)$
21. $(8a-4)(a^3+3a)$
22. $(3d^2-5d)(2d^2+9)$
23. $(5x^3+x^2)(4x^2-3)$
24. $(6x^3+1)(2x^2-7x)$
25. $(4a+b)(3a-5b)$
26. $(2a-11b)(5a-4b)$

For the following exercises, find the products.

27. $(x+9)^2$
28. $(y-10)^2$
29. $(a-7)^2$
30. $(x+11)^2$
31. $(2m-3)^2$
32. $(3a+5)^2$
33. $(5x+12)^2$
34. $(4x-1)^2$
35. $(x+2)(x-2)$
36. $(a-8)(a+8)$
37. $(b-6)(b+6)$
38. $(x+15)(x-15)$
39. $(4x+1)(4x-1)$
40. $(11x-3)(11x+3)$
41. $(2a+7b)(2a-7b)$
42. $(3a+4b)(3a-4b)$
43. $(x+2)(x^2-11x-9)$
44. $(x-8)(x^2+2x+3)$
45. $(3x-7)(2x^2+x-4)$
46. $(2x+9)(5x^2+6x+1)$

For the following exercises, use the greatest common factor to factor the polynomials.

47. $14x+4xy-18xy^2$
48. $49ab^2-35a^2b+77a$
49. $30x^3y-45x^2y^2+135xy^3$
50. $200k^3y^3-30k^2y^3+40y^3$
51. $36a^4b^2-18a^3b^3+54a^2b^4$
52. $6x^4-2x^3+3x^2-x$
53. $80x^7-16x^3y$
54. $44x^6y-55x^5y^2$

For the following exercises, factor the trinomial.

55. $a^2+9a-22$
56. $a^2+5a-24$
57. $x^2+12x+35$
58. $x^2+10x+24$
59. $x^2-9x+8$
60. $x^2-12x+27$
61. $x^2-4x+9$
62. $x^2+8x+10$
63. $x^2-x-30$
64. $x^2+x-20$
65. $t^2-9t-36$
66. $t^2-6t-7$

For the following exercises, factor the polynomial.

67. $7x^2 + 48x - 7$
68. $10h^2 - 9h - 9$
69. $2b^2 + 13b - 24$
70. $9d^2 - 73d + 8$

71. $5t^2 - 19t + 12$
72. $6x^2 + 17x + 7$
73. $-5x^2 - 50x - 135$
74. $-3x^2 - 15x - 12$

75. $-4x^3 - 20x^2 + 24x$
76. $-2x^3 + 28x^2 - 90x$

Factor the difference of squares.

77. $x^2 - 1$
78. $x^2 - 4$
79. $4m^2 - 9$

80. $16x^2 - 100$
81. $25x^2 - 196$
82. $121p^2 - 169$

83. $144a^2 - 49$
84. $36x^2 - 81$

For the following exercises, solve the quadratic equation by factoring.

85. $x^2 + 8x = 0$
86. $x^2 - 11x = 0$
87. $2a^2 - 14a = 0$
88. $3a^2 + 18a = 0$
89. $x^2 - 121 = 0$
90. $y^2 - 225 = 0$
91. $m^2 - 20m + 100 = 0$
92. $q^2 - 12q + 36 = 0$

93. $4x^2 - 25 = 0$
94. $9x^2 - 16 = 0$
95. $x^2 - 9x + 18 = 0$
96. $x^2 - 11x + 30 = 0$
97. $x^2 + 2x - 35 = 0$
98. $x^2 + 3x - 54 = 0$
99. $9x^2 + 30x + 16 = 0$
100. $8x^2 + 6x - 9 = 0$

101. $2x^2 + 9x - 5 = 0$
102. $6x^2 + 17x + 5 = 0$
103. $3x^2 + 38x + 80 = 0$
104. $4x^2 + 27x - 7 = 0$
105. $7x^2 - 19x - 6 = 0$
106. $5x^2 - 23x + 12 = 0$
107. $6x^2 - 23x + 20 = 0$
108. $12x^2 - 13x - 14 = 0$

For the following exercises, solve the equation by factoring. You will need to set the equation equal to 0 first, then find the greatest common factor, and finally see if further factoring is possible.

109. $4x^2 = 5x$
110. $2x^2 + 14x = 36$
111. $5x^2 = 5x + 30$
112. $4x^2 + 8 = 12x$
113. $3x^2 - 36 = 12x$

114. $2x^2 = 11x$
115. $3x^2 = 75$
116. $6x^2 = 54$
117. $x^3 + 5x^2 = -4x$
118. $x^3 + 7x^2 = -10x$

119. $2x^4 = -30x^3 - 88x^2$
120. $3x^4 = 27x^3 - 60x^2$

### Practice F — Answers

30. $(x - 6)(x + 1) = 0$;  $x = 6, x = -1$
31. $(x - 7)(x + 3) = 0$;  $x = 7, x = -3$

32. $(x + 5)(x - 5) = 0$;  $x = -5, x = 5$
33. $x(x + 10)(x + 1) = 0$; $x = 0, x = -10, x = -1$

# 4.2 Quadratic Functions in Standard Form

## Overview

Curved, dish-like antennas are commonly used to focus microwaves and radio waves to transmit television and telephone signals, as well as satellite and spacecraft communication. The cross-section of the antenna is in the shape of a parabola, which can be described by a quadratic function.

In this section, we will investigate quadratic functions, which frequently model problems involving area and projectile motion. You will learn to:

- Know the definition of a quadratic function in standard form
- Identify characteristics of parabolas
- Find the vertex and $y$-intercept of a parabola
- Determine a quadratic function's domain and range
- Solve problems involving a quadratic function's minimum or maximum value

## A. Quadratic Functions and their Graphs

In Section 4.1, we studied how to solve quadratic equations which are equations containing a polynomial in one variable where the highest power on the variable is 2. If we take the polynomial from a quadratic equation and use it as the formula for a function, the result is called a **quadratic function**. We will study quadratic functions in **standard form** as defined below.

> **Quadratic Function**
>
> A function that can be written in the form
>
> $$f(x) = ax^2 + bx + c$$
>
> where $a$, $b$, and $c$ are constants with $a \neq 0$, is called a **quadratic function**. We refer to this form as **standard form**.

Here are some examples of quadratic functions in standard form:

$f(x) = -4x^2 + 3x + 9$,    with constants $a = -4$, $b = 3$, and $c = 9$

$g(x) = x^2 - 5x$    with constants $a = 1$, $b = -5$, and $c = 0$

$h(x) = 2.71x^2 - 3.04$    with constants $a = 2.71$, $b = 0$, and $c = -3.04$

The $ax^2$ term is called the **quadratic term**. The $bx$ term is called the **linear term**. The $c$ term, which contains no variable, is simply called the **constant**.

In standard form, the constants $b$ and $c$ can equal 0, but the constant $a$ cannot equal 0. If $a$ were allowed to be 0, the $ax^2$ term would become 0, and the standard form would become $f(x) = bx + c$, which is a linear function, not a quadratic function.

The simplest quadratic function is given by

$$f(x) = x^2$$

In words, we say that this is a rule that squares the input to produce the output.

In the Figure 1 table, we represent the function $f(x) = x^2$ numerically with a table of selected input-output pairs. Notice the repetition of most output values. For example, the number 9 shows up twice as an output because when squared, the inputs 3 and −3 both result in the output 9.

| $x$ | $y = x^2$ |
|---|---|
| −3 | 9 |
| −2 | 4 |
| −1 | 1 |
| 0 | 0 |
| 1 | 1 |
| 2 | 4 |
| 3 | 9 |

Figure 1.

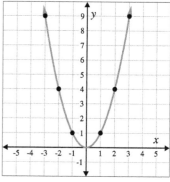

Figure 2.

When we plot the values in the table as ordered pairs and connect them with a smooth curve, we obtain the graph of $f(x) = x^2$ in Figure 2. The curve is called a **parabola**, and every quadratic function $y = ax^2 + bx + c$ has a parabola as the shape of its graph.

One important feature of the graph is that it has a turning point, called the **vertex**. For the function, $f(x) = x^2$, the vertex is at $(0, 0)$.

Another characteristic of the graph is that it's symmetric with respect to the $y$-axis. If we folded the graph along the $y$-axis, both halves of the graph would lie exactly on top of each other.

A vertical line that intersects a parabola at its vertex is called the **axis of symmetry** or **line of symmetry**. Figure 3 presents the graphs of a few more quadratic functions. Each has an axis of symmetry, which is shown here as a dashed line, that passes through its vertex.

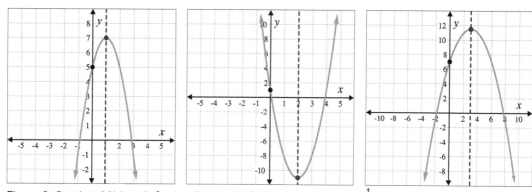

Figure 3. Graphs of $f(x) = -2x^2 + 4x + 5$, $g(x) = 3x^2 - 12x + 1$ and $h(x) = -\frac{1}{2}x^2 + 3x + 7$.

Some parabolas open upward in a U-shape, and some open downward in an upside-down U-shape. The sign of the coefficient $a$ has this effect on the orientation of the parabola. Below are the equations of the graphs in Figure 3.

For $f(x) = -2x^2 + 4x + 5$,     $a = -2$ and the parabola opens down.

For $g(x) = 3x^2 - 12x + 1$,     $a = 3$ and the parabola opens up.

For $h(x) = -\frac{1}{2}x^2 + 3x + 7$,     $a = -\frac{1}{2}$ and the parabola opens down.

In general, if the coefficient $a$ is positive, the parabola opens up. If the coefficient $a$ is negative, the parabola opens down.

The $y$-intercept of any function occurs when $x = 0$. Below are the $y$-intercepts for each function in Figure 3.

For $f(x) = -2x^2 + 4x + 5$,     $f(0) = 5$, so the $y$-intercept is $(0, 5)$.

For $g(x) = 3x^2 - 12x + 1$,     $f(0) = 1$, so the $y$-intercept is $(0, 1)$.

For $h(x) = -\frac{1}{2}x^2 + 3x + 7$,     $f(0) = 7$, so the $y$-intercept is $(0, 7)$.

In general, for a quadratic function of form $f(x) = ax^2 + bx + c$, the $y$-intercept of the graph is at $(0, c)$.

## Characteristics of Graphs of Quadratic Functions

For a quadratic function in standard form $f(x) = ax^2 + bx + c$, with $a \neq 0$:

- The shape of the graph is a parabola.
- If $a > 0$, then the parabola opens up. If $a < 0$, then the parabola opens down.
- The parabola has a turning point called the vertex.
- The parabola is symmetric about the axis of symmetry, a vertical line that intersects the vertex.
- The $y$-intercept is at $(0, c)$.

### Example 1

For the function $f(x) = -2x^2 + 8x + 1$, do the following:

a. Determine whether the parabola opens up or down.

b. Make a table of selected values using $x = -1, 0, 1, 2, 3, 4, 5$.

c. Graph the curve.

d. Identify the vertex.

e. Identify the line of symmetry and sketch it as a dashed line on the graph.

f. Identify the $y$-intercept.

## Solution

a. $a = -2$, so $a < 0$ and the parabola opens down.

b. Evaluate the function for the given $x$-values. For example:

$f(-1) = -2(-1)^2 + 8(-1) + 1 = -9$
$f(0) = -2(0)^2 + 8(0) + 1 = 1$
$f(1) = -2(1)^2 + 8(1) + 1 = 7$

and so on. See Figure 4.

c. Plot the points in Figure 4 and connect them with a smooth curve. The graph in Figure 5 includes most of these points.

| $x$ | $f(x)$ |
|---|---|
| -1 | -9 |
| 0 | 1 |
| 1 | 7 |
| 2 | 9 |
| 3 | 7 |
| 4 | 1 |
| 5 | -9 |

Figure 4.

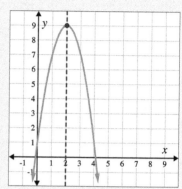

Figure 5.

d. From both the table and the graph, we can see the vertex is at (2, 9).

e. The line of symmetry is the vertical line that intercepts the vertex. See Figure 5.

f. The $y$-intercept of the graph is at (0, 1).

## Practice A

Try this problem, and then turn the page to check your work.

1. For the function $f(x) = x^2 + 6x + 2$

    a. Does the parabola open up or open down?

    b. Make a table of selected values using $x = -6, -5, -4, -3, -2, -1, 0$.

    c. Graph the curve. Draw the line of symmetry as a dashed line.

    d. Identify the vertex.

    e. Identify the $y$-intercept.

## B. The Vertex Formula

To find a formula for the $x$-coordinate of the vertex of a parabola $f(x) = ax^2 + bx + c$, we look at two cases.

If $b = 0$, the equation is of the form $f(x) = ax^2 + c$. The $y$-intercept is $(0, c)$ and is the vertex of the parabola. So the $x$-coordinate of the vertex is 0.

If $b \neq 0$, we can find the $x$-coordinate of the vertex in a few steps using the symmetry of parabolas. We start by finding the $x$-coordinate of the point on the parabola that is symmetric to the $y$-intercept.

This point has the same y-coordinate as the y-intercept, namely c.

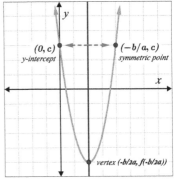

Figure 6.

| | |
|---|---|
| $c = ax^2 + bx + c$ | Substitute $c$ for $f(x)$. |
| $0 = ax^2 + bx$ | Subtract $c$ from both sides. |
| $0 = x(ax + b)$ | Factor the right-hand side. |
| $x = 0$ or $ax + b = 0$ | Apply the zero product property. |
| $x = 0$ or $ax = -b$ | Subtract $b$ from both sides. |
| $x = 0$ or $x = -\frac{b}{a}$ | Divide both sides by $a$. |

The x-coordinate of the y-intercept is 0. The x-coordinate of the symmetric point is $-\frac{b}{a}$. These points are equidistant from the line of symmetry., as you see in Figure 6.

We find the x-coordinate of the vertex by averaging the x-coordinate of the y-intercept and the x-coordinate of the symmetric point:

| | |
|---|---|
| $x = \frac{0 + \left(-\frac{b}{a}\right)}{2}$ | Use the definition of average. |
| $x = \frac{\left(-\frac{b}{a}\right)}{2}$ | Simplify the numerator. |
| $x = -\frac{b}{a} \cdot \frac{1}{2}$ | Dividing by 2 is equivalent to multiplying by the reciprocal, $\frac{1}{2}$. |
| $x = -\frac{b}{2a}$ | Multiply the fractions. |

The formula of the x-coordinate of the vertex is thus $x = -\frac{b}{2a}$. If $b = 0$, this formula gives $x = -\frac{0}{2a} = 0$, which agrees with our earlier statement, so the formula works for any value of $b$.

To find the y-coordinate of the vertex, we evaluate the function $f$ for the input $x = -\frac{b}{2a}$. That is, we substitute the x-coordinate of the vertex back into the original equation and compute the y-coordinate: $y = f\left(-\frac{b}{2a}\right)$

### Vertex Formula

To find the vertex of the graph of a quadratic function $f(x) = ax^2 + bx + c$:

1. Find the x-coordinate of the vertex using the **vertex formula**, $x = -\frac{b}{2a}$.
2. Find the y-coordinate of the vertex by evaluating $f$ at the value found in Step 1, $f\left(-\frac{b}{2a}\right)$.

### Example 2

Consider these functions:

1. $f(x) = -2x^2 + 4x + 5$
2. $f(x) = x^2 + 5x$

For each function, do the following:

a. Determine if the parabola opens up or down.

b. Find the $y$-intercept.

c. Find the vertex.

d. Verify your answers using the graph of the function on a graphing calculator.

**Solutions**

1. $f(x) = -2x^2 + 4x + 5$

a. $a = -2$, so the parabola opens down.

b. $c = 5$, so the $y$-intercept is $(0, 5)$.

c. Since $a = -2$ and $b = 4$, the $x$-coordinate of the vertex is: $x = -\frac{b}{2a} = -\frac{4}{2(-2)} = -\frac{4}{-4} = 1$.
The $y$-coordinate is found by evaluating $f(1)$: $y = f(1) = -2(1)^2 + 4(1) + 5 = 7$.
The vertex is $(1, 7)$.

d. Graph the function in the standard window (ZOOM 6) on the calculator. Make sure any plots are turned off. Use the trace feature on the calculator and enter the $x$-values of the $y$-intercept and the vertex to check your work. See figures 7 and 8.

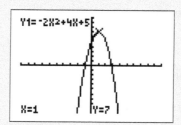

Figure 7.

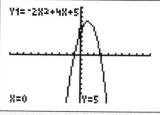

Figure 8.

2. $f(x) = x^2 + 5x$

a. $a = 1$, so the parabola opens up.

b. $c = 0$, so the $y$-intercept is $(0, 0)$.

c. Since $a = 1$ and $b = 5$, the $x$-coordinate of the vertex is: $x = -\frac{b}{2a} = -\frac{5}{2(1)} = -\frac{5}{2} = -2.5$. Evaluate $f(-2.5)$ for the $y$-coordinate: $y = f(-2.5) = (-2.5)^2 + 5(-2.5) = -6.25$

The vertex is $(-2.5, -6.25)$.

d. Verify the work as in problem 1 of this example. See Figures 9 and 10.

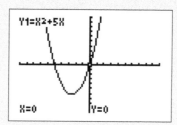

Figure 9.

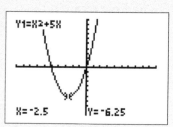

Figure 10.

## Practice B

Try this problem, and then turn the page to check your work.

2. For the function $f(x) = \frac{1}{2}x^2 - 4x + 3$

   a. Determine whether the parabola opens up or open down.
   
   b. Identify the y-intercept.
   
   c. Identify the vertex.
   
   d. Verify your work on a graphing calculator.

# C. Finding the Domain and Range of a Quadratic Function

Although many of the x-values we've worked with have been integer values, we can let x equal any real number in a quadratic function $f(x) = ax^2 + bx + c$. Therefore, the domain of any quadratic function is all real numbers, unless the context of a real-life application problem presents restrictions.

The range, however, depends on the location of the vertex and whether the parabola opens up or down - things affected by the values of the constants a, b, and c. Because the vertex is a turning point on the graph of the parabola, the range will consist of all y-values greater than or equal to the y-coordinate of the vertex or all y-values less than or equal to the y-coordinate of the vertex, depending on the orientation of the parabola.

## Practice A — Answers

$f(x) = x^2 + 6x + 2$

a. $a = 1$, so the parabola opens up.

b. 
| x | −6 | −5 | −4 | −3 | −2 | −1 | 0 |
|---|----|----|----|----|----|----|---|
| f(x) | 2 | −3 | −6 | −7 | −6 | −3 | 2 |

c.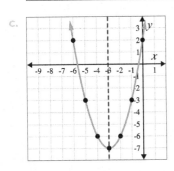

d. vertex: $(-3, -7)$

e. y-intercept: $(0, 2)$

If the parabola opens up, the y-coordinate of the vertex is the **minimum value** of the function, as in Figure 11. If the parabola opens down, as in Figure 12, the y-coordinate of the vertex is the **maximum value** of the function.

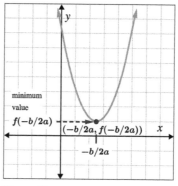

Figure 11.

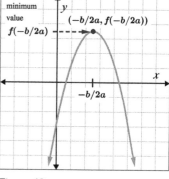

Figure 12.

## Domain and Range of a Quadratic Function

For a quadratic function $f(x) = ax^2 + bx + c$ with vertex $\left(-\frac{b}{2a}, f\left(-\frac{b}{2a}\right)\right)$:

- The domain is all real numbers.
- If $a > 0$, the parabola opens up and the range is: $y \geq f\left(-\frac{b}{2a}\right)$, or $\left[f\left(-\frac{b}{2a}\right), \infty\right)$.
- If $a < 0$, the parabola opens down and the range is: $y \leq f\left(-\frac{b}{2a}\right)$, or $\left(-\infty, f\left(-\frac{b}{2a}\right)\right]$.

### Example 3

Find the domain and range of $f(x) = -4x^2 - 12x - 3$.

### Solution

As with any quadratic function, the domain is all real numbers.

Because $a$ is negative, the parabola opens downward and the range will be y-values less than or equal to the y-coordinate of the vertex. We begin by finding the x-value of the vertex:

$$x = -\frac{b}{2a} = -\frac{-12}{2(-4)} = -\frac{-12}{-8} = -1.5$$

### Practice B — Answers

1. $f(x) = x^2 + 6x + 2$

   a. The parabola opens up.
   b. The y-intercept is (0, 3).
   c. The vertex is (4, –5).
   d. See figures 13 and 14.

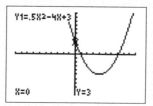

Figure 13.

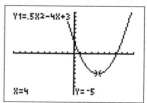

Figure 14.

We find the y-coordinate by evaluating $f(-1.5)$:

$$y = f(-1.5) = -4(-1.5)^2 - 12(-1.5) - 3 = 6$$

The vertex is (−1.5, 6). Since the parabola opens down, the y-coordinate of the vertex is the maximum value of the range. The range is: (−∞, 6].

## Example 4

Find the domain and range of the quadratic functions graphed below in Figures 15 to 17.

### Solution

For Figure 15, the domain is all real numbers or (− ∞, ∞). The y-coordinate of the vertex is the minimum value of the function. So, the range is $y \geq -4.5$ or we can write [−4.5, ∞).

For Figure 16, the domain is all real numbers or (− ∞, ∞). The y-coordinate of the vertex is the maximum value of the function. So, the range is $y \leq 8.63$ or we can write (−∞, 8.63].

For Figure 17, the domain is all real numbers or (− ∞, ∞). The y-coordinate of the vertex is the minimum value of the function. So, the range is $y \geq -7$ or we can write [−7, ∞).

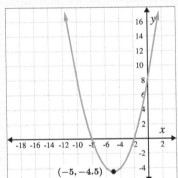

Figure 15.

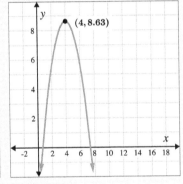

Figure 16.

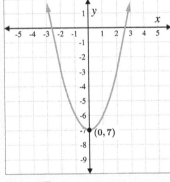

Figure 17.

## Practice C

Try the following problems. Turn the page to check your work.

3. Find the domain and range of $f(x) = -1.6x^2 - 4.8x + 3$.

4. Find the domain and range of $f(x) = 3x^2 + 12x + 14$.

5. Find the domain and range of a quadratic function with coefficient $a = 1.75$ and vertex at (5.25, −8.94)

## D. Applications Using Maximum or Minimum Value

As we've seen, the output of a quadratic function at the vertex is the maximum or minimum value of the function, depending on the orientation of the parabola. There are many real-world scenarios that involve finding the maximum or minimum value of a quadratic function, including applications involving area, projectile motion, and revenue.

### Example 5

A backyard farmer wants to enclose a rectangular space for a new garden within her fenced backyard. She has purchased 80 feet of wire fencing to enclose three sides, and she will use a section of the backyard fence as the fourth side.

a. Find a formula for the area of the garden if the sides of garden perpendicular to the existing backyard fence have length $L$.

b. What dimensions should the farmer make her garden to maximize the enclosed area?

c. What is the maximum area?

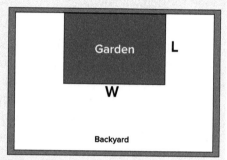

Figure 18.

### Solution

We use a diagram such as Figure 18 to record the given information. It is also helpful to introduce a temporary variable, $W$, to represent the width of the garden and the length of the fence section parallel to the backyard fence.

a. We have 80 feet of fence available, so $L + W + L = 80$, or more simply $2L + W = 80$. This equation allows us to represent the width, $W$, in terms of $L$.
Subtracting $2L$ from both sides, we get

$$W = 80 - 2L$$

Now we are ready to write an equation for the area the fence encloses. We know the area of a rectangle is length multiplied by width, so here is what we do:

$A = LW$      Use the area formula for rectangles.

$\phantom{A} = L(80-2L)$      Substitute $W = 80 - 2L$.

$A = 80L - 2L^2$      Apply the distributive property.

This formula represents the area of the fence in terms of the variable length $L$. It is a quadratic function and can be written in standard form:

$$A = f(L) = -2L^2 + 80L$$

b. The quadratic has a negative leading coefficient, $a = -2$, so the parabola will open downward, and the input value (the $L$-coordinate in this case) of the vertex will give the length that maximizes the area. Using the vertex formula:

$$L = -\frac{b}{2a} = -\frac{80}{2(-2)} = -\frac{80}{-4} = 20$$

So the length is 20 feet and the width is $W = 80 - 2L = 80 - 2(20) = 40$ feet.

The dimensions that will maximize the area of the garden, using 80 feet of fencing, are a length on the two shorter sides of 20 feet and a width of 40 feet.

c. We find the maximum area of the garden by evaluating the function at $L = 20$. This gives the output value of the vertex.

$$A = f(L) = -2(20)^2 + 80(20) = 800$$

The maximum area for the garden is 800 square feet.

This problem can be illustrated by graphing the quadratic function. We can see where the maximum area occurs on a graph of the quadratic function in Figure 19.

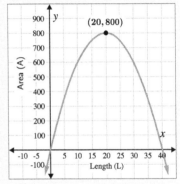

Figure 19.

## Example 6

A batter hits a baseball. The ball's height in feet after $t$ seconds is given by:

$$f(t) = -16t^2 + 88t + 3$$

With this in mind, answer the following:

a. What is the ball's height when the batter first hits it?
b. Find $f(2)$. What is its meaning in the context of this problem?
c. When does the ball reach its maximum height?
d. What is the maximum height of the ball?
e. Check your solutions using the graph and trace features on your calculator.

Practice C — Answers

2. The domain is all real numbers. The parabola opens down, and the vertex is at $(-1.5, 6.6)$, so the range is $y \leq 6.6$, or $(-\infty, 6.6]$.

3. The domain is all real numbers. The parabola opens up, and the vertex is at $(-2, 2)$, so the range is $y \geq 2$, or $[2, \infty)$.

4. The domain is all real numbers. The parabola opens up, so the range is $y \geq -8.94$, or $[-8.94, \infty)$.

## Solution

**a.** The batter hits the ball at time $t = 0$ seconds. We substitute $t = 0$ in the equation:

$$f(0) = -16(0)^2 + 88(0) + 3 = 3$$

The baseball is 3 ft high when the batter hits it.

**b.** $f(2) = -16(2)^2 + 88(2) + 3 = 115$

After $t = 2$ seconds, the height of the baseball is 115 ft.

**c.** Because $a = -16$, the graph of $f$ will be a parabola that opens down, and the vertex will be a maximum. We use the vertex formula to find the $t$-value of the vertex.

$$t = \frac{-b}{2a} = \frac{-88}{2(-16)} = 2.75$$

It takes 2.75 seconds for the baseball to reach its maximum height.

**d.** The maximum height of the ball can be found by evaluating the function at $t = 2.75$:

$$f(2.75) = -16(2.75)^2 + 88(2.75) + 3 = 124$$

The ball reaches a maximum hight of 124 feet.

**e.** To view the function on our calculator, we enter the equation and adjust the viewing window. Use common sense and the values already seen in the problem to help you choose a good window, such as $x$: [0, 8, 1] $y$: [0, 150, 10]. See Figure 20.

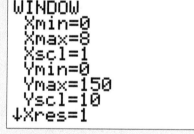

Figure 20.

Notice the $y$-intercept is at (0, 3) representing the initial height of the ball. Use the trace button to confirm the ball is at 115 feet after 2 seconds and that the vertex is at (2.75, 124). See Figure 21.

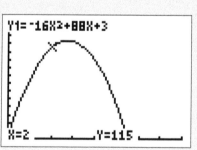 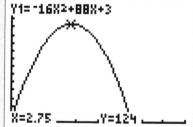

Figure 21. Confirming the work on a graphing screen.

When a business projects sales and revenue it often takes into account the fact that the unit price of an item affects its supply and demand. That is, if the unit price goes up, the demand for the item will usually decrease. For example, a local newspaper currently has 84,000 subscribers at a quarterly charge of $30. Market research has suggested that if the owners raise the price to $32, they would lose 5,000 subscribers.

Revenue is the amount of money a business brings in. In this case, the revenue can be found by multiplying the price per subscription times the number of subscribers, or quantity. We can introduce variables, $p$ for price per subscription and $Q$ for quantity, giving us the equation Revenue = $pQ$. We use this equation and scenario in the following example.

### Example 7

A local newspaper estimates that the quantity of subscriptions sold, $Q$, depends on the quarterly price $p$ in dollars of the subscription according to the equation $Q = -2{,}500p + 159{,}000$.

a. Find a formula for the revenue $R = f(p)$ of the newspaper if $Q$ number of subscriptions is sold at a quarterly price $p$.

b. Find $f(40)$ and interpret this value in the context of the problem.

c. What price should the newspaper charge in order to maximize revenue?

d. What is the maximum revenue?

### Solution

a. The revenue $R$ is found using $R = pQ$, so in this case, that means $R = p(-2{,}500p + 159{,}000)$. Applying the distributive property, we obtain a quadratic function in standard form that models the revenue of the newspaper as a function of the subscription price:

$$R = f(p) = -2500p^2 + 159{,}000p$$

b. To find $f(40)$, substitute $p = 40$ in the equation:

$$f(40) = -2500(40)^2 + 159{,}000(40) = 2{,}360{,}000$$

When the subscription price is $40, the newspaper's revenue is $2,360,000.

c. The leading coefficient of the model is negative so the vertex will provide a maximum value. To find the quarterly subscription price that will maximize revenue for the newspaper, we find the input value of the vertex, which in this case is the $p$-coordinate:

$$p = -\frac{b}{2a} = -\frac{159{,}000}{2(-2500)} = -\frac{159{,}000}{-5000} = 31.8$$

The model tells us that the maximum revenue will occur if the newspaper charges $31.80 for a quarterly subscription.

d. To find the maximum revenue, we evaluate the revenue function at $p = 31.8$:

$$f(31.8) = -2500(31.8)^2 + 159{,}000(31.8) = 2{,}528{,}100$$

The maximum revenue for the newspaper will be $2,528,100 if it charges $31.80 per subscription. We illustrate the situation by graphing the quadratic function in Figure 22.

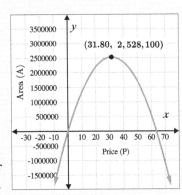

Figure 22.

## Exercises 4.2

For each of the following quadratic equations, do this:
- a. Determine if the parabola opens up or down.
- b. Make a table of selected values using $x = -3, -2, -1, 0, 1, 2, 3, 4$.
- c. Graph the curve.
- d. Identify the vertex.
- e. Identify the $y$-intercept.

1. $f(x) = 3x^2 + 6x - 5$
2. $f(x) = 2x^2 - 8x + 1$
3. $f(x) = -x^2 + 4x + 3$
4. $f(x) = -2x^2 + 4x + 6$

For each of the following quadratic functions, do the following and verify your answers with a graphing calculator.
- a. Determine if the parabola opens up or down.
- b. Find the $y$-intercept.
- c. Find the vertex.

5. $f(x) = x^2 - 4$
6. $f(x) = 9 - x^2$
7. $f(x) = x^2 + 8x + 13$
8. $f(x) = x^2 - 12x + 39$
9. $f(x) = -x^2 + 2x + 5$
10. $f(x) = x^2 - 8x + 16$
11. $f(x) = x^2 - x + \frac{5}{4}$
12. $f(x) = -\frac{1}{2}x^2 + 3x$

For the following exercises, find the vertex of the parabola and determine the domain and range of the function.

13. $f(x) = -4x^2 - 12x - 2$
14. $f(x) = -6x^2 - 24x - 25$
15. $f(x) = -2.4x^2 - 4.8x + 3.2$
16. $f(x) = x^2 + 5x + 9$
17. $f(x) = 2.2x^2 - 13.2x + 9.9$
18. $f(x) = -2x^2 + 10x + 7.5$

19.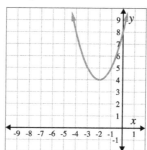

20.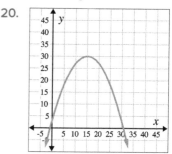

21. A farmer plans to use 130 feet of fencing and a side of his barn to enclose a rectangular pen for two "bummer lambs," which are lambs raised apart from their mothers for one reason or another. $L$, the length of the pen, can be written in terms of $W$, the width, as follows: $L = 130 - 2W$. The following equation shows the relationship between the width of the pen and $A$, the area of the pen, where the width is in feet and the area is in square feet: $A = W(130 - 2W)$.
- a. Write the equation for the area function in standard quadratic form.
- b. Find the width that will produce the maximum area of the pen.
- c. What is the maximum area?

22. A rancher plans to use 300 feet of fencing to enclose two adjacent rectangular corrals. $L$, the length of each corral, can be written in terms of $W$, the width of each corral, as follows: $L = 100 - \frac{4}{3}W$. The following equation shows the relationship between the width of each corral and $A$, the total area of both corrals, where the width is in feet and the area is in square feet: $A = W(100 - \frac{4}{3}W)$

   a. Write the equation for the area function in standard quadratic form.
   b. Find the width that will produce the maximum area of the corrals.
   c. What is the maximum area?

23. A manufacturer of garden hoses has daily production costs of $C = 800 - 10x + 0.25x^2$, where $C$ is the total cost in dollars, and $x$ is the number of hoses produced. How many garden hoses should be produced each day to yield a minimum production cost? What is the minimum production cost?

24. A manufacturer of sprinkler heads has daily production costs of $C = 745 - 12x + 0.12x^2$, where $C$ is the total cost in dollars, and $x$ is the number of sprinkler heads produced. How many sprinkler heads should be produced each day to yield a minimum production cost? What is the minimum production cost?

25. $y$, the height in feet of a ball thrown by a child, is given by $y = -\frac{1}{12}x^2 + 2x + 4$, where $x$ is the horizontal distance in feet from where the ball is thrown. What is the maximum height of the ball?

26. The path of a diver off diving platform is given by $y = -\frac{4}{9}x^2 + \frac{24}{9}x + 12$, where $y$ is the height in feet, and $x$ is the horizontal distance from the end of the diving board. What is the maximum height of the diver?

27. $P$, a company's daily profit in dollars, is given by $P = -2x^2 + 120x - 800$, where $x$ is the number of units produced per day. Find $x$ so that the daily profit is a maximum. What is the maximum profit?

28. A service industry has $P$, a daily profit in dollars, given by $P = -3.5x^2 + 329x - 700$, where $x$ is the number of people served per day. Find $x$ so that the daily profit is a maximum. What is the maximum profit?

29. The height in meters of a toy rocket $t$ seconds after it is launched is given by $f(t) = -4.9t^2 + 98t + 2$.

   a. What is the rocket's height when it is first launched?
   b. Find $f(7)$ and explain its meaning in the context of this problem.
   c. When does the rocket reach its maximum height?
   d. What is the maximum height of the rocket?

30. The height in meters of an arrow $t$ seconds after it is shot upwards from a crossbow is given by $f(t) = -4.9t^2 + 25t + 1$.

   a. What is the arrow's height when it is first shot from the bow?
   b. Find $f(1)$ and explain its meaning in the context of this problem.
   c. When does the arrow reach its maximum height?
   d. What is the maximum height of the arrow?

31. A person on the edge of a cliff throws a stone up and out so that after the stone reaches its maximum height, it falls to the beach below. The stone's height in feet above the beach after $t$ seconds is given by $f(t) = -16t^2 + 40t + 120$.

    a. Find the $y$-intercept of the graph of the function. What does it mean in this situation?

    b. Find the vertex of the graph of the function. What does it mean in this situation?

32. A person on the roof of a building tosses a penny up and out so that after the penny reaches its maximum height, it falls to the ground below. The penny's height in feet above the ground after $t$ seconds is given by $f(t) = -16t^2 + 32t + 68$.

    a. Find the $y$-intercept of the graph of the function. What does it mean in this situation?

    b. Find the vertex of the graph of the function. What does it mean in this situation?

33. If 65 apple trees are planted in a certain orchard, the average yield per tree will be 1500 apples per year. For each additional tree planted in the same orchard, the annual yield per tree drops by 20 apples. $P$, the number of apples produced per year, is dependent on $n$, the number of additional apple trees planted, and is given by $P = (65 + n)(1500 - 20n)$. Write the equation in standard form and use it to find how many more trees (above 65) should be planted to produce the maximum crop of apples. What is the maximum crop size?

34. It's estimated that an average of 14,000 people will attend a basketball game when the Portland Trailblazers play at the Moda Center, and the admission price is $100. For each $5 added to the price, the attendance will decrease by 280. $R$, the revenue per game, is dependent on $n$, the number of $5 increases in price, and is given by $R = (100 + 5n)(14{,}000 - 280n)$ Write the equation in standard form and use it to find the admission price that will produce the largest revenue per game. What is the largest revenue per game?

# 4.3 The Square Root Property

## Overview

In this section, we will study how to simplify expressions that have square roots. This skill allows us to solve certain quadratic equations using the square root property. We have already solved quadratic equations by factoring and applying the zero product property, but not all quadratic equations can be solved this way. It's also true that some quadratic equations do not have real number solutions, and we will introduce imaginary numbers in order to study these cases.

In this section, you will:

- Learn the product property and quotient property for square roots
- Simplify expressions with square roots
- Rationalize the denominator of a radical expression
- Use the square root property to solve quadratic equations
- Learn the definition of an imaginary number
- Solve equations with imaginary and complex solutions

## A. Evaluating Square Roots

When the square root of a number is squared, the result is the original number. Since the square root of 16 is 4, we know that $4^2 = 16$. The square root operation undoes squaring just as subtraction undoes addition. Here are some more examples:

If the square root of 100 is 10, then $10^2 = 100$.

If the square root of 49 is 7, then $7^2 = 49$.

Before we use the square root operation to solve quadratic equations, we need a solid understanding of the notation and rules for working with square roots. In general, if $k$ is a positive real number, then the square root of $k$ is a number that, when multiplied by itself (squared) gives back $k$.

We must be careful when working with square roots to distinguish between equations such as $x^2 = 25$ and $x = \sqrt{25}$.

The equation $x^2 = 25$ has two solutions: $x = 5$ and $x = -5$.

The equation $x = \sqrt{25}$ has one solution: $x = 5$.

For the second equation, we give only the positive value whose square is 25. The positive number whose square is the number under a square root symbol is defined to be the **principal square root**. The square root obtained using a calculator is always the principal or positive square root.

## Principal Square Root

The **principal square root** of $k$ is the nonnegative number that when multiplied by itself, equals $k$. It is written as a **radical expression**, $\sqrt{k}$.

If $k > 0$ and $\sqrt{k} = x$, then $x^2 = k$.

The principal square root of $k$ is written as $\sqrt{k}$. As Figure 1 illustrates, the symbol is called a **radical**, the term under the symbol is called the **radicand**, and the entire expression is called a **radical expression**.

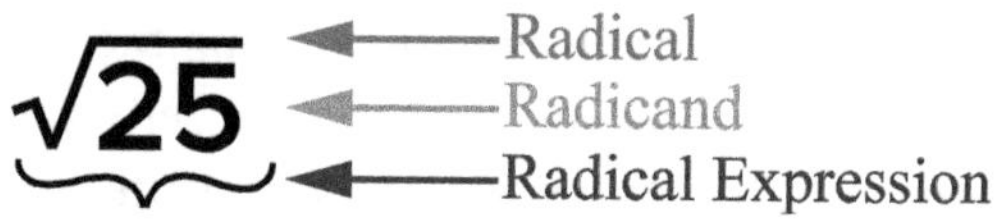

$\sqrt{25}$ ← Radical / Radicand / Radical Expression

Figure 1.

### Example 1

Evaluate each expression.

1. $\sqrt{100}$
2. $\sqrt{\sqrt{16}}$
3. $\sqrt{25 + 144}$
4. $\sqrt{49} - \sqrt{81}$

**Solution**

1. $\sqrt{100} = 10$ because $10^2 = 100$.
2. $\sqrt{\sqrt{16}} = \sqrt{4} = 2$ because $4^2 = 16$ and $2^2 = 4$.
3. $\sqrt{25 + 144} = \sqrt{169} = 13$ because $13^2 = 169$.
4. $\sqrt{49} - \sqrt{81} = 7 - 9 = -2$ because $7^2 = 49$ and $9^2 = 81$.

## ☠ Warning! Incorrect Approach! ☠

For problem 3 in the example above, $\sqrt{25 + 144}$, we *cannot* find the square roots of 25 and 144 separately before adding. Doing so would result in $\sqrt{25} + \sqrt{144} = 5 + 12 = 17$, which is *not* equivalent to $\sqrt{25 + 144} = \sqrt{169} = 13$.

The order of operations requires us to add the terms under the radical symbol before finding the square root. This is because the radical symbol acts as parentheses when the radical bar is extended. In general terms:

$$\sqrt{a + b} \neq \sqrt{a} + \sqrt{b}$$

## Practice A

Evaluate each expression. When you're finished, turn the page to check your work.

1. $\sqrt{225}$
2. $\sqrt{\sqrt{81}}$
3. $\sqrt{25 - 9}$
4. $\sqrt{36} + \sqrt{121}$

## B. Product and Quotient Properties for Square Roots

Let's investigate some properties of the square root operation. We have already stated something that is *not* true about square roots, namely $\sqrt{a + b} \neq \sqrt{a} + \sqrt{b}$. In words, the square root of a sum is not equal to the sum of the square roots. This is the case for differences as well: $\sqrt{a - b} \neq \sqrt{a} - \sqrt{b}$. However, there are useful properties of radicals that enable us to simplify radical expressions containing products and quotients.

### Using the Product Property to Simplify Square Roots

The first property we will look at is the **product property for square roots**, which allows us to separate the square root of a product of two numbers into the product of two separate radical expressions. For example, we can rewrite $\sqrt{15}$ as $\sqrt{3} \cdot \sqrt{5}$. We can also use the product rule to express the product of multiple radical expressions as a single radical expression.

### The Product Rule for Square Roots

If $a$ and $b$ are nonnegative, then

$$\sqrt{ab} = \sqrt{a} \cdot \sqrt{b}$$

In words, the square root of a product is the product of the square roots.

When asked to simplify a square root, we rewrite it such that there are no perfect squares as factors in the radicand. Remember, **perfect squares** are numbers whose square root is an integer. It is useful to memorize the first 15 perfect squares:

1, 4, 9, 16, 25, 36, 49, 64, 81, 100, 121, 144, 169, 196, 225

Given a square root expression, we can use the product rule to simplify the expression by following these steps:

1. Factor the largest perfect square from the radicand.
2. Write the radical expression as a product of radical expressions.
3. Simplify.

### Example 2

Using the product rule to simplify the following radical expressions.

1. $\sqrt{98}$  
2. $\sqrt{300}$

**Solutions**

1. The largest perfect square factor of 98 is 49 because 49 · 2 = 98.

$\sqrt{98} = \sqrt{49 \cdot 2}$  Factor the radicand.

$\phantom{\sqrt{98}} = \sqrt{49} \cdot \sqrt{2}$  Apply the product property, $\sqrt{ab} = \sqrt{a} \cdot \sqrt{b}$.

$\phantom{\sqrt{98}} = 7\sqrt{2}$  Simplify.

The expressions $\sqrt{98}$ and $7\sqrt{2}$ are equivalent irrational numbers where the latter is said to be in simplified form. We can check the equivalency on a calculator as shown in Figure 2. The radical expression is said to be in *exact* form. The expression of the number on the calculator in decimal form is *approximate* because the number has been rounded to nine decimal places.

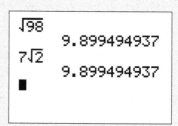

Figure 2. Verify that $\sqrt{98} = 7\sqrt{2}$.

2. The largest perfect square factor of 300 is 100. Note that 25 and 4 are also perfect square factors of 300, just not the largest one.

$\sqrt{300} = \sqrt{100 \cdot 3}$  Factor the radicand.

$\phantom{\sqrt{300}} = \sqrt{100} \cdot \sqrt{3}$  Apply the product property, $\sqrt{ab} = \sqrt{a} \cdot \sqrt{b}$.

$\phantom{\sqrt{300}} = 10\sqrt{3}$  Simplify.

If we had mistakenly used 25 as the largest perfect square factor in the first step, we would still arrive at the same simplified radical expression. It would just take a extra few steps.

$\sqrt{300} = \sqrt{25 \cdot 12}$  Factor the radicand using 25 as a factor.

$\phantom{\sqrt{300}} = \sqrt{25} \cdot \sqrt{12}$  Apply the product property, $\sqrt{ab} = \sqrt{a} \cdot \sqrt{b}$.

$\phantom{\sqrt{300}} = 5\sqrt{12}$  Simplify — and notice that $\sqrt{12}$ can be factored further.

$\phantom{\sqrt{300}} = 5\sqrt{4}\sqrt{3}$  Apply the product property again.

$\phantom{\sqrt{300}} = 5 \cdot 2\sqrt{3}$  Simplify.

$\phantom{\sqrt{300}} = 10\sqrt{3}$  Multiply.

**Practice A — Answers**

1. 15      2. 3      3. 4      4. 17

We can also use the product property to combine radical expressions into one radical expression. Sometimes, when the product is expressed as a single radical, it can be more easily simplified. Other times, it's useful to break a product of two radicals into a product of three or more radicals first. At those times, regrouping the factors may help in the simplification process.

### Example 3

Using the product rule to simplify the following radical expressions.

1. $\sqrt{12} \cdot \sqrt{3}$
2. $\sqrt{105} \cdot \sqrt{21}$

**Solutions**

1. Use the product rule to combine the radicals.

   $\sqrt{12} \cdot \sqrt{3} = \sqrt{12 \cdot 3}$     Express the product as a single radical expression: $\sqrt{ab} = \sqrt{a} \cdot \sqrt{b}$.

   $= \sqrt{36}$     Simplify.

   $= 6$

2. First factor the radical expressions into the product of smaller radicals.

   $\sqrt{105} \cdot \sqrt{21} = \sqrt{3}\sqrt{5}\sqrt{7} \cdot \sqrt{3}\sqrt{7}$     Apply the product property: $\sqrt{ab} = \sqrt{a} \cdot \sqrt{b}$.

   $= \sqrt{3}\sqrt{3} \cdot \sqrt{7}\sqrt{7} \cdot \sqrt{5}$     Reorder the factors.

   $= \sqrt{9} \cdot \sqrt{49} \cdot \sqrt{5}$     Express equivalent factors as a single radical expression.

   $= 3 \cdot 7 \cdot \sqrt{5}$     Simplify.

   $= 21\sqrt{5}$     Multiply.

### Using the Quotient Rule to Simplify Square Roots

Just as we can rewrite the square root of a product as a product of square roots, we can rewrite the square root of a quotient as a quotient of square roots by using the **quotient property for square roots**. It can be helpful to separate the numerator and denominator of a fraction under a radical so that we take their square roots separately. We can rewrite $\sqrt{\frac{25}{4}}$ as $\frac{\sqrt{25}}{\sqrt{4}}$. The second fraction is simplified to $\frac{5}{2}$.

### The Quotient Property for Square Roots

If $a$ and $b$ are nonnegative and $b \neq 0$, then

$$\sqrt{\frac{a}{b}} = \frac{\sqrt{a}}{\sqrt{b}}$$

In words, the square root of a quotient is the quotient of the square roots.

When given a radical expression, we can use the quotient property of square roots to simplify an expression with these steps:

1. Write the radical expression as the quotient of two radical expressions.
2. Simplify the numerator and denominator.

## Example 4

Use the quotient rule to simplify the following radical expressions.

1. $\sqrt{\frac{36}{121}}$    2. $\sqrt{\frac{13}{81}}$    3. $\sqrt{\frac{18}{49}}$

**Solutions**

1. $\sqrt{\frac{36}{121}}$

$\sqrt{\frac{36}{121}} = \frac{\sqrt{36}}{\sqrt{121}}$    Apply the quotient property: $\sqrt{\frac{a}{b}} = \frac{\sqrt{a}}{\sqrt{b}}$.

$= \frac{6}{11}$    Simplify.

2. $\sqrt{\frac{13}{81}}$

$\sqrt{\frac{13}{81}} = \frac{\sqrt{13}}{\sqrt{81}}$    Apply the quotient property: $\sqrt{\frac{a}{b}} = \frac{\sqrt{a}}{\sqrt{b}}$.

$= \frac{\sqrt{13}}{9}$    Simplify the denominator. The radical expression in the numerator can't be simplified.

3. $\sqrt{\frac{18}{49}}$

$\sqrt{\frac{18}{49}} = \frac{\sqrt{18}}{\sqrt{49}}$    Apply the quotient property: $\sqrt{\frac{a}{b}} = \frac{\sqrt{a}}{\sqrt{b}}$.

$= \frac{\sqrt{18}}{7}$    Simplify the denominator.

$= \frac{\sqrt{9}\sqrt{2}}{7}$    Factor the radicand and apply the product property: $\sqrt{ab} = \sqrt{a} \cdot \sqrt{b}$

$= \frac{3\sqrt{2}}{7}$    Simplify.

## Practice B

Simplify the radical expressions, and then turn the page to check your work.

5. $\sqrt{150}$

6. $\sqrt{32}$

7. $\sqrt{3} \cdot \sqrt{15}$

8. $\sqrt{\frac{16}{225}}$

9. $\sqrt{\frac{63}{64}}$

## C. Rationalizing the Denominator of a Radical Expression

When a radical expression is written in simplest form, mathematical convention says that it will not contain a radical in the denominator. We can remove radicals from the denominators of fractions using a process called **rationalizing the denominator**.

We know that multiplying by 1 does not change the value of an expression. Multiplying by $\frac{3}{3}$ or $\frac{\sqrt{2}}{\sqrt{2}}$ is equivalent to multiplying by 1, so we use this property of multiplication to rewrite expressions that contain radicals in the denominator. To remove radicals from the denominators of fractions, then, multiply by the form of 1 that will eliminate the radical.

For example, to rationalize the denominator of the expression $\frac{5}{\sqrt{2}}$, we multiply the fraction by $\frac{\sqrt{2}}{\sqrt{2}}$ and simplify the denominator:

$$\frac{5}{\sqrt{2}} \cdot \frac{\sqrt{2}}{\sqrt{2}} = \frac{5\sqrt{2}}{\sqrt{4}} = \frac{5\sqrt{2}}{2}$$

The last expression in the line above is said to have a rationalized denominator because 2 is a rational number. We can verify that the original expression is equivalent to the last expression by using a calculator, as in Figure 3.

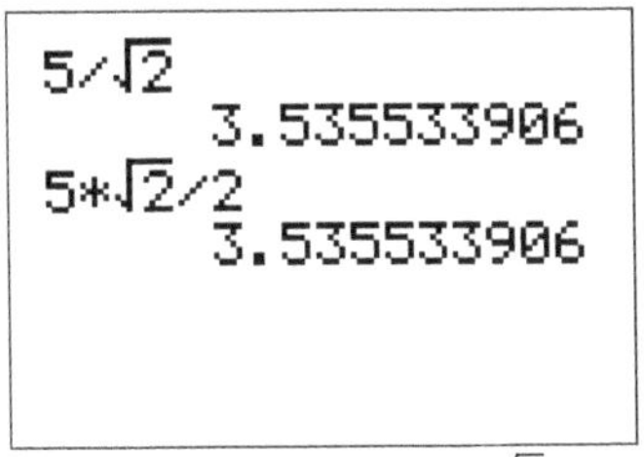

When you have an expression with a single square root radical term in the denominator, you can rationalize the denominator by following these steps:

1. Multiply the numerator and denominator by the radical in the denominator.
2. Simplify.

Figure 3 Verifying $\frac{5}{\sqrt{2}} = \frac{5\sqrt{2}}{2}$.

### Example 5

Write the expression in simplest form.

1. $\dfrac{3}{2\sqrt{7}}$
2. $\dfrac{2\sqrt{3}}{3\sqrt{10}}$
3. $\sqrt{\dfrac{169}{500}}$

**Solutions**

1. The radical in the denominator is $\sqrt{7}$, so multiply the fraction by $\dfrac{\sqrt{7}}{\sqrt{7}}$.

$$\frac{3}{2\sqrt{7}} = \frac{3}{2\sqrt{7}} \cdot \frac{\sqrt{7}}{\sqrt{7}} \quad \text{Multiply by 1: } \frac{\sqrt{7}}{\sqrt{7}} = 1.$$

$$= \frac{3\sqrt{7}}{2\sqrt{49}} \quad \text{Multiply the fractions. Apply the product property in the denominator: } \sqrt{ab} = \sqrt{a} \cdot \sqrt{b}.$$

$$= \frac{3\sqrt{7}}{2 \cdot 7} \quad \text{Simplify.}$$

$$= \frac{3\sqrt{7}}{14} \quad \text{Multiply.}$$

2. The radical in the denominator is $\sqrt{10}$. So multiply the fraction by $\frac{\sqrt{10}}{\sqrt{10}}$.

$\frac{2\sqrt{3}}{3\sqrt{10}} = \frac{2\sqrt{3}}{3\sqrt{10}} \cdot \frac{\sqrt{10}}{\sqrt{10}}$    Multiply by 1: $\frac{\sqrt{10}}{\sqrt{10}} = 1$.

$= \frac{2\sqrt{30}}{3\sqrt{100}}$    Multiply fractions. Apply the product property in the denominator: $\sqrt{ab} = \sqrt{a} \cdot \sqrt{b}$.

$= \frac{2\sqrt{30}}{30}$    Simplify the denominator: $3\sqrt{100} = 3 \cdot 10 = 30$.

$= \frac{\sqrt{30}}{15}$    Simplify the fraction: $\frac{2}{30} = \frac{1}{15}$.

3. Apply the quotient rule for exponents and simplify. Then rationalize the denominator.

$\sqrt{\frac{169}{500}} = \frac{\sqrt{169}}{\sqrt{500}}$    Apply the quotient property: $\sqrt{\frac{a}{b}} = \frac{\sqrt{a}}{\sqrt{b}}$.

$= \frac{13}{\sqrt{100}\sqrt{5}}$    Simplify the numerator. Apply the product property in the denominator: $\sqrt{ab} = \sqrt{a} \cdot \sqrt{b}$.

$= \frac{13}{10\sqrt{5}}$    Simplify the denominator.

$= \frac{13}{10\sqrt{5}} \cdot \frac{\sqrt{5}}{\sqrt{5}}$    Multiply by 1: $\frac{\sqrt{5}}{\sqrt{5}} = 1$.

$= \frac{13\sqrt{5}}{10\sqrt{25}}$    Multiply fractions. Apply the product property in the denominator: $\sqrt{ab} = \sqrt{a} \cdot \sqrt{b}$.

$= \frac{13\sqrt{5}}{50}$    Simplify the denominator: $10\sqrt{25} = 10 \cdot 5 = 50$.

## Practice C

Write the expression in simplest form and then turn the page to check your work.

10. $\frac{4}{\sqrt{3}}$      11. $\frac{12\sqrt{3}}{\sqrt{2}}$      12. $\sqrt{\frac{25}{99}}$

### Practice B — Answers

5. $5\sqrt{6}$      7. $3\sqrt{5}$      9. $\frac{3\sqrt{7}}{8}$

6. $4\sqrt{2}$      8. $\frac{4}{15}$

## D. Solving Quadratic Equations Using the Square Root Property

Now that we have a solid foundation for using and simplifying square roots, we're ready to use square rooting to solve some quadratic equations. To do so we introduce the **square root property**, which gives us an efficient way to solve quadratic equations containing an $x^2$ term (quadratic) but no $x$ term (linear). With this property, we isolate the $x^2$ and apply the square root property.

> **The Square Root Property**
>
> For $k \geq 0$, the square root property states:
>
> $x^2 = k$, then $x = \pm\sqrt{k}$.

Here's an example of how the square root property works: if we solve $x^2 = 25$ mentally, we know there are two answers, 5 and −5. The square root property gives us the answer as $x = \pm\sqrt{25}$, which simplifies to 5 and −5. Because $\sqrt{25}$ by itself is just the principal square root, 5, the $\pm$ sign is necessary to show both square roots.

When solving a quadratic equation that contains no $x$ term, we can use the square root property by following these steps:

1. Isolate $x^2$ on the left side of the equal sign.
2. Replace the Step 1 equation with $x =$ the positive or negative square root of the expression on the right side.
3. Simplify the radical expression on the right side, either keeping the $\pm$ to represent the two answers or writing the two answers out separately.

The most common student error in applying the square root property is to forget the $\pm$ sign. When you forget this symbol, you only obtain one of two possible solutions.

### Example 6

Use the square root property to solve this quadratic equation: $x^2 = 8$

**Solution**
Apply the square root property, and simplify the radical. Remember to use the $\pm$ sign.

$x^2 = 8$ — Original equation.
$x = \pm\sqrt{8}$ — Apply the square root property.
$x = \pm\sqrt{4}\sqrt{2}$ — Apply the product property.
$x = \pm 2\sqrt{2}$ — Simplify.

The solutions are $x = 2\sqrt{2}$ and $x = -2\sqrt{2}$. In Figure 4, we use a calculator to verify our work.

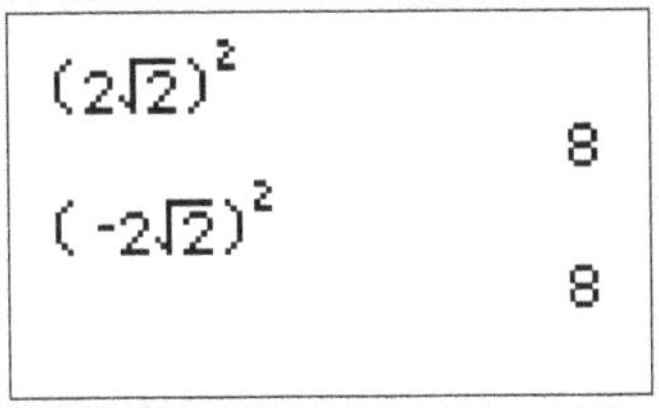

$(2\sqrt{2})^2$   8
$(-2\sqrt{2})^2$   8

Figure 4. Verify the solutions.

## Example 7

Solve the quadratic equation: $4x^2 + 1 = 8$

### Solution

First, we isolate $x^2$. Then we apply the square root property.

| | |
|---|---|
| $4x^2 + 1 = 8$ | Original equation. |
| $4x^2 = 7$ | Subtract 1 from both sides. |
| $x^2 = \frac{7}{4}$ | Divide both sides by 4. |
| $x = \pm\sqrt{\frac{7}{4}}$ | Apply the square root property. |
| $x = \pm\frac{\sqrt{7}}{2}$ | Simplify using the quotient property. |

The solutions are $x = \frac{\sqrt{7}}{2}$ and $x = -\frac{\sqrt{7}}{2}$. This time, we verify the solutions using the graph and intersect method on the graphing calculator. Note that $\frac{\sqrt{7}}{2} \approx 1.3229$ and $-\frac{\sqrt{7}}{2} \approx -1.3229$. See Figure 5.

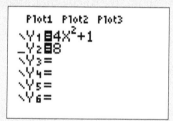 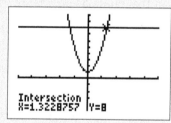 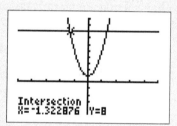

Figure 5. Verify by graph and intersect method.

The square root property allows us to undo the squaring of a variable. But it can also be used to undo the squaring of an expression. We can think of the square root property as

$\square^2 = k$ implies $\square = \pm\sqrt{k}$, where almost any expression might be in the box.

In words, if we have *something* squared equal to a number, that *something* is equal to plus or minus the square root of the number. So e can use the square root property to solve equations of the form $(ax + b)^2 = k$. For a quadratic equation in this form, the square root property states:

If $(ax + b)^2 = k$, then $ax + b = \pm\sqrt{k}$.

Recall that when solving, we reverse the order of operations for calculations. So we address the expression in parentheses last. This requires us to write the equation with the ± symbol as two separate equations to complete the solving process.

### Practice C — Answers

10. $\frac{4\sqrt{3}}{3}$    11. $6\sqrt{6}$    12. $\frac{5\sqrt{11}}{33}$

## Example 8

Solve: $(5x - 4)^2 = 121$

**Solution**

We start by applying the square root property where the squared expression is $5x - 4$.

| | | |
|---|---|---|
| $(5x - 4)^2$ | $= 121$ | Original equation. |
| $5x - 4$ | $= \pm\sqrt{121}$ | Apply the square root property. |
| $5x - 4$ | $= \pm 11$ | Simplify the radical expression. |
| $5x - 4 = -11$ | or $5x - 4 = 11$ | Write as two equations. |
| $5x = -7$ | or $5x = 15$ | Add 4 to both sides in both equations. |
| $x = -\frac{7}{5}$ | or $x = \frac{15}{5} = 3$ | Divide both sides by 5 in both equations. |

The two solutions are $x = -\frac{7}{5}$ and $x = 3$.

## Example 9

Solve: $3(x + 7)^2 - 15 = 0$

**Solution**

Start by isolating the squared expression, $(x + 7)^2$. Then use the square root property.

| | | |
|---|---|---|
| $3(x + 7)^2 - 15$ | $= 0$ | Original equation. |
| $3(x + 7)^2$ | $= 15$ | Add 15 to both sides. |
| $(x + 7)^2$ | $= 5$ | Divide both sides by 3. |
| $x + 7$ | $= \pm\sqrt{5}$ | Apply the square root property. |
| $x + 7 = -\sqrt{5}$ | or $x + 7 = \sqrt{5}$ | Write as two equations. |
| $x = -7 - \sqrt{5}$ | or $x = -7 + \sqrt{5}$ | Subtract 7 from both sides. |

When an expression is the combination of a rational and an irrational term, we typically write the rational term first. We can present the two solutions together as $x = -7 \pm \sqrt{5}$.

## Practice D

Solve for $x$. When you're finished, turn the page to check your solutions.

13. $x^2 = 20$
14. $7x^2 + 5 = 257$
15. $(4x + 9)^2 = 25$
16. $-10(x - 1)^2 + 270 = 0$

## E. Complex Numbers and Complex Solutions

In the last part of this section we will learn about an unusual type of numbers called *imaginary numbers*. While the description "imaginary" may make it sound as if these numbers are silly or useless, nothing could be further from the truth. Imaginary numbers and their cousins, *complex numbers*, have many practical applications.

Imaginary numbers are used in calculations involving alternating electrical currents. Spectrum analyzers — those cool color displays you see when music is playing — are programmed using imaginary numbers. And for mathematicians, one of the most fun and beautiful applications of complex numbers is the development of fractals, including the famous Mandelbrot Set. Look it up!

### The Square Root of a Negative Number

We know how to find the square root of a positive real number, so we've been able to solve equations of the form $x^2 = k$, where $k$ is nonnegative. But what if $k$ *is* negative? In section 2.4, we learned that an equation such as $x^2 = -1$ has no real-number solutions. So a new number $i$ is defined as the square root of $-1$:

$$i = \sqrt{-1}$$

When we square both sides of this identity, we find that $i^2 = -1$:

$$i = \sqrt{-1}$$
$$i^2 = (\sqrt{-1})^2$$
$$i^2 = -1$$

We name this new number, the **imaginary unit** *i*, which is the number whose square is $-1$.

### The Imaginary Unit *i*

The **imaginary unit**, written $i$, is the number whose square is $-1$. We write:

$$i^2 = -1 \quad \text{and} \quad i = \sqrt{-1}$$

When simplifying a radical expression, if the value of the radicand is negative, the square root is said to be an imaginary number. We can write the square root of any negative number as a multiple of $i$. Consider the square root of $-49$:

$$\sqrt{-49} = i\sqrt{49} = 7i$$

The number $7i$ is of the form $bi$. A number of the form $bi$, where $b$ is a real number, is called an **imaginary number**.

### Square Root of a Negative Number

If $k$ is a positive real number, then $\sqrt{-k} = i\sqrt{k}$.

## Example 10

Simplify the following expressions to $bi$ form, where $b$ is a real number.

1. $\sqrt{-100}$  2. $\sqrt{-12}$  3. $-5\sqrt{-121}$

Solutions

1. $\sqrt{-100} = i\sqrt{100} = 10i$
2. $\sqrt{-12} = i\sqrt{12} = i\sqrt{4}\sqrt{3} = 2i\sqrt{3}$
   In this case, $b = 2\sqrt{3}$ is a real number. The order of the factors is by convention.
3. $-5\sqrt{-121} = -5i\sqrt{121} = -5i(11) = -55i$

A **complex number** is the sum of a real number and an imaginary number. A complex number is expressed in standard form as $a + bi$, where $a$ is the real number part and $bi$ is the imaginary number part.

## Complex Number

A **complex number** in standard form is written

$$a + bi$$

where $a$ and $b$ are real numbers and $i$ is the imaginary unit.

Here are some examples of complex numbers:

$7 + 2i$,  $8 - 5i$,  $-6 - i$,  $1 + 4i\sqrt{2}$

Because the real numbers $a$ and $b$ can equal zero, we have $a + 0i = a$. So all real numbers are complex numbers. Further, $0 + bi = bi$ implies that all pure imaginary numbers are complex numbers, too. A complex number that is not a real number, meaning $b \neq 0$, is called an imaginary number.

Imaginary numbers differ from real numbers in that a squared imaginary number produces a negative real number. Remember that when a *positive* real number is squared, the result is a positive real number. When a *negative* real number is squared, the result is also a positive real number. Here are examples for each case:

$6^2 = 36$  The square of a positive real number is positive.
$(-6)^2 = 36$  The square of a negative real number is positive.
$(6i)^2 = 6^2 i^2 = 36(-1) = -36$  The square of an imaginary number is *negative*.

### Practice D — Answers

13. $x = \pm 2\sqrt{5}$  14. $x = \pm 6$  15. $x = -\frac{7}{2}$ and $x = -1$  16. $x = 1 \pm 3\sqrt{3}$

## Solving Quadratic Equations with Complex Solutions

We can now extend the square root property to include the case where $k$ is negative. Remember that if we begin with a radical expression then it simplifies to its *one* principal square root. But if we begin with an equation of form $x^2 = k$, then if $k \neq 0$, there will be *two* solutions.

### Example 11

Solve the following equations. Find all imaginary number solutions.

1. $x^2 = -81$
2. $x^2 = -48$
3. $6x^2 - 14 = -164$

**Solutions**

1. $x^2 = -81$

    | | |
    |---|---|
    | $x^2 = -81$ | Start with the original equation. |
    | $x = \pm\sqrt{-81}$ | Apply the square root property. |
    | $x = \pm i\sqrt{81}$ | Definition of the square root of a negative number. |
    | $x = \pm 9i$ | Simplify. |

2. $x^2 = -48$

    | | |
    |---|---|
    | $x^2 = -48$ | Original equation. |
    | $x = \pm\sqrt{-48}$ | Apply the square root property. |
    | $x = \pm i\sqrt{48}$ | Definition of the square root of a negative number. |
    | $x = \pm 4i\sqrt{3}$ | Simplify: $\sqrt{48} = \sqrt{16}\sqrt{3} = 4\sqrt{3}$ |

3. $6x^2 - 14 = -164$. We isolate the expression $x^2$ before applying the square root property.

    | | |
    |---|---|
    | $6x^2 - 14 = -164$ | Original equation. |
    | $6x^2 = -150$ | Add 14 to both sides. |
    | $x^2 = -25$ | Divide both sides by 6. |
    | $x = \pm\sqrt{-25}$ | Apply the square root property. |
    | $x = \pm i\sqrt{25}$ | Definition of the square root of a negative number. |
    | $x = \pm 5i$ | Simplify. |

If we try to verify our solutions to the last equation using the graph and intersect method on a graphing calculator, we discover in Figure 6 on the next page that the graphs do not intersect. This is because the equation has no real-number solutions.

We can verify the solution on the home screen of the calculator, as in Figure 7. The TI-83 or 84 calculator key pad has the imaginary unit $i$ above the decimal point.

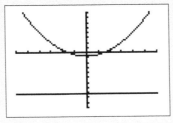

Figure 6.

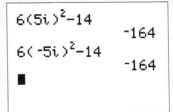

Figure 7. Verify the solutions.

### Example 12

Solve: $(x+5)^2 = -169$  Write the solution in standard complex form.

**Solution**

$(x+5)^2 = -169$      Original equation.

$x + 5 = \pm\sqrt{-169}$      Apply the square root property.

$x + 5 = \pm i\sqrt{169}$      Definition of the square root of a negative number.

$x + 5 = \pm 13i$      Simplify.

$x = -5 \pm 13i$      Write the solutions in standard complex form.

## Practice Set E

Simplify the following expressions. When you're finished, turn the page to check your solutions.

17. $\sqrt{-144}$

18. $\sqrt{-24}$

Solve the following equations.

19. $x^2 + 10 = -65$

20. $2(x-9)^2 + 8 = 0$

## Exercises 4.3

Simplify. Rationalize the denominators of fractions.

1. $\sqrt{196}$
2. $\sqrt{225}$
3. $\sqrt{28}$
4. $\sqrt{54}$
5. $\sqrt{30} \cdot \sqrt{6}$
6. $\sqrt{24} \cdot \sqrt{3}$
7. $\sqrt{\frac{16}{81}}$
8. $\sqrt{\frac{49}{121}}$
9. $\sqrt{\frac{17}{144}}$
10. $\sqrt{\frac{7}{25}}$
11. $\frac{6}{\sqrt{5}}$
12. $\frac{4}{\sqrt{13}}$
13. $\frac{3}{\sqrt{63}}$
14. $\frac{10}{\sqrt{125}}$
15. $\sqrt{\frac{3}{7}}$
16. $\sqrt{\frac{11}{20}}$

Solve using the square root property. The solutions are real numbers. Write the solutions in simplified form.

17. $x^2 = 36$
18. $x^2 = 49$
19. $x^2 - 14 = 0$
20. $x^2 - 23 = 0$
21. $p^2 = 40$
22. $t^2 = 288$
23. $9t^2 = 2$
24. $16p^2 = 29$
25. $6n^2 - 17 = 7$
26. $8n^2 + 43 = 243$
27. $(3x - 8)^2 = 49$
28. $(4x + 3)^2 = 121$
29. $(x + 6)^2 = 7$
30. $(x - 13)^2 = 5$
31. $(x + 9)^2 = 12$
32. $(x - 10)^2 = 27$
33. $5(x - 1)^2 + 42 = 447$
34. $4(x + 3)^2 - 19 = 125$

Simplify.

35. $\sqrt{-9}$
36. $\sqrt{-64}$
37. $\sqrt{-50}$
38. $\sqrt{-8}$
39. $\sqrt{-\frac{4}{25}}$
40. $\sqrt{-\frac{64}{169}}$
41. $2\sqrt{-48}$
42. $3\sqrt{-20}$

Find all complex-number solutions. Simplify the solutions and write in the standard form for complex numbers.

43. $x^2 = -4$
44. $x^2 = -49$
45. $x^2 = -45$
46. $x^2 = -18$
47. $8x^2 + 33 = 17$
48. $5x^2 + 41 = -14$
49. $(x - 9)^2 = -100$
50. $(x - 2)^2 = -25$
51. $3(x + 1)^2 + 58 = 34$
52. $3(x + 8)^2 + 261 = 111$
53. $\left(x - \frac{7}{4}\right)^2 = -\frac{9}{16}$
54. $\left(x + \frac{2}{3}\right)^2 = -\frac{1}{9}$

**Practice E — Answers**

17. $12i$
18. $2i\sqrt{6}$
19. $x = \pm 5i\sqrt{3}$
20. $x = 9 \pm 2i$

# 4.4 The Quadratic Formula

## Overview

Factoring and using the square root property are both solving methods that only work for some quadratic equations. In this section, we use an important equation called the quadratic formula, which can be used to solve *any* quadratic equation. We also study the connection between the solutions of a quadratic equation and the graph of the related quadratic function. In this section you will learn to:

- Solve quadratic equations by using the quadratic formula
- Find $x$-intercepts of quadratic functions using the quadratic formula
- Find $x$-intercepts of quadratic functions using a graphing calculator
- Use the discriminant to determine the number and type of solutions of a quadratic equation
- Make predictions with a quadratic model

## A. The Quadratic Formula

A third method of solving quadratic equations is by using the quadratic formula, which will solve all quadratic equations. We can derive the quadratic formula by yet another solving method called "completing the square." Since the method for completing the square is not covered in this course, we'll omit the derivation of the formula and just present the result below.

### The Quadratic Formula

Written in standard form, $ax^2 + bx + c = 0$, where $a$, $b$, and $c$ are real numbers and $a \neq 0$, any quadratic equation can be solved using the **quadratic formula**:

$$x = \frac{-b \pm \sqrt{b^2 - 4ac}}{2a}$$

We can solve a quadratic equation using the quadratic formula by following these steps:

1. Make sure the equation is in standard form: $ax^2 + bx + c = 0$.
2. Make note of the values of the coefficients and constant term, $a$, $b$, and $c$.
3. Carefully substitute the values noted in step 2 into the formula. To avoid careless errors, use parentheses around each number input.
4. Calculate and solve.

Although the quadratic formula works on any quadratic equation in standard form, it's easy to make errors in substituting the values into the formula and simplifying the resulting expression. Pay close attention when substituting, and use parentheses, especially when inserting a negative number. You will also need to make sure both terms in the numerator are divided by the denominator.

## Example 1

For $x^2 + 2x - 15 = 0$, do the following:

a. Solve for $x$ using the quadratic formula.

b. Graph the related quadratic function, $y = x^2 + 2x - 15$, and note the coordinates of the $x$-intercepts.

### Solution

a. Identify the coefficients: $a = 1, b = 2, c = -15$. Then use the quadratic formula.

$x = \dfrac{-b \pm \sqrt{b^2 - 4ac}}{2a}$     The quadratic formula.

$x = \dfrac{-(2) \pm \sqrt{(2)^2 - 4(1)(-15)}}{2(1)}$     Substitute: $a = 1, b = 2,$ and $c = -15$.

$x = \dfrac{-2 \pm \sqrt{64}}{2}$     Simplify: $(2)^2 - 4(1)(-15) = 4 - {-60} = 64$.

$x = \dfrac{-2 \pm 8}{2}$     Simplify further: $\sqrt{64} = 8$.

$x = \dfrac{-2 + 8}{2}$ or $x = \dfrac{-2 - 8}{2}$     Write as two equations.

$x = 3$ or $x = -5$     Compute.

Notice in the second to last step, the fraction bar is extended below both terms in the numerator. *Both* terms are divided by 2. In this problem, the calculation is easy to do in your head, but if we use a calculator to simplify an expression such as this, we must use two separate steps or put parentheses around the numerator. See Figure 1.

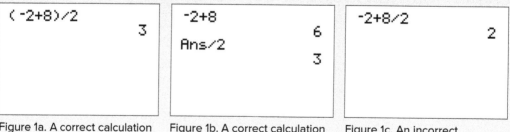

Figure 1a. A correct calculation of $x = \dfrac{-2+8}{2}$ in one step.

Figure 1b. A correct calculation of $x = \dfrac{-2+8}{2}$ in two steps.

Figure 1c. An incorrect calculation of $x = \dfrac{-2+8}{2}$.

b. For $x^2 + 2x - 15 = 0$, the related quadratic function is $f(x) = x^2 + 2x - 15$. Enter the equation into the calculator and find a good viewing window. Enter the solutions to the quadratic equation, $x = 3$ and $x = -5$, as values in TRACE mode. See Figure 2.

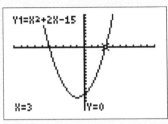

Figure 2a. An x-intercept of the graph.

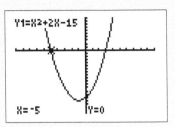

Figure 2b. An x-intercept of the graph

Notice that the solutions to the quadratic equation are the $x$-coordinates of the $x$-intercepts of the graph of the quadratic function, namely $(3, 0)$ and $(-5, 0)$. In other words, the input values $x = 3$ and $x = -5$ both give an output of $y = 0$.

We can also solve the equation used in Example 1 by factoring and applying the zero product property: $x^2 + 2x - 15 = 0$ becomes $(x - 3)(x + 5) = 0$, so either

$x - 3 = 0$ or $x + 5 = 0$ and

$x = 3$ or $x = -5$.

We obtain the same solutions using either method. The next example has a quadratic equation that cannot be factored, so we use the quadratic formula to solve.

## Example 2

For $3x^2 = 9x - 4$, do the following:

a. Solve for $x$ using the quadratic formula.

b. Graph the related quadratic function and verify that the solutions are the $x$-coordinates of the $x$-intercepts.

### Solution

a. We rewrite the equation in standard form by setting the equation equal to 0.

$3x^2 = 9x - 4$  Original equation.

$3x^2 - 9x = -4$  Subtract $9x$ from both sides.

$3x^2 - 9x + 4 = 0$  Add 4 to both sides.

Identify the coefficients: $a = 3$, $b = -9$, $c = 4$. Then use the quadratic formula.

$x = \dfrac{-(-9) \pm \sqrt{(-9)^2 - 4(3)(4)}}{2(3)}$  Substitute $a = 3$, $b = -9$, and $c = 4$ in the quadratic formula.

$x = \dfrac{9 \pm \sqrt{33}}{6}$  Simplify: $(-9)^2 - 4(3)(4) = 81 - 48 = 33$

The solutions to the equation are $x = \frac{9 + \sqrt{33}}{6}$ and $x = \frac{9 - \sqrt{33}}{6}$. Rounded to three decimal places the approximate solutions are: $x = \frac{9 + \sqrt{33}}{6} \approx 2.457$ and $x = \frac{9 - \sqrt{33}}{6} \approx 0.543$.

b. The related quadratic function is $f(x) = 3x^2 - 9x + 4$. Once again, the solutions to the quadratic equation are the $x$-coordinates of the $x$-intercepts of the graph of the parabola, as you see in Figure 3. Notice the parentheses used when entering the $x$ value.

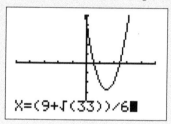
Figure 3a. Entering a radical expression for x.

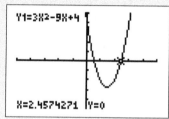
Figure 3b. An x-intercept of the graph

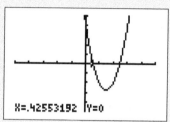
Figure 3c. An x-intercept of the graph.

## Example 3

For $4x^2 - 20x + 25 = 0$, do the following:

a. Solve for $x$ using the quadratic formula.

b. Graph the related quadratic function and verify that the solution is the $x$-coordinate of the $x$-intercept.

### Solution

a. Identify the coefficients: $a = 4$, $b = -20$, $c = 25$. Then use the quadratic formula.

$x = \frac{-(-20) \pm \sqrt{(-20)^2 - 4(4)(25)}}{2(4)}$  Substitute $a = 4$, $b = -20$, and $c = 25$ in the quadratic formula.

$x = \frac{20 \pm \sqrt{0}}{8}$  Simplify: $(-20)^2 - 4(4)(25) = 400 - 400 = 0$

$x = \frac{20}{8}$  $\sqrt{0} = 0$

$x = \frac{5}{2} = 2.5$  Reduce the fraction.

This time we found only one solution, $x = 2.5$.

b. The related quadratic function is $f(x) = 4x^2 - 20x + 25$. When we look at the graph of the parabola this time, we see in Figure 4 that it has only one $x$-intercept at $(2.5, 0)$. We found one solution rather than two because when we evaluated the radical in the quadratic formula we obtained $\pm\sqrt{0} = 0$. Adding or subtracting 0 to a number does not change the number.

The original equation, $4x^2 - 20x + 25 = 0$, can be written in factored form as $(2x - 5)(2x - 5) = (2x - 5)^2 = 0$. Because the factors are identical, we only need to set one of factors equal to zero to find the solution. Solving $2x - 5 = 0$ gives $x = \frac{5}{2}$, which is sometimes referred to as a double solution.

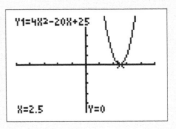

Figure 4.

In Example 4, we use the quadratic formula to solve a quadratic equation that has imaginary-number solutions.

### Example 4

For $-5x^2 - 5 = -8x$, do the following:

a. Solve for $x$ using the quadratic formula.
b. Graph the related quadratic function and note any $x$- or $y$-intercepts.

#### Solution

a. We rewrite the equation in standard form by setting the equation equal to 0.

$-5x^2 - 5 = -8x$     Original equation.

$-5x^2 + 8x - 5 = 0$     Add $8x$ to both sides and write in standard form: $ax^2 + bx + c = 0$.

Identify the coefficients: $a = -5$, $b = 8$, $c = -5$. Then use the quadratic formula.

$x = \dfrac{-(8) \pm \sqrt{(8)^2 - 4(-5)(-5)}}{2(-5)}$     Substitute $a = -5$, $b = 8$, and $c = -5$ in the quadratic formula.

$x = \dfrac{-8 \pm \sqrt{-36}}{-10}$     Simplify: $(8)^2 - 4(-5)(-5) = 64 - 100 = -36$.

$x = \dfrac{-8 \pm 6i}{-10}$     Square root of a negative number: $\sqrt{-k} = i\sqrt{k}$.

$x = \dfrac{4 \pm 3i}{5}$     Reduce the fraction. Divide each term by $-2$.

In the last step, we divide the expression $\pm 6i$ by $-2$, which results in $-3i$ and $3i$. These two results could be shortened to $\mp 3i$, but since the order of the two solutions does not matter, we write the more traditional form, $\pm 3i$.

The solutions are two complex numbers: $x = \dfrac{4 + 3i}{5}$ and $x = \dfrac{4 - 3i}{5}$. Recall that complex numbers in standard form, $a + bi$, have a real and an imaginary part. We write the solutions in standard complex form by dividing both terms of the numerator by 5: $x = \dfrac{4}{5} + \dfrac{3}{5}i$ and $x = \dfrac{4}{5} - \dfrac{3}{5}i$.

**b.** Graph the quadratic function, $f(x) = -5x^2 + 8x - 5$. The $y$-intercept is $(0, -5)$. There are no $x$-intercepts. See Figure 5. This is related to the fact that we found no real-numbered solutions to the equation $-5x^2 + 8x - 5 = 0$.

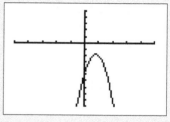

Figure 5. $f(x) = -5x^2 + 8x - 5$.

## Quadratic Formula Program

You should solve a few quadratic equations by hand using the quadratic formula. Once you understand how to apply the formula, your professor may want you to download a **quadratic formula program** to your calculator. This will allow you to solve more complex or multifaceted problems quickly without getting bogged down in one computation. Ask your professor for more information.

## Practice A

Solve the following quadratic equations using the quadratic formula. When you're finished, turn the page to check your work.

1. $9x^2 + 3x - 2 = 0$
2. $3x^2 + 2 = 12x$
3. $2x^2 - 4x + 6 = 0$

## B. Finding x-Intercepts of Quadratic Functions

Examples 1 through 4 in the previous section demonstrate that the $x$-intercepts of a quadratic function can be found by setting the function $y = f(x)$ equal to 0 and solving for $x$. This is true for a quadratic function in any form, no matter what solving method is used. However, if the quadratic function is in standard form, $f(x) = ax^2 + bx + c$, then we can always solve the equation $ax^2 + bx + c = 0$ for $x$ using the quadratic formula. The solutions given by the quadratic formula are the $x$ values of the $x$-intercepts of the graph of the related quadratic function. Recall from Section 4.2 that for a quadratic function in standard form, $f(x) = ax^2 + bx + c$, the $y$-intercept is at $(0, c)$.

### Example 5

Find the $y$ and $x$-intercepts of the quadratic function $f(x) = 3x^2 + 5x - 2$. Verify your work by graphing the function on a graphing calculator.

**Solution**

The function is in standard form with $c = -2$. The $y$-intercept is at $(0, -2)$.

For the *x*-intercepts, we find all solutions of $f(x) = 0$:

$$0 = 3x^2 + 5x - 2$$

Apply the quadratic formula with $a = 3$, $b = 5$, and $c = -2$:

$$x = \frac{-(5) \pm \sqrt{(5)^2 - 4(3)(-2)}}{2(3)} = \frac{-5 \pm \sqrt{49}}{6} = \frac{-5 \pm 7}{6}$$

The last expression simplifies to the solutions $x = \frac{1}{3}$ and $x = -2$.
The *x*-intercepts are $(\frac{1}{3}, 0)$ and $(-2, 0)$.

We verify our work by graphing $f(x)$ on a calculator and entering the values $x = 0$, $x = \frac{1}{3}$, and $x = -2$ while in TRACE mode. Figure 6 shows one of these checks.

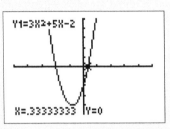

Figure 6. Verify the work.

To find the *x*-intercepts of the function in Example 5, we used the quadratic formula to solve the equation $0 = 3x^2 + 5x - 2$. This equation can be solved quickly using a quadratic formula program on a calculator as demonstrated in Figure 7. Notice the program is set to round solutions to three decimal places. The results, of course, are the same: $x \approx .333$ and $x = -2$ give the first coordinates of the *x*-intercepts of the graph of the related function.

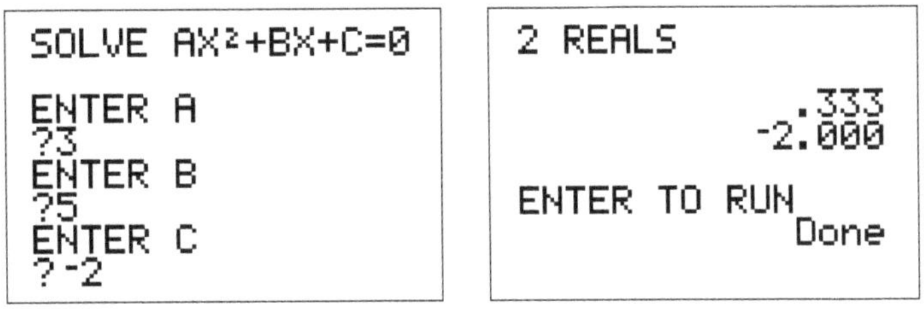

In Example 6, we estimate the *x*-intercepts of a quadratic function using the graph-and-intersect method first demonstrated in Section 3.2.

Figure 7.

### Example 6

Use a graphing calculator and the graph of the function to estimate the *x*-intercepts of the quadratic function $f(x) = -0.5x^2 - 3x + 4$. Round values to three decimal places.

#### Solution

Since the *y*-coordinate of the *x*-intercepts is 0, we are solving this equation:

$$0 = -0.5x^2 - 3x + 4$$

Enter the left side of the equation as $Y_1$, and enter the right side of the equation as $Y_2$. Next, use the inter-

Figure 8.

sect feature under the [CALC] menu above the [TRACE] button.

The approximate coordinates of the *x*-intercepts are (-7.123, 0) and (1.123, 0).

Notice in Figure 9 that one of the *y*-coordinates is written in scientific notation. Remember that $y = 1\text{E-}12$ implies $y = 1 \times 10^{-12} = 0.000000000001 \approx 0$. So the *y*-coordinate of the *x*-intercept is approximately 0.

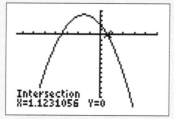 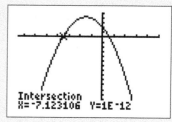

Figure 9.

When using a calculator to find or verify *x*-intercepts, the displayed *y*-value may differ slightly depending on the "guess" entered during the intersection procedure, but the *y*-value will always round to zero.

In the next example, we interpret the intercepts of a graph in the context of a realistic situation.

### Example 7

A batter hits a fast pitch at a baseball game. The height of the baseball (in feet) *t* seconds after it has been hit can be modeled by the function $f(t) = -16t^2 + 120t + 3.5$. With that in mind, do the following:

a. Graph the function on a calculator in the window [-1, 10, 1] by [-20, 240, 20].

b. Find and interpret $f(0)$.

c. Find the *t*-intercepts of the graph of the function and interpret their meaning in the context of the problem.

d. When is the baseball at a height of 80 feet?

Solution

a. Figure 10 shows the graph of the function.

b. $f(0) = -16(0)^2 + 120(0) + 3.5 = 3.5$. The *y*-intercept of the parabola is (0, 3.5). In the context of this problem, that means that at time $t = 0$, when the batter hits the ball, the height of the baseball is 3.5 feet.

c. We find the *t*-intercepts of the parabola by solving the following equation for *t*:

$$0 = -16t^2 + 120t + 3.5$$

This is most easily solved with a quadratic formula program on a calculator where $a = -16$, $b = 120$, and $c = 3.5$. The solutions, rounded to three decimal places, are $t \approx -0.029$ and $t \approx 7.529$, so the *t*-intercepts are approximately (−0.029, 0) and (7.529, 0).

The first solution, $t \approx -0.029$, is a negative value that would suggest going back in time before the baseball was hit by the batter. This value makes no sense in the context of the problem and is an example of model breakdown.

The second solution, $t \approx 7.529$, means that the height of the ball will be 0 feet after approximately 7.529 seconds. Or more plainly, it will take 7.529 seconds for the baseball to land.

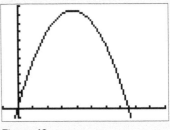

Figure 10.

d. To find when the baseball is at a height of 80 feet, we substitute 80 for $f(t)$ in the equation:

$$80 = -16t^2 + 120t + 3.5$$

We choose to solve this equation using the graph-intersect method on a graphing calculator, where $Y_1 = 80$ and $Y_2 = -16t^2 + 120t + 3.5$.

One solution is $t \approx 0.703$ and is shown in Figure 11. The second solution is $t \approx 6.797$.

We can also set the equation equal to 0 by subtracting 80 from both sides to obtain standard form:

$$0 = -16t^2 + 120t - 76.5$$

We can quickly solve this equation using a quadratic formula program, and of course, we obtain the same solutions. The baseball will be at a height of 80 at approximately 0.703 seconds on its way up and at 6.797 seconds on its way back down.

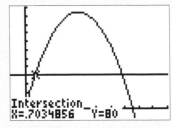

Figure 11.

## Practice B

Try the following problems. Then turn the page to check your solutions.

4. Find the x- and y-intercepts of $f(x) = -2x^2 + 6x - 8$.

5. Use a graphing calculator to estimate the x-intercepts of $f(x) = -1.8x^2 + 13.2x + 24.9$. Round values to three decimal places.

6. A small rocket is launched so that the height of the rocket (in meters) $t$ seconds after it has been launched can be modeled by the function $f(t) = -4.9t^2 + 51t + 2$. Find and interpret the t-intercepts in the context of the problem.

Practice A — Answers

1. $x = -\frac{2}{3}, x = \frac{1}{3}$
2. $x = \frac{6 \pm \sqrt{30}}{3}$
3. $x = \frac{4 \pm i\sqrt{32}}{4} = \frac{4 \pm 4i\sqrt{2}}{4} = 1 \pm i\sqrt{2}$

## C. Determining the Number and Type of Solutions

The quadratic formula does more than generate the solutions to a quadratic equation. It also tells us about the nature of the solutions. To determine the nature of the solutions we need only consider the **discriminant**, which is the expression under the radical, $b^2 - 4ac$:

$$x = \frac{-b \pm \sqrt{b^2 - 4ac}}{2a}$$

### The Discriminant

For $ax^2 + bx + c = 0$, where $a$, $b$, and $c$ are real numbers, the **discriminant** is the expression under the radical in the quadratic formula:

$$b^2 - 4ac$$

It tells us whether the solutions are real numbers or imaginary numbers and how many solutions of each type to expect.

If the discriminant is positive, its square root is a real number, and the quadratic formula gives two real-numbered solutions. If the discriminant is negative, its square root is an imaginary number, and the quadratic formula gives two imaginary-number solutions. Finally, if the discriminant is 0, the quadratic formula gives:

$$x = \frac{-b \pm \sqrt{0}}{2a} = \frac{-b}{2a}$$

Therefore, there is one real-number solution. The table below relates the value of the discriminant to the solutions of a quadratic equation.

| Value of Discriminant | Results |
| --- | --- |
| $b^2 - 4ac = 0$ | one real-number solution |
| $b^2 - 4ac > 0$ | two real-number solutions |
| $b^2 - 4ac < 0$ | two imaginary-number solutions |

### Example 8

Use the discriminant to find the number and type of solutions to the following quadratic equations:

1. $x^2 + 4x + 4 = 0$
2. $5x^2 - 11x - 3 = 0$
3. $3x^2 - 10x + 15 = 0$

**Solutions**

Calculate the discriminant $b^2 - 4ac$ for each equation and state the expected type of solutions.

1. $b^2 - 4ac = (4)^2 - 4(1)(4) = 0$. There will be one real-number solution.
2. $b^2 - 4ac = (-11)^2 - 4(5)(-3) = 181 > 0$. There will be two real-number solutions.
3. $b^2 - 4ac = (-10)^2 - 4(3)(15) = -80 < 0$. There will be two imaginary-number solutions.

We've seen that the number of x-intercepts of a parabola can vary depending upon the location of the graph. There is a direct correlation between the value of the discriminant of a quadratic equation and the number of x-intercepts of the corresponding quadratic function. The table below presents this correlation.

| Discriminant | Solutions to $ax^2 + bx + c = 0$ | x-intercepts of $f(x) = ax^2 + bx + c$ |
| --- | --- | --- |
| $b^2 - 4ac = 0$ | one real-number solution | one x-intercept (the same point as the vertex) |
| $b^2 - 4ac > 0$ | two real-number solutions | two x-intercepts (the graph crosses the x-axis twice) |
| $b^2 - 4ac < 0$ | two imaginary-number solutions | no x-intercepts (the graph never crosses the x-axis) |

Figures 12, 13, and 14 provide examples of each case.

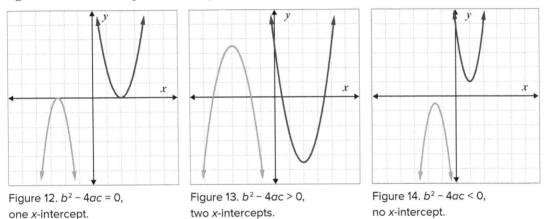

Figure 12. $b^2 - 4ac = 0$, one x-intercept.

Figure 13. $b^2 - 4ac > 0$, two x-intercepts.

Figure 14. $b^2 - 4ac < 0$, no x-intercept.

## Practice C

For each quadratic equation, do the following: find the discriminant, state the number and type of solutions, and state the number of x-intercepts of the corresponding quadratic function. Turn the page to check your work.

7. $6x^2 + 5x - 2 = 0$
8. $0 = -3x^2 + 2x - 8$
9. $x^2 - 10x + 25 = 0$

### Practice B — Answers

4. The y-intercept is $(0, -8)$. The x-intercepts are $(-4, 0)$ and $(1, 0)$.
5. $(-1.556, 0)$ and $(8.889, 0)$
6. The approximate values for the t-intercepts are $(-0.039, 0)$ and $(10.447, 0)$. The first point has a negative value and makes no sense in the context of the problem (model breakdown). The second point is interpreted to mean the rocket will land after approximately 8.889 seconds.

## D. Deciding Which Solving Method to Use

In this chapter, we've discussed three ways to solve quadratic equations: factoring, the square root property, and the quadratic formula. In addition, we can also solve an equation graphically with the graph-intersect method. How do we decide which method to use?

Here are some guidelines to help you decide which method to use to solve a quadratic equation:

- Use **factoring** for equations that can be put in the form $ax^2 + bx + c = 0$ and can be easily factored. If $(x - p)(x - q) = 0$, then $x = p$ or $x = q$. Most quadratic equations are not factorable, so only a small percentage can be solved by factoring. However, if an equation *can* be solved by simple factoring, this is often the easiest method.
- Use the **square root property** for equations that can easily be put into the form $x^2 = k$ or $(ax + b)^2 = k$ and then $x = \pm k$ or $ax + b = \pm k$.
- Use **graphing methods** for finding real solutions only. This is most useful when other related questions are asked that can be solved with the graphing features of the calculator.
- Use the **quadratic formula** for any quadratic equation — especially those that cannot be easily solved by another method.

### Example 9

Solve the following equations using the method of your choice.

1. $2x^2 - 29 = 5$
2. $x^2 - 8x + 12 = 0$
3. $(x + 6)(x - 7) = 3(x + 2)$

**Solutions**

1. The equation can easily be put into the form $x^2 = k$.

   | | |
   |---|---|
   | $2x^2 - 29 = 5$ | Original equation. |
   | $2x^2 = 34$ | Add 29 to both sides. |
   | $x^2 = 17$ | Divide both sides by 2. |
   | $x = \pm\sqrt{17}$ | Apply the square root property. |

### Practice C — Answers

7. discriminant = 73; two real-number solutions; two x-intercepts
8. discriminant = -92; two imaginary-number solutions; no x-intercepts
9. discriminant = 0; one real-number solution; one x-intercept

2. The equation is easily factorable, so $x^2 - 8x + 12 = 0$ becomes $(x - 6)(x - 2) = 0$. Therefore, $x - 6 = 0$ or $x - 2 = 0$ and $x = 6$ or $x = 2$.

3. First we write the equation in $ax^2 + bx + c = 0$ form:

$(x + 6)(x - 7) = 3(x + 2)$    Original equation.
$x^2 - x - 42 = 3x + 6$    Apply the distributive property.
$x^2 - 4x - 48 = 0$    Set equal to 0 by subtracting $3x$ and 6 from both sides.

The trinomial $x^2 - 4x - 48$ is a prime polynomial, so we can't solve by factoring. Because there's a quadratic term and a linear term, it also can't be put into the form $x^2 = k$ or easily put into the form $(ax + b)^2 = k$. We must therefore use the quadratic formula with $a = 1$, $b = -4$, and $c = -48$. The exact solutions are $x = \frac{4 \pm \sqrt{208}}{2} = \frac{4 \pm 4\sqrt{13}}{2} = 2 \pm 2\sqrt{13}$. Using a quadratic formula program, we find approximate solutions: $x \approx 9.211$ and $x \approx -5.211$.

You will have the opportunity to use all of the methods we have discussed for solving quadratic equations as you work the exercises.

# Exercises 4.4

Use the quadratic formula to solve the equation for all real and non-real solutions. Write the solutions in exact form.

1. $-2x^2 + 7x + 4 = 0$
2. $3x^2 + 5x - 8 = 0$
3. $5x^2 - 6x = 3$
4. $x^2 - 9x = -13$
5. $6x^2 = 2x - 2$
6. $4x^2 = 28x - 49$
7. $9x^2 + 24x + 16 = 0$
8. $-4x^2 + x - 11 = 0$

For the following quadratic functions, find the $x$- and $y$-intercepts. Round your answers to three decimal places.

9. $f(x) = 2x^2 + 5x + 3$
10. $f(x) = 2x^2 - 12x - 14$
11. $f(x) = -2x^2 - 8x - 5$
12. $f(x) = -3x^2 - 5x + 1$
13. $f(x) = x^2 - 4x - 2$
14. $f(x) = x^2 - 5x - 2$

For exercises 15 to 18, enter the expressions into your graphing calculator and find the $x$-intercepts of the function using the graph-intersect method — [2nd] [CALC] 5:intersect. Round your answers to the nearest thousandth.

15. $Y_1 = 4x^2 + 3x - 2$,
    $Y_2 = 0$
16. $Y_1 = -3x^2 + 8x - 1$,
    $Y_2 = 0$
17. $Y_1 = -0.5x^2 + 4.9x - 2.6$,
    $Y_2 = 0$
18. $Y_1 = 2.4x^2 + 3.7x - 5.1$,
    $Y_2 = 0$

19. To solve the quadratic equation $x^2 + 5x - 7 = 4$, graph these two equations and find the points of intersection. Round solutions to the nearest thousandth.
    $Y_1 = x^2 + 5x - 7$
    $Y_2 = 4$

20. To solve the quadratic equation $0.3x^2 + 2x - 4 = 2$, graph the following two equations and find the points of intersection. Round solutions to the nearest thousandth.
$Y_1 = 0.3x^2 + 2x - 4$
$Y_2 = 2$

21. The value of Abercrombie and Fitch stock had a price $P$ (in dollars) that can be modeled by $P = 0.2t^2 - 5.6t + 50.2$, where $t$ is the time in months from 1999 to 2001 ($t = 1$ is January 1999). Find the two months in which the price of the stock was $30.

22. A formula for the normal systolic blood pressure for a man age $A$, measured in mmHg, is given as $P = 0.006A^2 - 0.02A + 120$. Find the age to the nearest year of a man whose normal blood pressure measures 125 mmHg.

23. A flower pot falls off a balcony. It's height (in feet) is given by the formula $f(t) = 100 + 7t - 16t^2$, where $t$ is measured in seconds. How long will it take for the flower pot to hit the ground?

24. A rock is tossed off a cliff and falls to the beach below. It's height (in feet) is given by the formula $f(t) = 122 + 12.5t - 16t^2$, where $t$ is measured in seconds. How long will it take for the rock to hit the beach?

For the following exercises, find the discriminant of the quadratic equation and use it to determine the number and the type of solutions. Then, state the number of $x$-intercepts of the parabola for the corresponding quadratic function.

25. $0 = -6x^2 - 5x + 2$
26. $0 = 16x^2 - 24x + 9$
27. $0 = 3x^2 - 2x + 8$
28. $0 = 2x^2 + 7x - 4$
29. $0 = x^2 - 10x + 25$
30. $0 = -5x^2 + 3x - 6$
31. $0 = -2x^2 + 5x - 3$
32. $0 = 6x^2 - 2x + 1$

For the following exercises, find all solutions, real or imaginary, using the method of your choice. When needed, round solutions to three decimal places.

33. $2(x - 5)^2 - 10 = -82$
34. $0 = 2x^2 - 10x + 4$
35. $x^2 + 7x - 18 = 0$
36. $4x^2 + x = 1$
37. $-3x^2 + 9x = -2$
38. $\frac{1}{2}x^2 + 3x + 1 = 0$
39. $4x^2 - 3x - 1 = 0$
40. $(x - 3)^2 + 2 = 0$
41. $-2(x + 3)^2 - 6 = 0$
42. $-4 = x^2 + 6x$
43. $(2x - 1)(4x + 3) + 7 = -5(x^2 - 8)$
44. $(3x - 2)(x + 5) + 24 = -x^2 - 6$
45. $5x^2 + 48 = 13$
46. $7x^2 + 100 = 37$
47. $0 = x^2 + x - 30$
48. $x^2 + 6x - 55 = 0$
49. $(x + 3)(x - 2) = 4(x + 1)$
50. $3(x + 4)^2 = 10$

# 4.5 Modeling with Quadratic Functions

## Overview

Outside of the college math classroom, you will rarely be asked to find the vertex or the $x$-intercepts of a quadratic function. In the real world, however, you might very well be asked to find the conditions that produce maximum revenue or minimum cost for a business. You might also want to determine when a projectile will reach its maximum height or when an engine is at the highest point on its power curve. In this section, we focus on the application of quadratic models to real-world situations. We'll use several features of a graphing calculator to help us solve problems.

In this section, you will learn to:

- Find maximum and minimum values using a graphing calculator
- Find and interpret both input and output values of a quadratic function in the context of a real-world problem
- Find a quadratic regression equation using a graphing calculator

## A. Using a Graphing Calculator to Find Maximum or Minimum Values

In Section 4.2, for a quadratic function in standard form, $y = f(x) = ax^2 + bx + c$, we used an algebraic method to determine the coordinates of the vertex of a parabola. Recall that $x = \frac{-b}{2a}$ determines the $x$-coordinate of the vertex. The $y$-coordinate is found when we substitute $x = \frac{-b}{2a}$ in the equation and evaluate $f\left(\frac{-b}{2a}\right)$.

For quadratic functions, the vertex represents the point on the curve where either the maximum or minimum value of the function occurs. We can also use the maximum and minimum features on a graphing calculator to find the coordinates of a vertex. Example 1 illustrates this process.

In a real-world problem, the coordinates of the vertex often represent very useful information. The independent variable tells us *when* or *where* the maximum or minimum value of the function occurs. The dependent variable tells us the maximum or minimum value of the function.

### Example 1

A company manufactures solar panels. The business manager projects that the profit (in dollars) from making $x$ solar panels per week can be modeled by this formula:

$P = f(x) = -2.5x^2 + 198x + 10{,}100$

How many solar panels should be produced each week in order to maximize profit? What is the maximum profit?

## Solution

The leading coefficient, $a = -2.5$, is negative, so the parabola opens downward. The maximum value of the function will be at the vertex. We use a graphing calculator to find the coordinates of the vertex.

First, enter the equation $y = -2.5x^2 + 198x + 10100$ into the calculator and select a good viewing window. Try $[-10, 120, 10]$ by $[-50, 16000, 1000]$. See Figure 1.

To approximate the coordinates of the vertex, enter the keystrokes [2nd], [CALC], and [4]. See Figure 2a.

The "maximum" under the [CALC] menu first asks for the left bound. Use your $x$-axis scaling to help you determine an $x$-value well to the left of the vertex, say 30, and [ENTER]. It then asks for the right bound. Choose an $x$-value well to the right of the vertex, say 50, and enter. Finally, the calculator asks for a guess. For the "guess", choose any number between your left and right bounds, say 40, and [ENTER]. See Figure 2b.

In Figure 2c, you see that the calculator provides the coordinates of the vertex, $(39.6, 14020.4)$. The value of the independent variable, $x = 39.6$, represents 39.6 solar panels. For practical reasons, we round this to 40. The value of the dependent variable, $y = 14{,}020.4$ represents the profit of the company in dollars.

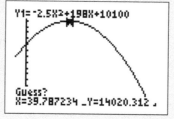
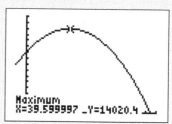

Figure 1. Viewing window.

Figures 2a and 2b. Steps for calculating the coordinates of a vertex.

Figure 2c. The coordinates of the vertex.

Manufacturing 40 solar panels per week will maximize profit. The maximum profit per week is about \$14,020.

## Practice A

Practice finding maximum and minimum values using the graphing calculator feature. Round values to three decimal places. Then turn the page to check your solutions.

1. Use a graphing calculator to find the maximum value of $y = -0.4x^2 - 5x - 7$.
2. Use a graphing calculator to find the minimum value of $y = \frac{1}{3}x^2 - \frac{6}{5}x + \frac{1}{2}$. In the Calc Menu on your calculator, choose "3: minimum" instead of "4: maximum."

## Section 4.5: Modeling with Quadratic Functions

## B. Using and Interpreting a Quadratic Model

In the next two examples, we apply what we've learned about quadratic functions to make estimates and predictions with realistic quadratic models. We'll use the graph of a function to help us visualize the relationship between the variables in the model. It's important to be able to interpret solutions correctly within the context of the real-life situations that are modeled.

### Example 2

A man stands at the edge of a high cliff near the ocean. He throws a rock upward so that it first goes higher than the cliff and then lands on the beach below. The height of the rock, in feet above the beach, is given by $f(t) = -16t^2 + 56t + 124$, where $t$ is time in seconds after the rock is thrown.

a. Graph the function in a good viewing window.
b. Find $f(0)$ and $f(3)$ and interpret these values.
c. When does the rock reach its maximum height? What is the maximum height?
d. When does the rock hit the beach?
e. Find $t$, when $f(t) = 140$. Interpret the solutions.

### Solution

a. Enter the equation in the calculator. To choose a good viewing window, we first think about possible values for the input variable $t$, which represents the time in seconds that the rock is in motion. From personal experience, we estimate that the rock will fall to the beach in a matter of seconds, so we set the x-window to [0, 8, 1].

The output values represent the height of the rock, so this will include 0 and higher. The value of the constant in the function is 124, so we go above that to ensure we can see the whole graph. We also go below 0 because we want to be able to see the key points on the graph when we find values. Set the y-window to [-20, 200, 20] and graph the function. See Figures 3a and 3b.

b. Because we already have $f$ graphed, we find $f(0)$ and $f(3)$ using the TRACE mode on our calculator. Press TRACE, type 0, and ENTER. Repeat this process for the input value 3. We find that $f(0) = 124$ and $f(3) = 148$. See Figure 3c.

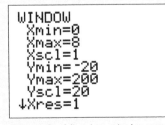
Figure 3a. Viewing window.

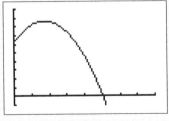
Figure 3b. Graph.

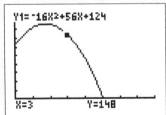
Figure 3c. Evaluating $f(3)$ on the graphing screen.

We interpret $f(0) = 124$ to mean the rock was first thrown from a height of 124 feet. We interpret $f(3) = 148$ to mean that after 3 seconds, the rock is 148 feet high. Additionally, we see from the graph that at this time, the rock has already reached its maximum height and is falling to the beach.

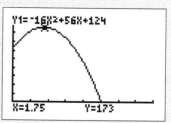

Figure 4. Maximum height of rock.

c. The rock reaches the maximum height at the vertex of the parabola. Instead of using the maximum feature on our calculator we choose to use the vertex formula. The input value, $t$, of the vertex is:

$$t = -\frac{b}{2a} = -\frac{56}{2(-16)} = 1.75$$

The rock reaches a maximum height after 1.75 seconds.

We find the maximum height, which is the $y$-coordinate of the vertex, using the [TRACE] mode on the calculator, as in Figure 4.

The rock reaches a maximum height of 173 feet.

d. To find when the rock hits the beach, we need to determine when the height is zero. Substituting $f(t) = 0$ in the original equation, we have $0 = -16t^2 + 56t + 124$.

We've learned that there is more than one way to solve this equation. To be efficient, this time we use the quadratic formula program on the calculator to solve: $a = -16$, $b = 56$, and $c = 124$. The solutions are $t \approx -1.538$ and $t \approx 5.038$ seconds.

The first solution is negative, which doesn't make sense in the context of the problem. The second solution tells us it takes about 5.038 seconds for the rock to land on the beach. You can verify this on the graphing screen as well.

e. To find $t$, when $f(t) = 140$, we substitute in the original equation: $140 = -16t^2 + 56t + 124$.

We could set this equation equal to zero and solve using the quadratic formula, but since we already have the function graphed, let's use the graph-intersect method. Enter $Y_2 = 140$ along with the original function and find the intersection points of the two graphs. Figures 5a and 5b show that the input values are $t \approx 0.314$ and $t \approx 3.186$.

The rock will be 140 feet above the beach at about 0.314 seconds on its way up and again at about 3.186 seconds on its way down.

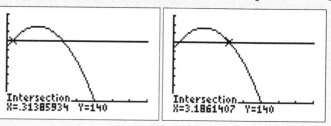

Figures 5a and 5b. Solutions to the equation $140 = -16x^2 + 56x + 124$.

## Practice A — Answers

1. The vertex is at $(-6.25, 8.625)$. The maximum value is 8.625.
2. The vertex is at $(2.80, -3.46)$. The minimum value is $-3.46$.

The next example requires the same set of skills in a slightly different context.

## Example 3

A small family farm supplements its income from grain and vegetable crops by farming and selling tilapia fish. Each month, there are overhead costs, and if the fish harvest increases, these costs also rise. This is accounted for in the following function that models the monthly profit in dollars as a function of the number of pounds of fish $x$ the farm sells: $P = f(x) = -0.0001x^2 + 0.6x - 500$.

a. Find $f(0)$ and interpret this value.
b. Find the $x$-intercepts of the graph of this function, and interpret these values.
c. Find the vertex of the graph of the function and interpret the meaning of both coordinates.

### Solution

a. Evaluate the function at $x = 0$:

$$f(0) = -0.0001(0)^2 + 0.6(0) - 500 = -500$$

If the farm sells 0 lbs of fish, it will lose $500. Because of overhead costs, the farm will lose money if it does not sell any tilapia fish.

b. After entering the function into $Y_1$, it may take some trial and error to come up with a reasonable window setting. [0, 6000, 1000] by [−500, 500, 100] works well. Enter $Y_2 = 0$ and use the [CALC] 5:intersect feature to find the $x$-intercepts of the graph. Figures 6a and 6b show that the $x$-intercepts are (1000, 0) and (5000, 0).

If the farm sells 1,000 lbs or 5,000 lbs of fish, it will break even. If it sells between 1,000 and 5,000 lbs, it will make a profit.

c. Use the [CALC] 4:maximum feature on the calculator to find the coordinates of the vertex. The vertex is (3000, 400). See Figure 6c.

If the farm sells 3,000 lbs of fish per month they will make a maximum profit of $400.

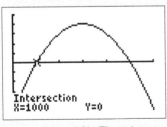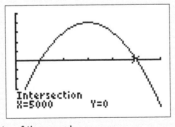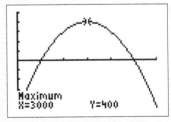

Figures 6a and 6b. The x-intercepts of the graph.  Figure 6c. The vertex.

## Practice B

3. A stone is thrown upward from the top of a 112-foot high cliff overlooking the ocean. The stone's height in feet above the ocean $t$ seconds after it is thrown can be modeled by the equation $h(t) = -16t^2 + 96t + 112$. Use the model to answer the following questions.

   a. Find and interpret $h(2)$.

   b. When does the stone reach the maximum height? What is the maximum height?

   c. When does the stone land in the ocean?

When you're finished, turn the page to check your work.

## C. Finding a Model Using Data in a Table and Quadratic Regression

In Section 1.4, we found linear regression equations on the calculator to model data that had the characteristics of a linear function. In Section 2.5, we used exponential regression equations to model data that appeared to grow or decay exponentially. Now we look at data that can be modeled with a quadratic regression equation.

### Example 4

A study was done to compare the speed $x$ in miles per hour with the mileage $y$ in miles per gallon of an automobile. The table below provides the results of the study.

| Speed, mph (x) | 15 | 25 | 35 | 45 | 55 | 65 | 75 |
|---|---|---|---|---|---|---|---|
| Mileage, mpg (y) | 22.3 | 27.5 | 29.2 | 30.2 | 30.1 | 27.4 | 23.3 |

Using the information from the table, do the following:

a. Use a graphing calculator to create a scatter plot of the data.

b. Use the regression feature of the calculator to find a model that fits the data.

c. Approximate the speed at which the car gets the greatest mileage.

d. Predict the mileage when the speed is 85 mph.

### Solution

a. Use [STAT] and EDIT to enter the data into Lists 1 and 2 on the calculator. Turn on Plot 1 and graph the data in a good viewing window. The Figure 7a scatter plot shows that the data has a parabolic trend.

b. Use the regression feature — [STAT] and [CALC] [5]:QuadReg — to find the quadratic model. See Figure 7b. By rounding the coefficients, we have $y = -0.0084x^2 + 0.771x + 12.87$.

c. Graph the data and the model in the same viewing window as shown in Figure 7c. Use the [CALC] 4:maximum feature of the calculator to find the coordinates of the vertex. The vertex is approximately (46, 31). The speed at which the mileage is greatest, then, is about 46 mph. At that speed, the car's mileage is about 31 miles per gallon.

d. Using the [TRACE] feature — and making sure that we're tracing the regression model, not the scatter plot — enter "$x = 85$." (You may have to adjust the $x$-max in the viewing window to 85 or greater.) At 85 miles per hour, the mileage is about 17.5 miles per gallon.

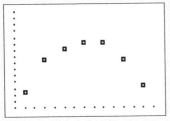
Figure 7a.

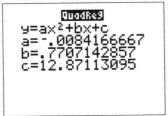
Figure 7b.

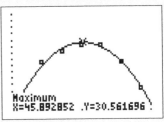
Figure 7c.

## Example 5

A basketball is dropped from a height of 10 meters. The height of the basketball is recorded at intervals of about 0.1 second. The table below shows the results.

| Time, seconds ($x$) | 0 | 0.11 | 0.19 | 0.30 | 0.39 | 0.52 | 0.61 | 0.68 | 0.80 | 0.92 | 1.0 |
|---|---|---|---|---|---|---|---|---|---|---|---|
| Height, meters ($y$) | 10 | 9.952 | 9.801 | 9.560 | 9.215 | 8.775 | 8.235 | 7.594 | 6.864 | 6.033 | 5.107 |

Use a graphing calculator to find a regression model that fits the data. Then use the model to predict the time when the basketball will hit the ground.

### Solution

Begin by entering the data into Lists 1 and 2 on the calculator. Choose a good viewing window and display the scatter plot, as shown in Figure 8a.

From the scatter plot, we see the data has a parabolic trend, so using the regression feature of the calculator, we find the quadratic model, as shown in Figure 8b.

The quadratic model is given by
$y = -4.754x^2 - 0.085x + 10.005$.
Notice that the $y$-intercept of the model is approximately (0, 10) which represents the initial height of the basketball.

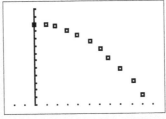
Figure 8a.

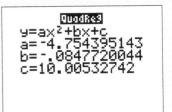
Figure 8b.

To predict when the basketball will hit the ground, we substitute 0 for $y$ in the model and solve the resulting equation for $x$: $0 = -4.754x^2 - 0.085x + 10.005$

We use a quadratic formula program on the calculator to solve: $a = -4.754$, $b = -0.085$, and $c = 10.005$. The solutions are $x = -1.460$ and $x = 1.442$. We disregard the negative solution as it is outside the practical domain of the model.

The basketball will hit the ground after about 1.442 seconds.

When fitting an equation model to a set of data, we need to pay close attention to the trend of the data. Sometimes it's obvious from a scatter plot that the trend is linear or quadratic or exponential, but other times it may not be easy to distinguish between models. We can find several models for the data and graph them with the data. Then we visually determine the model that best fits the data. Remember, the best model doesn't have to contain any of the actual data points, but it should come *close* to the data points and be the best overall representation of the trend seen in the data.

## Example 6

The percentage of American college students who are minorities has been increasing since 1970. The table below comes from the National Center for Education Statistics. It shows percentages of minority students for various years.

| Year | 1976 | 1980 | 1990 | 2000 | 2005 | 2008 |
|---|---|---|---|---|---|---|
| Percent | 16.2 | 16.9 | 20.7 | 29.6 | 32.3 | 34.7 |

Using this table, do the following:

a. Let $y = f(x)$ represent the percentage of American college students who are minorities $x$ years since 1970. Find a linear equation, a quadratic equation, and an exponential equation to model the data. Compare how well the models fit the data.

b. Compare the $y$-intercepts of the three models and interpret these values.

c. For years before 1970, which of the three models most likely gives the best estimates of the percentages of minority students?

d. Use each model to predict the percentage of American college students who are minorities in 2020.

### Practice B — Answer

3. a. $h(2) = 240$. At 2 seconds, the stone is 240 feet above the ocean. b. The stone reaches its maximum height at 3 seconds. The maximum height is 256 feet. c. In 7 seconds, the stone will land in the ocean.

## Solution

**a.** Enter the data into Lists 1 and 2 in the graphing calculator and graph the scatter plot of the data in a good viewing window. Remember, $x = 0$ corresponds to 1970.

Next, find linear, quadratic, and exponential regression equations and graph each with the scatter plot. Figure 9 presents the regression equations.

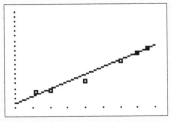 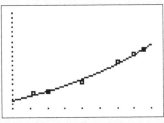

Figure 9a. Linear model.    Figure 9b. Quadratic model.    Figure 9c. Exponential model.

With the coefficients rounded, the regressions are:

Linear:        $y = f(x) = 0.6024x + 11.1$
Quadratic:     $y = g(x) = 0.0108x^2 + 0.1281x + 14.7$
Exponential:   $y = h(x) = 13.4(1.0255)^x$

It appears that the quadratic and exponential models are each a slightly better fit than the linear model.

**b.** When we round the $y$-intercept of each model, we find:

Linear:        $(0, 11.1)$ In 1970, the model estimates 11.1% minority students.
Quadratic:     $(0, 14.7)$ In 1970, the model estimates 14.7% minority students.
Exponential:   $(0, 13.4)$ In 1970, the model estimates 13.4% minority students.

Each of these estimates seems reasonable within the context of the problem.

**c.** To look at the models for the years before 1970, we adjust the viewing window to include negative $x$ and $y$ values. Try: $[-40, 40, 10]$ by $[-10, 50, 10]$. In Figure 10, we graph all three models on the same screen.

The linear model predicts a negative percentage of minority students before 1950, which does not make sense.

The quadratic model predicts that the percentage of minority students was higher before 1970 and that this percentage decreased before it increased. This seems unlikely given common knowledge about minority access to education in the history of the United States.

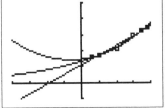

Figure 10. Comparing years before 1970.

The exponential model gives us the most reasonable estimation. Due to the horizontal asymptote at $y = 0$, this model predicts that there were always some — if very few — minorities attending American colleges and that this percentage has always been increasing.

d. To predict the percentage of American college students who are minorities in 2020, we let $x = 2020 - 1970 = 50$ and evaluate each model at this input value:

Linear: $f(50) = 0.6024(50) + 11.1 \approx 41.2$
Quadratic: $g(50) = 0.0108(50)^2 + 0.1281(50) + 14.7 \approx 48.1$
Exponential: $h(50) = 13.4(1.0255)^{50} \approx 47.2$

The linear, quadratic, and exponential models predict 41.2%, 48.1%, and 47.2% minority students respectively. They all seem like reasonable predictions with the linear model showing the smallest growth over time.

## Exercises 4.5

Use a graphing calculator to solve the following problems. Approximate values to three decimal places where needed.

1. Suppose that the price per unit in dollars of a cell phone production is modeled by $p = 45 - 0.0125x$, where $x$ is in thousands of phones produced, and the revenue represented by thousands of dollars is $R = x \cdot p = x(45 - 0.0125x)$. Write this equation in standard quadratic form, and find the production level that will maximize revenue. What is the maximum revenue?

2. Dr. Evil launches a rocket from a submarine. The rocket's height, in meters above sea level, as a function of time in seconds, is given by $h(t) = -4.9t^2 + 229t + 234$. Find the time it takes for the rocket to reach its maximum height. What is the maximum height?

3. A ball is thrown from the top of a building. The ball's height, in meters above ground, as a function of time in seconds, is given by $h(t) = -4.9t^2 + 24t + 8$. How long does it take to reach maximum height and what is that height?

4. Safeco Field in Seattle holds 62,000 Mariners fans. With a ticket price of $11, the average attendance has been 26,000. When the price dropped to $9, the average attendance rose to 31,000. This pattern is accounted for in the following equation which models revenue, $R$, as a function of the drop in ticket price, $x$, both in dollars. $R = (11 - x)(26000 + 2500x)$ Write the equation model in standard quadratic form and use it to predict the ticket price that would maximize revenue.

5. A power boat engine has a power curve approximated by $y = -\frac{x^2}{5000} + \frac{21x}{25} - 32$, where $x$ is the number of revolutions per minute (rpm) and $y$ is the horsepower generated. At what number of revolutions per minute is the engine putting out maximum horsepower? What is the maximum horsepower?

6. An engine in a stunt plane has a power curve approximated by $y = -\frac{x^2}{15,000} + \frac{2x}{5} - 12$, where $x$ is the number of revolutions per minute (rpm) and $y$ is the horsepower generated. At what number of revolutions per minute is the engine putting out maximum horsepower? What is the maximum horsepower?

7. The height of a softball is given by $f(t) = -16t^2 + 45t + 6.5$, where $y$ represents the height of the ball in feet and $x$ is the number of seconds that have elapsed since the softball was thrown.
   a. Find $f(0)$ and $f(2)$ and interpret these values.
   b. Find the time it takes for the softball to reach the maximum height. What is the maximum height?
   c. When will the softball hit the ground?
   d. When will the softball be 25 feet in the air?

8. The height of a golf ball is given by $f(t) = -16t^2 + 85t + 0.2$, where $y$ represents the height of the ball in feet and $t$ is the number of seconds that have elapsed since the golf ball was teed off.
   a. Find $f(0)$ and $f(3)$ and interpret these values.
   b. Find the time it takes for the golf ball to reach the maximum height. What is the maximum height?
   c. When will the golf ball hit the ground?
   d. When will the golf ball be 50 feet in the air?

9. The following table shows data collected relating the maximum load in kilograms that a beam can support to the depth of the beam in centimeters.

   | Depth, x (cm) | 10 | 12 | 14 | 16 | 18 | 20 | 22 | 24 |
   |---|---|---|---|---|---|---|---|---|
   | Load, y (kg) | 4,900 | 7,200 | 9,900 | 12,500 | 16,000 | 20,000 | 24,000 | 28,500 |

   a. Use a graphing calculator to find a quadratic regression model for this data.
   b. Use the model to estimate the load that the beam can support if its depth is 15 cm.

10. The table below shows the number $y$ of U.S. Supreme Court cases waiting to be tried for the years 1995 to 2000. (Source: Office of the Clerk, Supreme Court of the U.S.) Let $x$ represent years since 1990, so that $x = 5$ corresponds to 1995.

    | Year | 1995 | 1996 | 1997 | 1998 | 1999 | 2000 |
    |---|---|---|---|---|---|---|
    | Cases, y | 7,567 | 7,602 | 7,695 | 8,083 | 8,445 | 8,965 |

    a. Use a graphing calculator to find a quadratic regression model for this data.
    b. Use the model to predict the year in which there will be 15,000 U.S. Supreme Court cases waiting to be tried.

11. A new company recorded its sales units versus its profit (or loss) for eight consecutive quarters. See the table below. Note the units sold are in hundreds and the profit is in thousands of dollars.

| Units Sold x (hundreds) | 0 | 25 | 40 | 50 | 60 | 75 | 80 | 100 |
|---|---|---|---|---|---|---|---|---|
| Profit y (thous. $) | -30 | 50 | 70 | 70 | 75 | 40 | 30 | -60 |

a. Using a graphing calculator, make a scatter plot of the data to verify the data has a quadratic trend. Find a quadratic regression model for this data.

b. Use the model to predict the number of units that should be sold to maximize profit.

c. Find the $x$-intercepts of the graph of the model and interpret these values.

12. A new company recorded its production units per day versus its profit (or loss) for those days. See the table below.

| Units Sold x (hundreds) | 0 | 10 | 20 | 30 | 40 | 50 | 60 | 70 |
|---|---|---|---|---|---|---|---|---|
| Profit y (thous. $) | -70 | 70 | 140 | 210 | 230 | 170 | 100 | 50 |

a. Using a graphing calculator, make a scatter plot of the data to verify the data has a quadratic trend. Find a quadratic regression model for this data.

b. Use the model to predict the number of units that should be produced to maximize profit.

c. Find the $x$-intercepts of the graph of the model and interpret these values.

13. The table below shows the amount $y$ in billions of dollars spent on books and maps in the United States for the years 1990 to 2000. Let $x$ represent years since 1990.

| Year | 1990 | 1991 | 1992 | 1993 | 1994 | 1995 | 1996 | 1997 | 1998 | 1999 | 2000 |
|---|---|---|---|---|---|---|---|---|---|---|---|
| Amount, y (billion $) | 16.5 | 16.9 | 17.7 | 18.8 | 20.8 | 23.1 | 24.9 | 26.3 | 28.2 | 30.7 | 33.9 |

a. Use a graphing calculator to find *both* a linear and a quadratic regression model for this data.

b. Choose the model that seems to best fit the data. Explain your answer.

c. Use each model to estimate the sales of books and maps in 2016.

CHAPTER 5
# Further Topics in Algebra

In this chapter, we cover several different topics that will be useful to the student who intends to enroll in more math or science classes. First, we look at direct and inverse variation relationships. In a sealed container, for example, if the pressure on a gas increases, its volume decreases. This is just one case of an inverse variation relationship important to chemists and physicists.

We'll then take an introductory look at the patterns found in lists of numbers called sequences. We will also find that we can write formulas to describe these patterns and then use the formulas to solve problems.

We conclude with a section on dimensional analysis, a widely-used technique for converting units of measure. The problem-solving method introduced here is fundamental to the study of chemistry and physics, but is also useful in a variety of practical situations. If you hear your favorite baseball pitcher threw a fastball that traveled 150 feet per second, for example, you can use dimensional analysis to calculate this speed in miles per hour.

5.1  Variation ............................................................................................. page 264

5.2  Arithmetic Sequences .......................................................................page 276

5.3  Geometric Sequences ....................................................................... page 288

5.4  Dimensional Analysis ....................................................................... page 299

# 5.1 Variation

## Overview

Wally's Used Cars has just offered Shayla a position in sales. The position offers 16% commission on her car sales. If she sells a vehicle for $4,600, for example, she earns 16% of $4,600, which is $736. Shayla's earnings depend solely on the amount of her sales. As her sales increase, so will her earnings. We say her earnings vary *directly* with her sales.

Each morning before work, Jacob runs 4 miles as part of a healthy lifestyle. The speed at which he runs determines the time it takes to complete the run. If Jacob increases his speed, the time it takes to run 4 miles decreases. We say the time it takes to complete the run varies *inversely* with the speed.

In this section, we'll look at variation relationships such as the ones described above. You will learn how to:

- Use a single point to find a direct variation equation or an inverse variation equation
- Solve direct variation problems
- Solve inverse variation problems
- Distinguish between direct and inverse variation models

## A. Direct Variation

In the example above, Shayla's earnings per vehicle sale can be found by multiplying the sales price of the vehicle, $s$, by 16%, her commission. The formula $E = 0.16s$ tells us her earnings, $E$, come from the product of her commission and the sales price of the vehicle.

The table below presents some possible sales and observe that as the sales price increases, the earnings increase as well. The earnings increase in a predictable way. If we double the sales price of the vehicle from $4,600 to $9,200, we double the earnings from $736 to $1,472.

| $s$, sales price | $E = 0.16s$ | Interpretation |
|---|---|---|
| $4,600 | $E = 0.16(4,600) = 736$ | A sale of a $4,600 vehicle results in $736 earnings. |
| $9,200 | $E = 0.16(9,200) = 1,472$ | A sale of a $9,200 vehicle results in $1472 earnings. |
| $18,400 | $E = 0.16(18,400) = 2,944$ | A sale of a $18,400 vehicle results in $2944 earnings. |

On the next page, Figure 1 offers a graphical representation of the equation $E = 0.16s$. As one variable increases, the second variable increases as a multiple of the first. A relationship in which one quantity is a constant multiplied by another quantity is called **direct variation**.

# Direct Variation

For $k \neq 0$, if $x$ and $y$ are related by an equation of the form $y = kx$, then we say $y$ **varies directly** with $x$, or $y$ is proportional to $x$. We call $k$ the **variation constant** or the constant of proportionality. The equation $y = kx$ is called a **direct variation equation**.

The formula we used to represent Shayla's earnings, $E = 0.16s$, is a direct variation equation of the form $y = kx$. We say that earnings vary directly with the sales price of the car. The value $k = 0.16$ is the variation constant in this situation. Here are a few more examples:

1. For $y = 2.5x$, we say $y$ varies directly with $x$, with variation constant $k = 2.5$.

2. For $p = 10t$, we say $p$ varies directly with $t$, with variation constant $k = 10$.

3. For $N = \frac{7}{11}h$, we say $N$ varies directly with $h$, with variation constant $k = \frac{7}{11}$.

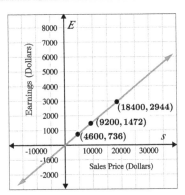

Figure 1. The direct variation equation $E = 0.16s$.

If one variable varies directly with another variable, and if we know one point that lies on the graph, we can find the variation constant $k$. Once we know the value of $k$, we can write a direct variation equation.

## Example 1

If $y$ varies directly with $x$, and $y$ is 28 when $x$ is 7, find an equation relating $x$ and $y$.

### Solution

The first part of the sentence gives the general relationship between $x$ and $y$. The statement "$y$ varies directly with $x$" is equivalent to the equation $y = kx$.

The second part of the sentence gives a point on the graph that we use to find the variation constant $k$. Substitute $x = 7$ and $y = 28$ into the equation and solve for $k$.

$y = kx$   Start with the general direct variation form.
$28 = k(7)$   Substitute $x = 7$ and $y = 28$.
$4 = k$   Divide both sides by 7.

The variation constant is $k = 4$. The specific direct variation equation is $y = 4x$.

Notice that the equation $y = 4x$ from Example 1 is a linear equation. The graph of $y = 4x$ in Figure 2 is a line with slope $m = 4$ and a y-intercept of $(0, 0)$. In general, the direct variation equation $y = kx$ represents a linear function whose graph is a line that passes through the origin $(0, 0)$. The variation constant $k$ is the slope of this line, or the rate of change in $y$ with respect to $x$.

In Figure 2, we graph three direct variation equations where $k = \frac{1}{2}$, $k = 2$, and $k = 4$. If the value of $k$ is positive, giving us a positive slope, then $y = kx$ is an increasing function. If the value of $x$ increases, the value of $y$ increases.

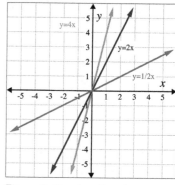

Figure 2. Graphs of $y = kx$.

In general, if we assume $y$ varies directly with $x$ with a positive variation constant $k$, then:

- If the value of $x$ increases, the value of $y$ increases.
- If the value of $x$ decreases, the value of $y$ decreases.

### Finding and Using a Direct Variation Equation

We're often asked to not only write a direct variation equation but to answer further questions using the equation. First we use the given data point to find the variation constant $k$ and write the equation. Then we use the equation to find the requested value of the variable.

### Example 2

The variable $p$ varies directly with $q$. If $p = 15$ when $q = 35$, find the value of $p$ when $q = 70$. What is the value of $q$ when $p = 54$?

#### Solution

Represent the direct variation equation as $p = kq$. Next, substitute $p = 15$ and $q = 35$ in the equation and solve for the variation constant $k$.

$p = kq$  Start with the general direct variation form.
$15 = k(35)$  Substitute $p = 15$ and $q = 35$.
$\frac{3}{7} = k$  Divide both sides by 35 and reduce the fraction: $\frac{15}{35} = \frac{3}{7}$

The variation constant is $k = \frac{3}{7} \approx 0.429$. To maintain accuracy, especially for the follow-up questions, we use the fraction, rather than the rounded value, to represent $k$ in our equation:

$p = \frac{3}{7}q$

Use the equation to find the value of $p$ when $q = 70$. Substitute $q = 70$ in the equation:

$p = \frac{3}{7}(70) = 30$

So $p = 30$ when $q = 70$.

Now use the equation again to find the value of $q$ when $p = 54$. Substitute $p = 54$ in the equation and solve for $q$:

$$54 = \frac{3}{7}q \quad \text{Substitute } p = 54 \text{ in the equation.}$$

$$\frac{7}{3} \cdot 54 = \frac{7}{3} \cdot \frac{3}{7}q \quad \text{Multiply both sides by the reciprocal of } \frac{3}{7} \text{ to solve for } q.$$

$$126 = q$$

So $q = 126$ when $p = 54$.

Variation models are quite common in the sciences, particularly in chemistry and physics. To find and use a direct variation model, follow these steps:

1. Write a general direct variation equation using the variables given.
2. Substitute the given values of the variables in the equation and solve for the constant of variation, $k$.
3. Use the constant of variation to write the specific direct variation model.
4. Use the equation from Step 3 to make estimates for requested values of the variable.

### Example 3

The weight of an object on Mars, $M$, varies directly with its weight on Earth, $E$. If an object weighs 100 lbs on Earth, then it only weighs 40 lbs on Mars. With that in mind, do the following:

a. *Curiosity* is a car-sized robotic rover exploring Gale Crater on Mars as part of NASA's Mars Science Laboratory mission. This rover weighs almost 2 tons, 4000 lbs., on Earth. How much does it weigh on Mars?

b. *Curiosity* has retrieved and analyzed many rock samples while exploring Mars. If a rock sample weighs 7.2 kg on Mars, how much does it weigh on Earth?

### Solution

a. A general direct variation model for this situation will have the form $M = kE$. Given $E = 100$ when $M = 40$, we can solve for $k$.

$$M = kE \quad \text{Start with the direct variation form.}$$
$$40 = k(100) \quad \text{Substitute } E = 100 \text{ and } M = 40.$$
$$0.4 = k \quad \text{Divide both sides by 100: } \frac{40}{100} = 0.4$$

The constant of variation is $k = 0.4$. This time we will use the decimal form of the constant since we do not have to round the value. So the direct variation model is $M = 0.4E$.

To find $M$ when $E = 4000$, we substitute in the equation: $M = 0.4(4000) = 3200$. The rover, *Curiosity*, weighs 3200 lbs on Mars.

**b.** Substitute $M = 7.2$ in the equation and solve for $E$.

$7.2 = 0.4E$      Substitute $M = 7.2$.

$18 = E$      Divide both sides by 0.4.

▸ A rock that weighs 7.2 kg on Mars will weigh 18 kg on Earth.

## Practice A

For each of the following problems, write a direct variation equation, and then find the requested variable values. When you are done, turn the page and check your solutions.

1. If $y$ varies directly with $x$, and $y = 6.75$ when $x = 3$, find $y$ when $x = 7$.
2. If $C$ varies directly with $n$, and $C = \frac{28}{3}$ when $n = 14$, find $n$ when $C = 18$.
3. The temperature of a certain gas, $T$, varies directly with its pressure, $p$. A temperature of 190 K produces a pressure of 50 pounds per square inch (psi). What is the temperature of the gas when the pressure is 65 psi? What pressure will the gas have at 290 K?

## B. Inverse Variation

Water temperature in an ocean varies inversely with the water's depth. The formula $T = \frac{14{,}000}{d}$ gives us the temperature in degrees Fahrenheit at a depth, $d$, of feet below the ocean's surface at a certain location in the Pacific Ocean. Below we have a table of some possible depths and observe that as the depth increases, the water temperature decreases.

| $d$, depth in feet | $T = \frac{14{,}000}{d}$ | Interpretation |
| --- | --- | --- |
| 500 | $\frac{14{,}000}{500} = 28$ | At a depth of 500 feet, the water temperature is 28° F. |
| 1000 | $\frac{14{,}000}{1000} = 14$ | At a depth of 1,000 feet, the water temperature is 14° F. |
| 2000 | $\frac{14{,}000}{2000} = 7$ | At a depth of 2,000 feet, the water temperature is 7° F. |

We notice in the relationship between these variables that as one quantity increases, the other decreases. The two quantities are said to be inversely proportional and each term varies inversely with the other. Inversely proportional relationships are also called **inverse variations**.

## Inverse Variation

For $k \neq 0$, if $x$ and $y$ are related by an equation of the form $y = \frac{k}{x}$, then we say $y$ **varies inversely** with $x$, or $y$ is **inversely proportional** to $x$. We call $k$ the **variation constant** or the **constant of proportionality**. The equation $y = \frac{k}{x}$ is called an **inverse variation equation**.

The formula we used to represent the temperature of the ocean, $T = \frac{14{,}000}{d}$, is an inverse variation equation with variation constant $k = 14{,}000$.

We say the water temperature varies inversely with the depth of the water because, as the depth increases, the temperature decreases. Figure 3 offers a graphical representation of this inverse variation.

Here are a few more examples of inverse variation equations:

1. For $y = \frac{5}{x}$, we say $y$ varies inversely with $x$ with variation constant $k = 5$.

2. For $p = \frac{0.22}{w}$, we say $p$ varies inversely with $w$ with variation constant $k = 0.22$.

3. For $P = \frac{7.2}{V}$, we say $P$ varies inversely with $V$ with variation constant $k = 7.2$.

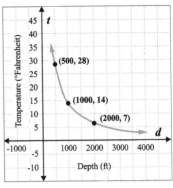

Figure 3.

### Finding and Using an Inverse Variation Equation

If one variable varies inversely with another variable and we know one data point, we can find the variation constant $k$. Once we know the value of $k$, we can write an inverse variation equation and use it to find requested values.

#### Example 4

If $y$ varies inversely with $x$, and if $y = 9$ when $x = 4$, find $y$ when $x = 7.2$.

**Solution**

The first part of the sentence gives the general relationship between $x$ and $y$. The statement "$y$ varies inversely with $x$" is equivalent to this equation:

$$y = \frac{k}{x}$$

The second part of the sentence gives a data point that we use to find the variation constant $k$. Substitute $x = 4$ and $y = 9$ into the equation and solve for $k$.

$y = \frac{k}{x}$    Start with the general inverse variation form.

$9 = \frac{k}{4}$    Substitute $x = 4$ and $y = 9$.

$36 = k$    Multiply both sides by 4.

The variation constant is $k = 36$. The specific inverse variation equation is $y = \frac{36}{x}$. To use the equation to find $y$ when $x = 7.2$, substitute $x = 7.2$ in the equation:

$$y = \frac{36}{7.2} = 5$$

So $y = 5$ when $x = 7.2$.

Notice in Example 4 that as the *x*-value *increased* from $x = 4$ in the first part of the example, to $x = 7.2$, the *y*-value *decreased* from $y = 9$ to $y = 5$ in the second part.

If we assume that *y* varies inversely with *x* with a positive variation constant *k*, then for positive values of *x*:

- If the value of *x* increases, the value of *y* decreases.
- If the value of *x* decreases, the value of *y* increases.

## Example 5

If *m* varies inversely with *b*, and if $m = 15$ when $b = 7$, find *b* when $m = 5.25$.

### Solution

The general inverse variation equation is $m = \frac{k}{b}$. Substitute $m = 15$ and $b = 7$ into the equation and solve for *k*.

$15 = \frac{k}{7}$   Substitute $m = 15$ and $b = 7$.

$105 = k$   Multiply both sides by 7.

The variation constant is $k = 105$. The specific inverse variation equation is $m = \frac{105}{b}$. To use the equation to find *b* when $m = 5.25$, substitute $m = 5.25$ in the equation and solve for *b*.

$5.25 = \frac{105}{b}$   Substitute $m = 5.25$.

$5.25b = 105$   Multiply both sides by *b*.

$b = 20$   Divide both sides by 5.25.

So $b = 20$ when $m = 5.25$.

Inverse variation models, like direct variation models, are common in many fields of science. To find and use an inverse variation model, follow these steps:

1. Write a general inverse variation equation using the variables given.
2. Substitute the given values of the variables in the equation and solve for the constant of variation, *k*.
3. Use the constant of variation to write the specific inverse variation model.
4. Use the equation from step 3 to make estimates for requested values of the variable.

### Practice A — Answers

1. $y = 2.25x, y = 15.75$
2. $C = \frac{2}{3}n, n = 27$
3. $T = 3.8p, T = 247$ K, and $p \approx 76.3$ psi

## Example 6

$V$, the volume of a gas varies inversely with $P$, the pressure that the gas exerts on its container. If the volume is 50 cubic feet when the pressure is 48 pounds per square inch (psi), what is the volume if the pressure increases to 80 psi? What pressure would correspond to a volume of 100 cubic feet?

### Solution

The general inverse variation equation is $V = \frac{k}{P}$. Substitute $V = 50$ and $P = 48$ in the equation and solve for $k$.

$\quad\quad 50 = \frac{k}{48}\quad$ Substitute $V = 50$ and $P = 48$.

$\quad 2400 = k\quad$ Multiply both sides by 48.

The variation constant is $k = 2400$. The specific inverse variation equation is $V = \frac{2400}{P}$. To find the volume when the pressure is 80 psi, substitute $P = 80$ in the equation:

$$V = \frac{2400}{80} = 30$$

The volume of the gas is 30 cubic feet when the pressure is 80 pounds per square inch.

To find the pressure when the volume is 100 cubic feet, substitute $V = 100$ in the equation and solve for $P$.

$\quad\quad 100 = \frac{2400}{P}\quad$ Substitute $V = 100$.

$\quad 100P = 2400\quad$ Multiply both sides by $P$.

$\quad\quad P = 24\quad$ Divide both sides by 100.

The pressure is 24 pounds per square inch when the volume is 100 cubic feet.

## Practice B

For each of the following problems, write an inverse variation equation, and then find the requested variable values. When you are done, turn the page and check your solutions.

4. If $y$ varies inversely with $x$, and $y = 24$ when $x = 3$, find $y$ when $x = 4$.
5. If $C$ varies inversely with $F$, and $C = 1.6$ when $F = 16$, find $F$ when $C = 2$.
6. $h$, the average heart rate in beats per minute (bpm) of most mammals varies inversely with $L$, the mammal's average life span in years. A lion has an average heart rate of 63 beats per minute and an average life span of 25 years. What is the average heart rate of an elephant if its life span is about 70 years? Estimate the life span of a mouse whose heart rate is 634 bpm.

## C. Direct and Inverse Variation of a Power

So far we've described functions in which the dependent variable varies directly or inversely with the independent variable. Some quantities vary directly or inversely with the square of another quantity, or with the cube of another quantity, or with the square root of another quantity, and so on. Here are some examples:

1. $y = kx^4$    $y$ varies directly with the fourth power of $x$.
2. $r = k\sqrt{A}$    $r$ varies directly with the square root of $A$.
3. $p = \dfrac{k}{q^3}$    $p$ varies inversely with the cube of $q$.

We use "varies directly" to mean the dependent variable is equal to a constant multiplied by a power of the independent variable. We use "varies inversely" to mean the dependent variable is equal to a constant divided by a power of the independent variable.

### Example 7

The quantity $y$ varies directly with the cube of $x$. If $y = 25$ when $x = 2$, find $y$ when $x$ is 6.

**Solution**

The general formula for direct variation with a cube is $y = kx^3$. Substitute $x = 2$ and $y = 25$, and solve for $k$.

$$y = kx^3 \quad \text{Start with the general formula.}$$
$$25 = k(2^3) \quad \text{Substitute } x = 2 \text{ and } y = 25.$$
$$25 = k(8) \quad \text{Simplify.}$$
$$3.125 = k \quad \text{Divide both sides by 8.}$$

The variation constant $k = 3.125$. The specific direct variation equation is $y = 3.125 x^3$. To find $y$ when $x = 6$, substitute $x = 6$ in the equation and solve for $y$:

$$y = 3.125(6)^3 = 675$$

So $y = 675$ when $x = 6$.

### Practice B — Answers

1. $y = \dfrac{72}{x}$; $y = 18$
2. $C = \dfrac{25.6}{F}$; $F = 12.8$
3. $h = \dfrac{1575}{L}$; $h = 23$ bpm; $L = 2.5$ years

### Example 8

$I$, the illumination provided by a car's headlight, varies inversely as the square of $d$, the distance from the headlight. A headlight produces 60 footcandles (fc) at a distance of 10 ft. With this in mind, do the following:

a. Write an equation that models the relationship between the illumination provided by the headlight and the distance from the headlight.

b. Use this model to create a table of values displaying the illumination in 5-ft increments, starting at 10 ft.

c. What will be the illumination at 40 ft?

**Solution**

a. The general formula for inverse variation with a square is $I = \frac{k}{d^2}$. Substitute $I = 60$ and $d = 10$ in the equation and solve for $k$.

$\quad I = \frac{k}{d^2}$    Start with the general formula.

$\quad 60 = \frac{k}{10^2}$    Substitute $I = 60$ and $d = 10$.

$\quad 6000 = k$    Multiply both sides by $10^2 = 100$.

The variation constant is $k = 6000$. So the following equation models this situation:

$$I = \frac{6000}{d^2}$$

b. We'll use a graphing calculator to create a table of values. Enter the equation and set the table to begin at 10 with the change in the independent variable equal to 5. See Figure 4b. We observe that as the distance from the headlight increases, the intensity of illumination decreases.

c. We use the graph or table to find that at a distance of 40 ft, the illumination is just 3.75 fc.

Figure 4a. Table setup.

Figure 4b. Table for $y = \frac{6000}{x^2}$.

## Exercises 5.1

For the following exercises, write an equation describing the relationship of the given variables.

1. $y$ varies directly with $x$ and when $x = 4$, $y = 80$.
2. $y$ varies directly with $x$ and when $x = 36$, $y = 12$.
3. $p$ varies directly with the cube of $q$ and when $q = 2$, $p = 24$.
4. $w$ varies directly with the square of $n$ and when $n = 5$, $w = 15$.
5. $V$ varies directly with the fourth power of $t$ and when $t = 2.5$, $V = 234.375$.
6. $p$ varies directly with the square root of $c$ and when $c = 36$, $p = 30$.
7. $y$ varies inversely with $x$ and when $x = 4$, $y = 2$.
8. $y$ varies inversely with $x$ and when $x = 3.7$, $y = 46.25$.
9. $h$ varies inversely with the square of $d$ and when $h = 8$, $d = 0.4$.
10. $C$ varies inversely with the cube of $L$ and when $L = 3$, $C = 2$.
11. $r$ varies inversely with the square root of $A$ and when $A = 25$, $r = 3$.
12. $Z$ varies inversely with the cube root of $n$ and when $n = 64$, $Z = 5$.

For the following exercises, use the given information to find the unknown value.

13. $y$ varies directly with $x$. When $x = 3$, then $y = 12$. Find $y$ when $x = 8.5$.
14. $p$ varies directly with the square of $x$. When $x = 10$, then $p = 25$. Find $p$ when $x = 6$.
15. $D$ varies directly with the cube of $t$. When $t = 2$, then $D = 40$. Find $t$ when $D = 135$.
16. $B$ varies directly with the square root of $q$. When $q = 16$, then $B = 28$. Find $q$ when $B = 700$.
17. $y$ varies inversely with $x$. When $x = 3$, then $y = 12$. Find $y$ when $x = 4.5$.
18. $M$ varies inversely with the square of $n$. When $n = 4$, then $M = 3$. Find $M$ when $n = 2$.
19. $W$ varies inversely with the cube of $t$. When $t = 5$, then $W = 2$. Find $t$ when $W = 31.25$.
20. $y$ varies inversely with the square root of $x$. When $x = 64$, then $y = 12$. Find $y$ when $x = 36$.

For the following exercises, assume the variation constant $k$ is a positive number.

21. The variable $Q$ varies directly with $b$. Describe what happens to the value of $Q$ as $b$ increases.
22. The variable $p$ varies directly with $t$. Describe what happens to the value of $p$ as $t$ decreases.
23. The variable $T$ varies inversely with $m$. For positive values of $m$, what happens to the value of $T$ as $m$ decreases?
24. The variable $g$ varies inversely with $h$. For positive values of $h$, what happens to the value of $g$ as $h$ increases?
25. The electrical resistance of a cable of specified length varies inversely with the square of the diameter of the cable. Will a thicker cable have more or less resistance than a narrow cable?
26. The electrical resistance of a wire of specified diameter varies directly with the length of the wire. Will a longer wire have more or less resistance than a short wire?

For the following exercises, use the given information to answer the questions.

27. $S$, the distance that an object falls, varies directly with the square of $t$, the time of the fall. If an object falls 16 feet in one second, how long will it take for it to fall 144 feet?

28. $V$, the velocity of a falling object, varies directly with $t$, the time of the fall. If after 2 seconds, the velocity of the object is 64 feet per second, what is the velocity after 5 seconds?

29. The rate of vibration of a string under constant tension varies inversely with the length of the string. If a string is 24 inches long and vibrates 128 times per second, what is the length of a string that vibrates 64 times per second?

30. The volume of a gas held at constant temperature varies inversely with the pressure of the gas. If the volume of a gas is 1200 cubic centimeters when the pressure is 200 millimeters of mercury, what is the volume when the pressure is 300 millimeters of mercury?

31. The weight of an object above the surface of Earth varies inversely with the square of the distance from the center of Earth. If an object weighs 50 pounds when it is 3960 miles from Earth's center, what would it weigh it were 3970 miles from Earth's center?

32. The intensity of light measured in foot-candles varies inversely with the square of the distance from the light source. Suppose the intensity of a light bulb is 0.08 foot-candles at a distance of 3 meters. Find the intensity level at 8 meters.

33. $C$, the electrical current in a circuit, varies inversely with $R$, its resistance measured in ohms. When the current in a circuit is 40 amperes, the resistance is 10 ohms. Find the current if the resistance is 12 ohms.

34. $F$, the force needed on a wrench handle to loosen a certain bolt, varies inversely with $L$, the length of the handle. A force of 35 lbs is needed when the handle is 6 inches long. If a person needs only 20 lbs of force to loosen the bolt, estimate the length of the wrench handle.

# 5.2 Arithmetic Sequences

## Overview

In this section, we examine patterns in lists of numbers. There is a *wide* variety of pattern types:

2, 4, 8, 16, 32, 64, 128      The numbers in the list double.
7, 13, 19, 25, 31, 37, 43     The numbers in the list increase by 6.
1, 4, 9, 16, 25, 36, 49       The numbers in the list are increasing perfect squares.

We will focus on the patterns in lists of numbers that have a correlation to linear functions. In this section, you'll learn to do the following:

- Know the meaning of sequence, term, and term number
- Identify arithmetic sequences
- Find a formula for the general term of an arithmetic sequence
- Find a term or term number of an arithmetic sequence
- Use an arithmetic sequence to make estimates and predictions

## A. Introduction to Sequences

A section in a movie theater has 10 rows. The first row has 15 seats and each other row has 4 more seats than the row in front of it. So the numbers of seats in the 10 rows are:

15, 19, 23, 27, 31, 35, 39, 43, 47, 51

We call this and any other list of numbers "a sequence."

A **finite sequence** is a sequence that has a last term, such as 51 in the example of the movie theater section. A sequence that does not have a last term is called an **infinite sequence**. For example, the list of positive even numbers is an infinite sequence:

2, 4, 6, 8, 10, …

We use the ellipsis, three periods in a row, to mean "and so forth." It indicates that we are omitting some numbers in the sequence. When we don't put a final term after the ellipsis, it means that the pattern of numbers continues infinitely.

We also use specific notation to represent the terms in a sequence. The variable $a$ with a number subscript represents a term in a sequence. The value of the subscript is a positive integer that indicates the position of the term in the sequence.

Consider the finite sequence from the movie theater rows above:

We write the first term of the sequence as $a_1$, so $a_1 = 15$.
We write the second term of the sequence as $a_2$, so $a_2 = 19$.
We write third term of the sequence as $a_3$, so $a_3 = 23$. And so on.

We use the variable $n$ to represent term numbers when working with sequences. Because we can't have a fractional term position, the values of $n$ are positive integers.

Consider the infinite sequence of positive even numbers:

$$2, 4, 6, 8, 10, \ldots$$

The numbers in this sequence can be found from a formula that defines a function, $f(n) = 2n$. We generate the terms of the sequence by finding $f(1) = 2, f(2) = 4, f(3) = 6$, and so on. The inputs are positive integers, and the outputs are the terms of the sequence. This justifies a formal definition of a sequence.

## Sequence

A **sequence** is a function whose domain is the set of positive integers: $n = 1, 2, 3, 4, \ldots$

The notation for sequences uses a subscript to indicate the value of the input variable, while the function notation uses parentheses:

Instead of $f(1)$, we write $a_1$ for the *first term* of a sequence.
Instead of $f(2)$, we write $a_2$ for the *second term* of a sequence.
Instead of $f(3)$, we write $a_3$ for the *third term* of a sequence.
Instead of $f(n)$, we write $a_n$ for the *nth term* of a sequence.

The term $a_n$ is called the ***n*th term of a sequence** or the **general term** of a sequence. The general term is used to define the other terms of a sequence. When we write a formula for $a_n$, it defines the $n$th term of a sequence using the position of the term. If we know the formula for the general term, $a_n$, we can find any other term in the sequence.

### ▶ Example 1

Write the first four terms and the 25th term of the sequence defined by the formula $a_n = -3n + 8$.

**Solution**

| | | |
|---|---|---|
| $n = 1$ | $a_1 = -3(1) + 8 = 5$ | Substitute $n = 1$ into the formula. |
| $n = 2$ | $a_2 = -3(2) + 8 = 2$ | Repeat with values 2 through 4 and then 25 for $n$. |
| $n = 3$ | $a_3 = -3(3) + 8 = -1$ | |
| $n = 4$ | $a_4 = -3(4) + 8 = -4$ | |
| ... | ... | |
| $n = 25$ | $a_{25} = -3(25) + 8 = -67$ | |

The first five terms are 5, 2, −1, −4, and −7. The 25th term is −67.

In the table below, we list the sequence values.

| $n$   | 1 | 2 | 3  | 4  | 5  |
|-------|---|---|----|----|----|
| $a_n$ | 5 | 2 | −1 | −4 | −7 |

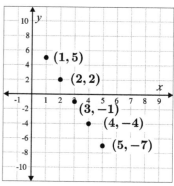

Figure 1.

Figure 1 is a graph of those table values.

We can see that this sequence has a linear relationship. However, notice that the points on the graph are not connected by a curve because the domain is the positive integers only.

### Example 2

Write the first five terms and the 50th term of the sequence defined by the formula $a_n = n^2 - 1$.

**Solution**

Substitute $n = 1$ into the formula. Repeat with values 2 through 5, and 50 for $n$.

$n = 1 \quad a_1 = (1)^2 - 1 = 0$

$n = 2 \quad a_2 = (2)^2 - 1 = 3$

$n = 3 \quad a_3 = (3)^2 - 1 = 8$

$n = 4 \quad a_4 = (4)^2 - 1 = 15$

$n = 5 \quad a_5 = (5)^2 - 1 = 24$

...     ...

$n = 50 \quad a_{50} = (50)^2 - 1 = 2499$

The first five terms are 0, 3, 8, 15, and 24. The 50th term is 2499.

### Practice A

Try the following problems. When you are done, turn the page and check your solutions.

1. Write the first five terms and the 30th term of the sequence defined by the formula $a_n = 5n - 4$.

2. Write the first five terms and the 99th term of the sequence defined by the formula $a_n = \dfrac{n}{n+1}$.

## B. Definition of Arithmetic Sequences

Now let's return to the sequence that represents the number of seats in successive rows in a movie theater:

15, 19, 23, 27, 31, 35, 39, 43, 47, 51

Notice that the difference between any term and the preceding term is 4:

19 − 15 = 4
23 − 19 = 4
27 − 23 = 4
...
51 − 47 = 4

With this sequence, we say that 4 is the common difference and that the sequence is arithmetic.

**Arithmetic sequences** are sequences whose terms change by a constant amount. Each term increases or decreases from the preceding term by the same constant value. This constant value is called the **common difference** of the sequence. Figure 2 shows a sequence with a common difference of −10.

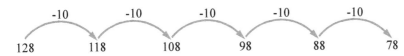

Figure 2.

Figure 3.

Figure 3 is another example of an arithmetic sequence. In this case, the common difference is 3. You can choose any term of the sequence and add 3 to find the next term.

### Arithmetic Sequence

An **arithmetic sequence** is a sequence where the difference between every pair of consecutive terms is a constant, $d$. The constant $d$ is called the **common difference**.

### Example 3

Determine whether the sequence is arithmetic. If it is, find the common difference $d$.

1. $1, 10, 19, 28, 37, \ldots$
2. $27, 22, 17, 12, 7, \ldots$
3. $2, 6, 18, 54, 162, \ldots$

**Solutions**

Subtract each term from the subsequent term to determine whether a common difference exists.

1. The sequence is arithmetic, because it has a common difference, $d = 9$:
   $10 - 1 = 9$
   $19 - 10 = 9$
   $28 - 19 = 9$
   $\ldots$

2. The sequence is arithmetic, because it has a common difference, $d = -5$.
   $72 - 77 = -5$
   $67 - 72 = -5$
   $62 - 67 = -5$
   $\ldots$

3. The sequence is *not* arithmetic. We can see from the first two differences that the sequence does not have a common difference: $6 - 2 = 4$, and $18 - 6 = 12$.

The graph of each sequence in Example 2 is shown in Figure 4. We can see from the graphs that Figures 4a and 4b are linear, while Figure 4c is *not* linear. Arithmetic sequences have a constant rate of change so their graphs will always be points on a line with slope $m$ equal to the common difference $d$.

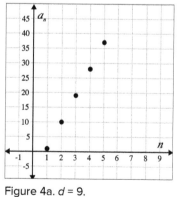
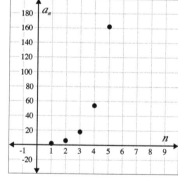

Figure 4a. $d = 9$.    Figure 4b. $d = -5$.    Figure 4c. No common difference.

### Practice A — Answers

1. The first five terms are 1, 6, 11, 16, 21. The 30th term is 146.
2. The first five terms are $\frac{1}{2}, \frac{2}{3}, \frac{3}{4}, \frac{4}{5}$, and $\frac{5}{6}$. The 99th term is $\frac{99}{100}$.

## Practice B

Determine whether the sequence is arithmetic. If it is, find $d$, the common difference. When you're done, turn the page to check your work.

3. 11, 8, 5, 2, –1, ...
4. 1, 3, 6, 10, 15, ...
5. –6, 14, 34, 54, 74, ...

## C. Formula for an Arithmetic Sequence

Now that we recognize an arithmetic sequence, we can find a formula for the sequence that gives the value of any term when the input is the term number. In general, we find the terms of an arithmetic sequence by beginning with the first term and adding the common difference repeatedly.

Let's return to the movie theater sequence:

15, 19, 23, 27, 31, 35, 39, 43, 47, 51

As you've seen, the common difference is 4. We add 4 to the first term, 15, to find the second term, 19. If we add 4 to the first term *two* times, we find the third term, 23. If we add 4 to the first term *three* times, we find the fourth term, 27. And so on.

For an arithmetic sequence with $d$, a common difference, the terms can also be stated like this:

$$a_1,\ a_1 + d,\ a_1 + d + d,\ a_1 + d + d + d, \ldots$$

We can simplify the general sequence like this:

$$a_1,\ a_1 + d,\ a_1 + 2d,\ a_1 + 3d, \ldots$$

We can use this pattern to find a formula that describes any term, $a_n$, of an arithmetic sequence:

$a_1 = a_1$

$a_2 = a_1 + d,$  Add $d$ once to the first term to get the second term.

$a_3 = a_1 + 2d,$  Add $d$ twice to the first term to get the third term.

$a_4 = a_1 + 3d$  Add $d$ three times to the first term to get the fourth term.

$a_n = a_1 + (n-1)d$  Add $d$ a total of $(n-1)$ times to the first term to get the $n$th term.

The $n$th term of an arithmetic sequence is equal to the first term plus $n-1$ times the common difference.

### Formula for an Arithmetic Sequence

A formula for the $n$th term of an arithmetic sequence with first term $a_1$ and common difference $d$ is given by $a_n = a_1 + (n-1)d$.

## Example 4

Find a formula for the sequence 15, 19, 23, 27, 31, 35, 39, 43, 47, 51.

### Solution

To find a formula for the sequence, we identify the first term, $a_1 = 15$, and the common difference $d = 4$. We substitute these values in the formula $a_n = a_1 + (n - 1)d$:

$a_n = 15 + (n - 1)(4)$   Substitute $a_1 = 15$ and $d = 4$.
$a_n = 15 + 4n - 4$   Apply the distributive property.
$a_n = 4n + 11$   Simplify.

The formula $a_n = 4n + 11$ describes the sequence. Notice that the form of the equation is linear. The variable is to the first power, and there is a constant rate of change ($m = 4$).

It's always a good idea to check your formula by verifying one or more terms in the sequence:

$a_4 = 4(4) + 11 = 27$, so the fourth term in the sequence is 27.
$a_{10} = 4(10) + 11 = 51$, so the tenth term in the sequence is 51.

The formula checks out.

## Example 5

Find the 30th term in the sequence 11, 5, –1, –7, –13, ...

### Solution

The sequence has a common difference $d = -6$, so the sequence is arithmetic. We use the common difference and the first term, $a_1 = 11$, in the formula $a_n = a_1 + (n - 1)d$ to write the formula for the sequence.

$a_n = 11 + (n - 1)(-6)$   Substitute $a_1 = 11$ and $d = -6$.
$a_n = 11 - 6n + 6$   Apply the distributive property.
$a_n = -6n + 17$   Simplify.

To find the 30th term, substitute $n = 30$ in the formula: $a_{30} = -6(30) + 17 = -163$. The 30th term of the sequence is –163.

### Practice B — Answers

3. The sequence is arithmetic. The common difference $d = -3$.

4. The sequence is not arithmetic because $3 - 1 \neq 6 - 3$.

5. The sequence is arithmetic. The common difference $d = 20$.

## Example 6

The number 88 is a term in the sequence 4, 7, 10, 13, ... What is its term number?

**Solution**

The sequence has a common difference $d = 3$, so the sequence is arithmetic. We use the common difference and the first term, $a_1 = 4$, in the formula $a_n = a_1 + (n-1)d$ to write the formula for the sequence.

| | |
|---|---|
| $a_n = 4 + (n-1)(3)$ | Substitute $a_1 = 4$ and $d = 3$. |
| $a_n = 4 + 3n - 3$ | Apply the distributive property. |
| $a_n = 3n + 1$ | Simplify. |

To find the term number $n$ for the term 88, substitute $a_n = 88$ in the formula and solve for $n$.

| | |
|---|---|
| $88 = 3n + 1$ | Substitute $a_n = 88$. |
| $87 = 3n$ | Subtract 1 from both sides. |
| $29 = n$ | Divide by 3. |

88 is the 29th term in the sequence.

Sometimes we're asked to determine whether a number is a term in a particular sequence. In this case, the process is the same as the one used in the last example. We write a formula for the sequence, set the formula equal to the number in question, and solve for $n$. If a number is a term in a sequence, then by definition, $n$ must be a positive whole number. If we discover through solving that $n$ is not a positive integer, then we know the number in question is *not* a term in the sequence.

## Example 7

Is the number 296 a term in the sequence 12, 19, 26, 33, 40, ... ?

**Solution**

The sequence has a common difference, $d = 7$, so the sequence is arithmetic. We use the common difference and the first term, $a_1 = 12$, in the formula $a_n = a_1 + (n-1)d$ to write the formula for the sequence.

| | |
|---|---|
| $a_n = 12 + (n-1)(7)$ | Substitute $a_1 = 12$ and $d = 7$. |
| $a_n = 12 + 7n - 7$ | Apply the distributive property. |
| $a_n = 7n + 5$ | Simplify. |

To determine whether the number 296 is a term in this sequence, substitute $a_n = 296$ in the formula and solve for $n$.

| | |
|---|---|
| $296 = 7n + 5$ | Substitute $a_n = 296$. |
| $291 = 7n$ | Subtract 5 from both sides. |
| $41.5714 \approx n$ | Divide by 7. |

The variable $n$ must be a positive whole number, so we know 296 is *not* a term in the sequence.

## Practice C

Now it's your turn. When you are done, turn the page and check your solutions.

6. Find a formula for the sequence 27, 20, 13, 6, −1, …

7. Find the 25th term of the sequence −50, −10, 30, 70, 110, …

8. Find the term number of the last term in the sequence 2, 11, 20, 29, 38, …, 281.

9. Is the number 367 a term in the sequence 19, 22, 25, 28, 31, …?

## D. Modeling with an Arithmetic Sequence

Some real-world situations involve a constant rate of change and an independent variable limited to positive integers. We can model these situations with arithmetic sequences. The starting value in problem is the first term of the sequence. The constant rate of change is the value of the common difference of the sequence.

As you will see in the next example, it can be tricky to align the first term of the sequence with the starting value of the problem. We have to be careful as we define the variable $n$ so that the formula gives the initial value when $n = 1$.

### Example 8

A five-year-old child begins receiving an allowance of $1 each week. Her parents promise her an annual increase of $2.50 per week.

  a. Write a formula for the child's weekly allowance.
  b. What will her allowance be when she is 13 years old?
  c. When will the child receive an allowance of $36 per week?

### Solution

We can model the situation with an arithmetic sequence with a first term of 1 and a common difference of 2.5.

a. Let $a_n$ be the amount of the allowance, and let $n$ be the number of years after age 4. Notice that we define $n$ so that $n = 1$ represents the girl at age 5. Now we use the formula for an arithmetic sequence to develop the allowance formula.

$a_n = 1 + (n - 1)(2.5)$   Substitute $a_1 = 1$ and $d = 2.5$.

$a_n = 1 + 2.5n - 2.5$   Apply the distributive property.

$a_n = 2.5n - 1.5$   Simplify.

The allowance formula is $a_n = 2.5n - 1.5$.

b. To find the child's allowance when she is 13 years old, we use $n = 13 - 4 = 9$. This will be the ninth year she receives an allowance from her parents:

$$a_n = 2.5(9) - 1.5 = 21$$

The child's allowance at age 13 will be $21.

c. To determine when the child will receive an allowance of $36 per week, we substitute $a_n = 36$ in the formula and solve for $n$.

| | |
|---|---|
| $36 = 2.5n - 1.5$ | Substitute $a_n = 36$. |
| $37.5 = 2.5n$ | Add 1.5 to both sides. |
| $15 = n$ | Divide both sides by 2.5 |

Recall that $n$ represents the number of years after age 4, so the child will be 19 years old — and no longer a child, actually — when she receives an allowance of $36 per week.

### Example 9

Wooden planks are stacked in layers in a lumber yard so that each layer has 3 less planks than the previous layer. The bottom layer has 52 planks.

a. Write a formula for the number of planks in the $n$th layer of the stack.

b. If the top layer of the stack has 10 planks, how many layers are there in the stack?

**Solution**

a. The situation can be modeled with an arithmetic sequence where the first term is $a_1 = 52$ and the common difference is $-3$.

| | |
|---|---|
| $a_n = 52 + (n-1)(-3)$ | Substitute $a_1 = 52$ and $d = -3$. |
| $a_n = 52 - 3n + 3$ | Apply the distributive property. |
| $a_n = -3n + 55$ | Simplify. |

The formula is $a_n = -3n + 55$.

b. The top layer has 10 planks, so we substitute $a_n = 10$ in the formula and solve for $n$.

| | |
|---|---|
| $10 = -3n + 55$ | Substitute $a_n = 10$. |
| $-45 = -3n$ | Subtract 55 from both sides. |
| $15 = n$ | Divide both sides by $-3$. |

The stack of planks has 15 layers.

## Exercises 5.2

1. What is an arithmetic sequence?
2. How is the common difference of an arithmetic sequence found?
3. How do we determine whether a sequence is arithmetic?
4. Describe how linear functions and arithmetic sequences are similar. How are they different?

In the following exercises, write the first five terms of the sequence. Determine whether or not the sequence is arithmetic. If it is, find the common difference.

5. $a_n = 5 + 12n$
6. $a_n = 8n - 3$
7. $a_n = \frac{1}{n+1}$
8. $a_n = 3(2)^n$
9. $a_n = -2 + (n - 1)15$
10. $a_n = 5(n + 2)$
11. $a_n = 100 - n^2$
12. $a_n = n^3$

In the following exercises, determine whether the sequence is arithmetic. If so, find the common difference $d$.

13. $0, \frac{1}{2}, 1, \frac{3}{2}, 2, \ldots$
14. $11.4, 9.3, 7.2, 5.1, 3, \ldots$
15. $4, 16, 64, 256, 1024, \ldots$
16. $3, 15, 75, 375, 1875, \ldots$
17. $55, 34, 13, -8, -29, \ldots$
18. $-31, -18, -5, 8, 21, \ldots$
19. $11.4, 9.5, 7.6, 5.7, 3.8, \ldots$
20. $\frac{7}{6}, \frac{6}{5}, \frac{5}{4}, \frac{4}{3}, \frac{3}{2},$

In the following exercises, write a formula for the arithmetic sequence. Then find the 25th term of the sequence.

21. $8, 13, 18, 23, 28, \ldots$
22. $2, 9, 16, 23, 30, \ldots$
23. $27, 20, 13, 6, -1, \ldots$
24. $358, 347, 336, 325, 314, \ldots$
25. $-12, -37, -62, -87, -112, \ldots$
26. $-7, -23, -39, -55, -71, \ldots$
27. $23.5, 25.25, 27, 28.75, 30.5, \ldots$
28. $14.3, 17.1, 19.9, 22.7, 25.5, \ldots$

In the following exercises, find the term number $n$ of the last term of the finite sequence.

29. $7, 12, 17, 22, 27, \ldots, 102$
30. $2, 9, 16, 23, 30, \ldots, 184$
31. $4, 10, 16, 22, 28, \ldots, 1426$
32. $5, 13, 21, 29, 37, \ldots, 533$
33. $-27, -39, -51, -63, -75, \ldots, -999$
34. $-3, -12, -21, -30, -39, \ldots, -390$
35. $675, 650, 625, 600, 575, \ldots -175$
36. $508, 476, 444, 412, 380, \ldots, -772$

### Practice C — Answers

6. $a_n = -7n + 34$
7. $a_n = 40n - 90;\ a_{25} = 910$
8. $a_n = 9n - 7;\ n = 32$ so 281 is the 32nd term in the sequence.
9. Yes, $n = 117$, a whole number, so 367 is the 117th term in the sequence.

With the following exercises, use the information provided to answer the questions.

37. Is 172 a term in the sequence 2, 9, 16, 23, 30, …? Explain.
38. Is 1888 a term in the sequence 5, 14, 23, 32, 41, …? Explain.
39. An auditorium has rows of seating that increase in length the further the row is from the stage. The first row has 21 seats, the second row has 24 seats, the third row has 27 seats, the fourth row has 30 seats, and so on.
   a. Let $a_n$ be the number of seats in the $n$th row. Write a formula to model this situation.
   b. Find $a_{12}$ and interpret this value.
   c. If the last row has 138 seats, how many rows are in the auditorium?
40. A marching band makes a triangle formation during a half-time show at a football game. In this formation there are 3 band members in the first row, 5 members in the second row, 7 members in the third row, 9 members in the fourth row, and so on.
   a. Let $a_n$ be the number of band members in the $n$th row. Write a formula to model this situation.
   b. Find $a_7$ and interpret this value.
   c. If there are 49 band members in the last row of the marching formation, how many rows are there?
41. A welder's starting salary is $32,500. Each year she receives a $950 raise.
   a. Let $a_n$ be the welder's salary in the $n$th year. Write a formula to model this situation.
   b. Find $a_5$ and interpret this value.
   c. In what year will her salary first exceed $50,000?
42. A teacher's starting salary is $38,000. Each year he receives a $1100 raise.
   a. Let $a_n$ be the teacher's salary in the $n$th year. Write a formula to model this situation.
   b. Find $a_6$ and interpret this value.
   c. In what year will his salary be $60,000?
43. A team of 8 people are conducting an experiment at a research station in Antarctica. The team arrived at the station with 209 daily food units. Each team member uses one food unit per day.
   a. Let $a_n$ be the number of food units remaining after the $n$th day. Write a formula to model this situation.
   b. Find $a_{14}$ and interpret this value.
   c. How many days will the food supply last?
44. Marco has an expense account to use on his world travels. He begins with $19,000. Every Friday, he withdraws $700 to help with his travel expenses.
   a. Let $a_n$ be the balance of the expense account after $n$ weeks. Write a formula to model this situation.
   b. Find $a_{15}$ and interpret this value.
   c. How many weeks will the money last?

# 5.3 Geometric Sequences

## Overview

Many jobs offer an annual cost-of-living increase to keep salaries consistent with inflation. Jamie, for example, is a recent college graduate and sales manager who earns an annual salary of $26,000. She's been promised a 2% cost of living increase each year.

We can find her annual salary in any given year by multiplying her salary from the previous year by 102% (1.02). Her salary will be $26,520 after one year; $27,050.40 after two years; $27,591.41 after three years, and so on. When a salary increases by a constant rate each year, the salary grows by a constant factor.

In this section, we will study sequences that grow in this way. You will learn how to:

- Identify a geometric sequence
- Find a formula for the general term of a geometric sequence
- Find a term or term number of a geometric sequence
- Use a geometric sequence to make estimates and predictions

## A. Definition of Geometric Sequence

In Section 5.2, we worked with arithmetic sequences whose terms increased or decreased by a constant value called the common difference. The common difference was *added* to a term to find the next term. In a geometric sequence, each term increases or decreases by a constant factor called the common ratio. The common ratio is *multiplied* by a term to find the next term. Consider this sequence:

$$1, 6, 36, 216, 1296, \ldots$$

Notice that the ratio of any term to its preceding term is 6:

$$\frac{6}{1} = 6, \quad \frac{36}{6} = 6, \quad \frac{216}{36} = 6, \quad \frac{1296}{6} = 6, \ldots$$

We call 6 the common ratio, and we call the sequence a geometric sequence. Multiplying any term of the sequence by the common ratio 6 generates the next term, as you see in Figure 1.

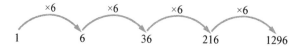

Figure 1. Geometric sequence with common ratio 6.

## Geometric Sequence

A **geometric sequence** is one in which any term divided by the previous term is a constant $r$. This constant $r$ is called the **common ratio** of the sequence.

We can determine if a set of numbers represents a geometric sequence by following these steps:

1. Divide each term by the previous term.
2. Compare the quotients. If they're the same, a common ratio exists, and the sequence is geometric.

### Example 1

Determine whether the sequence is geometric. If so, find the common ratio $r$.

1. $7, 21, 63, 189, 567, \ldots$
2. $1, 2, 6, 24, 120, \ldots$
3. $3072, 768, 192, 48, 12, \ldots$

**Solutions**

Divide each term by the previous term to determine whether a common ratio exists.

1. The sequence is geometric because it has a common ratio of 3:

   $\frac{21}{7} = 3, \quad \frac{63}{21} = 3, \quad \frac{189}{63} = 3, \quad \frac{567}{189} = 3$

2. The sequence is not geometric because there is not a common ratio.

   $\frac{2}{1} = 2, \quad \frac{6}{2} = 3, \quad \frac{24}{6} = 4$

3. The sequence is geometric because it has a common ratio of $\frac{1}{4}$:

   $\frac{768}{3072} = \frac{1}{4}, \quad \frac{192}{768} = \frac{1}{4}, \quad \frac{48}{192} = \frac{1}{4}, \quad \frac{48}{12} = \frac{1}{4}$

For the third sequence in Example 1, we can write the common ratio in decimal form:

$r = \frac{1}{4} = 0.25$

In fraction form, the common ratio reminds us that multiplying a term by $\frac{1}{4}$ gives the same result as dividing a term by 4.

## Practice A

Determine whether the sequence is geometric. If so, find $r$, the common ratio. When you're done, turn the page and check your solutions.

1. $3, 15, 75, 375, 1875, \ldots$
2. $2.4, 6, 15, 37.5, 93.75, \ldots$
3. $3, 9.6, 30.72, 70.656, 162.5088, \ldots$
4. $720, 360, 180, 90, 45, \ldots$

## B. Formula for a Geometric Sequence

Now that we can identify a geometric sequence, we can find the terms of a geometric sequence by beginning with the first term and multiplying by the common ratio repeatedly. For example, if the first term of a geometric sequence is $a_1 = 2$ and the common ratio is $r = 4$, we find the second term by multiplying $2 \cdot 4$ to get 8, we find the third term by multiplying $2 \cdot 4 \cdot 4$ to get 32, and so on.

$$a_1 = 2$$
$$a_2 = 2 \cdot 4 = 8$$
$$a_3 = 2 \cdot 4 \cdot 4 = 32$$
$$a_4 = 2 \cdot 4 \cdot 4 \cdot 4 = 128$$

The first four terms are 2, 8, 32, 128.

In general, if $a_1$ is the first term of a geometric sequence and $r$ is the common ratio, the sequence looks like this:

$$a_1,\ a_1 \cdot r,\ a_1 \cdot r \cdot r,\ a_1 \cdot r \cdot r \cdot r, \ldots$$

Using exponents, we have:

$$a_1,\ a_1 r,\ a_1 r^2,\ a_1 r^3, \ldots$$

We use the pattern to find a formula that describes the general term $a_n$ of a geometric sequence.

$a_1 = a_1$
$a_2 = a_1 r$ — Multiply $a_1$ by $r$ once to get the second term.
$a_3 = a_1 r^2$ — Multiply $a_1$ by $r$ twice to get the third term.
$a_4 = a_1 r^3$ — Multiply $a_1$ by $r$ three times to get the fourth term.
$a_n = a_1 r^{n-1}$ — Multiply $a_1$ by $r$ a total of $(n-1)$ times to get the $n$th term.

### Formula for a Geometric Sequence

The $n$th term of a geometric sequence is given by the formula:

$$a_n = a_1 r^{n-1}$$

where $a_1$ is the first term, and $r$ is the common ratio of the sequence.

### Practice A — Answers

1. geometric, $r = 5$
2. geometric, $r = 2.5$
3. not geometric
4. geometric, $r = \frac{1}{2}$

# Example 2

Find a formula that describes the terms of the sequence.

1. $2, 6, 18, 54, 162, \ldots$
2. $32, 16, 8, 4, 2, \ldots$

## Solutions

1. The sequence has a common ratio of $r = 3$, so the sequence is geometric. Substitute $a_1 = 2$ and $r = 3$ into the formula for geometric sequences, $a_n = a_1 r^{n-1}$:

$$a_n = 2(3)^{n-1}$$

We can verify the formula by entering $y = 2(3)^{x-1}$ in a graphing calculator and checking with the table feature that the first five terms are 2, 6, 18, 54, and 162. See Figure 2.

2. The sequence has a common ratio of $r = \frac{1}{2}$, so the sequence is geometric. Substitute $a_1 = 32$ and $r = \frac{1}{2}$ into the formula $a_n = a_1 r^{n-1}$:

$$a_n = 32\left(\frac{1}{2}\right)^{n-1}$$

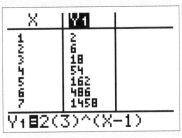

Figure 2. Verify the formula.

Notice that an explicit formula for a geometric sequence is in a form that is similar to the form of the exponential functions that we studied in Chapter 2. For both forms, the base is a constant number and the variable is in the exponent:

$$a_n = a_1 r^{n-1} \quad \text{and} \quad y = ab^x$$

One difference is we subtract 1 from the input in the geometric sequence formula to ensure the initial value is the first term of the sequence corresponding to $n = 1$. The exponential function formula has an initial value, $a$, that corresponds to $x = 0$.

The other main difference between the two forms is that the domain of the formula for geometric sequences is limited to positive integers: $n = 1, 2, 3, 4, \ldots$

The first formula from Example 2, $a_n = 2(3)^{n-1}$, describes an exponential function whose only inputs are positive integers and whose outputs are the terms of the sequence. The points $(1, 2)$, $(2, 6)$, $(3, 18)$, $(4, 54)$, and $(5, 162)$ are on the graph in Figure 3a on the next page. Because the base is 3, which is greater than 1, the graph has the shape of an exponential growth curve. We don't connect the dots because we can't have fractional term numbers.

The second formula in Example 2, $a_n = 32\left(\frac{1}{2}\right)^{n-1}$, describes an exponential function with positive integer inputs. The points (1, 32), (2, 16), (3, 8), (4, 4), and (5, 2) are on the graph in Figure 3b. Because the base is $\frac{1}{2}$, which is less than 1, the graph has the shape of an exponential decay curve. Again, we do not connect the dots.

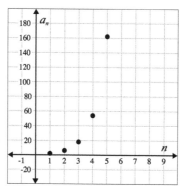

Figure 3a. The first five terms of the sequence $a_n = 2(3)^{n-1}$.

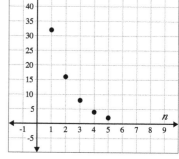

Figure 3b. The first five terms of the sequence $a_n = 32\left(\frac{1}{2}\right)^{n-1}$.

## Practice B

Write a formula for each geometric sequence. When you're done, turn the page and check your solutions.

5. $\frac{1}{8}, \frac{1}{4}, \frac{1}{2}, 1, 2, \ldots$
6. $2000, 400, 80, 16, 3.2, \ldots$
7. $1, 3, 9, 27, 81, \ldots$
8. $90, 9, 0.9, 0.09, 0.009, \ldots$

## C. Finding a Term or a Term Number of a Geometric Sequence

For a geometric sequence, we can use the formula $a_n = a_1(r)^{n-1}$ to find a term of the sequence. We input the term number for $n$ in the formula and compute the value of the term. We can also use the formula to solve for a term number, $n$, when we know a term in the sequence.

### Example 3

1. Find the 15th term of the sequence $5, 10, 20, 40, 80, \ldots$
2. Find the 12th term of the sequence $15625, 3125, 625, 125, 5, \ldots$

Solution

1. The sequence has a common ratio, $r = 2$, so the sequence is geometric. The first term is $a_1 = 5$. We substitute these values in the formula for geometric sequences to obtain $a_n = 5(2)^{n-1}$. To find the 15th term of the sequence, we substitute $n = 15$ in the formula:

   $a_{15} = 5(2)^{15-1} = 5(2)^{14} = 81{,}920$

   To verify our work, we enter $y = 5(2)^{x-1}$ in a graphing calculator and check that the first 5 terms are 5, 10, 20, 40, and 80. Then we check that the 15th term is 81,920. See Figures 4a and 4b.

2. The sequence has a common ratio, $r = \frac{1}{5}$, so it's geometric. The first term is $a_1 = 15{,}625$. We substitute these values in the formula for the sequence: $a_n = 15{,}625 \left(\frac{1}{5}\right)^{n-1}$.

To find the 12th term of the sequence, we substitute $n = 12$ in the formula:

$$a_{12} = 15{,}625 \left(\frac{1}{5}\right)^{12-1} = 15{,}625 \left(\frac{1}{5}\right)^{11} = 3.2 \times 10^{-4}$$

The 12th term is expressed in scientific notation. See Figure 4c for a calculator display of this value. In decimal form the 12th term of this sequence is 0.00032.

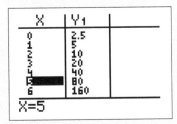 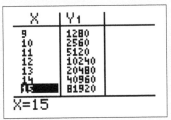 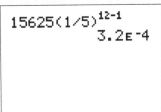

Figures 4a. and 4b. Verify the work.

Figure 4c. Finding the 12th term of the sequence.

In the next example, we are given a term in a geometric sequence and asked to find its term number. That requires us to solve an exponential equation for the variable $n$. We use the properties of logarithms we studied in Chapter 3 to help us solve for $n$.

## Example 4

The number 885,735 is a term in the sequence 5, 15, 45, 135, 405, .... What is its term number?

### Solution

The sequence has a common ratio, $r = 3$, so the sequence is geometric. We note that the first term is $a_1 = 5$ and write the formula for the sequence $a_n = 5\,(3)^{n-1}$.

| | | |
|---|---|---|
| $885735 = 5\,(3)^{n-1}$ | Substitute $a_n = 885{,}735$. |
| $177{,}147 = 3^{n-1}$ | Divide both sides by 5. |
| $\log(177{,}147) = \log(3^{n-1})$ | Take the logarithm of both sides. |
| $\log(177{,}147) = (n-1)\log(3)$ | Apply the power rule for logarithms. |
| $\dfrac{\log(177{,}147)}{\log(3)} = n - 1$ | Divide both sides by $\log(3)$. |
| $\dfrac{\log(177{,}147)}{\log(3)} + 1 = n$ | Add 1 to both sides. |
| $12 = n$ | Compute. |

885,735 is the 12th term of the sequence. We can use a graphing calculator to verify our work.

## Practice C

Try the following problems and then when you are done, turn the page and check your solutions.

9. Find the 9th term of the sequence 1536, 384, 96, 24, ...

10. Find the 11th term of the sequence 5, 30, 180, 1080, 6480, ...

11. The number 32 is a term in the sequence 19531.25, 7812.5, 3125, 1250, ... What's its term number?

## D. Modeling with a Geometric Sequence

Some real world situations involve quantities that increase or decrease by a constant factor and an independent variable limited to positive integers. These situations can be modeled with a geometric sequence. The starting value in the problem is the first term of the sequence. The constant factor is the value of the common ratio of the sequence.

### Example 5

A pendulum swings 15 feet left to right on its first swing. On each swing following the first, the pendulum swings $\frac{4}{5}$ of the distance of the previous swing. Let $a_n$ represent the distance the pendulum travels on the $n$th swing.

a. Find a formula that describes $a_n$.

b. Find $a_5$ and interpret this value.

c. How many swings until the distance the pendulum travels is less than 1 foot?

**Solution**

a. The distance of any swing of the pendulum is multiplied by a factor of $\frac{4}{5}$ to find the distance of the next swing. So $a_n$ is a geometric sequence with a common ratio $r = \frac{4}{5}$ and first term $a_1 = 15$. We substitute these values in the formula $a_n = a_1 r^{n-1}$:

$$a_n = 15\left(\frac{4}{5}\right)^{n-1}$$

b. To find $a_5$, we substitute $n = 5$ in the formula:

$$a_5 = 15\left(\frac{4}{5}\right)^{5-1} = 15\left(\frac{4}{5}\right)^4 = 6.144$$

The pendulum travels 6.144 feet on the fifth swing.

### Practice B — Answers

5. $a_n = \frac{1}{8}(2)^{n-1}$

6. $a_n = 2000\left(\frac{1}{5}\right)^{n-1}$ or $a_n = 2000(0.2)^{n-1}$

7. $a_n = (3)^{n-1}$

8. $a_n = 90\left(\frac{1}{10}\right)^{n-1}$ or $a_n = 90(0.1)^{n-1}$

c. To find when the distance is less than 1 foot, we let $a_n = 1$ in the formula and solve for $n$.

$$1 = 15\left(\frac{4}{5}\right)^{n-1} \qquad \text{Substitute } a_n = 1.$$

$$\frac{1}{15} = \left(\frac{4}{5}\right)^{n-1} \qquad \text{Divide both sides by 15.}$$

$$\log\left(\frac{1}{15}\right) = \log\left(\left(\frac{4}{5}\right)^{n-1}\right) \qquad \text{Take the logarithm of both sides.}$$

$$\log\left(\frac{1}{15}\right) = (n-1)\log\left(\frac{4}{5}\right) \qquad \text{Apply the power rule for logarithms.}$$

$$\frac{\log\left(\frac{1}{15}\right)}{\log\left(\frac{4}{5}\right)} = n - 1 \qquad \text{Divide both sides by } \log\left(\frac{4}{5}\right).$$

$$\frac{\log\left(\frac{1}{15}\right)}{\log\left(\frac{4}{5}\right)} + 1 = n \qquad \text{Add 1 to both sides.}$$

$$13.1359 \approx n \qquad \text{Compute.}$$

In this situation, $n$ is the number of swings, so we round our answer to the least integer greater than 13.1359. Then $n = 14$ represents 14 swings of the pendulum.

To verify our work, we use the formula $a_n = 15\left(\frac{4}{5}\right)^{n-1}$ to evaluate $a_{13}$ and $a_{14}$.

$a_{13} = 15\left(\frac{4}{5}\right)^{13-1} \approx 1.03$  On the 13th swing, the pendulum travels about 1.03 feet.

$a_{14} = 15\left(\frac{4}{5}\right)^{14-1} \approx 0.82$  On the 14th swing, the pendulum travels about 0.82 feet.

This verifies that on the 14th swing, the pendulum travels less than 1 foot.

## Example 6

Sarah has a new job with a large company, and her employer offers her two choices for structuring her yearly salary:

- **Plan A:** A starting salary of $30,000 with a 3.2% raise in her salary each year.
- **Plan B:** A starting salary of $34,000 with an $800 raise in her salary each year.

Let $a_n$ be Sarah's salary in dollars in the $n$th year. For both salary plans, complete the following:

a. Find a formula that describes $a_n$.
b. Predict her salary, to the nearest whole dollar, in the 5th year and in the 10th year.
c. If she stays with the same company for 32 years, what will her salary be the year she retires?
d. What advice could you give Sarah regarding which plan to choose?

### Solution

**a.** Start by finding a formula that describes $a_n$ for both plans.

*Plan A:* The sequence is geometric with first term $a_1 = 30{,}000$ and common ratio $r = 1 + 0.032 = 1.032$. We substitute these values in the formula $a_n = a_1 r^{n-1}$:

$$a_n = 30000(1.032)^{n-1}$$

| X | Y1 | Y2 |
|---|---|---|
| 26 | 65935 | 54000 |
| 27 | 68045 | 54800 |
| 28 | 70222 | 55600 |
| 29 | 72469 | 56400 |
| 30 | 74788 | 57200 |
| 31 | 77181 | 58000 |
| 32 | 79651 | 58800 |

X=32

Figure 5.

*Plan B:* The sequence is arithmetic with first term $a_1 = 34{,}000$ and common difference $d = 800$. We substitute these values in the formula $a_n = a_1 + (n - 1)d$:

$a_n = 34{,}000 + (n - 1)800$, which is equivalent to $a_n = 800n + 33{,}200$

| X | Y1 | Y2 |
|---|---|---|
| 6 | 35117 | 38000 |
| 7 | 36241 | 38800 |
| 8 | 37401 | 39600 |
| 9 | 38597 | 40400 |
| 10 | 39833 | 41200 |
| 11 | 41107 | 42000 |
| 12 | 42423 | 42800 |

X=12

Figure 6.

**b.** Enter the formulas in a graphing calculator. Use a table to find $a_5$ and $a_{10}$:

$$Y_1 = 30000(1.032)^{x-1}$$
$$Y_2 = 800x + 33{,}200$$

*Plan A:* As Figure 5 shows, in the 5th year, Sarah's salary is \$34,028. In the 10th year, it's \$39,833.

*Plan B:* In the 5th year, her salary is \$37,200. In the 10th year, it's \$41,200.

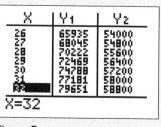

| X | Y1 | Y2 |
|---|---|---|
| 8 | 37401 | 39600 |
| 9 | 38597 | 40400 |
| 10 | 39833 | 41200 |
| 11 | 41107 | 42000 |
| 12 | 42423 | 42800 |
| 13 | 43780 | 43600 |
| 14 | 45181 | 44400 |

X=12

Figure 7.

**c.** To find Sarah's salary in the 32nd year, use the calculator table to find $a_{32}$ for both plans. See Figure 6.

*Plan A:* In the 32nd year, Sarah's salary is \$79,651.

*Plan B:* In the 32nd year, her salary is \$58,800.

**d.** We notice from part b that Sarah's salary will be higher in the 5th year and the 10th year if she chooses Plan B. From part c we see that her salary in the 32nd year is quite a bit higher with Plan A. As we scroll through the table on the calculator, we can see that Plan B has higher yearly salaries than Plan A until the 12th year. After that, Plan A always has higher salaries. See Figure 7.

We advise Sarah to choose Plan B if she only plans to stay with the company for 12 years or less. We advise her to choose Plan A if she plans to stay with the company for the rest of her working life.

## Exercises 5.3

1. What is a geometric sequence?
2. How is the common ratio of a geometric sequence found?
3. Describe how exponential functions and geometric sequences are similar. How are they different?
4. What is the difference between an arithmetic sequence and a geometric sequence?

For the following exercises, determine whether the sequence is arithmetic, geometric, or neither. If the sequence is arithmetic, write the common difference. If the sequence is geometric, write the common ratio.

5. 1, 3, 9, 27, 81, …
6. 14, 27, 40, 53, 66, …
7. 200, 180, 160, 140, 120, …
8. 2, 10, 60, 420, 3360, …
9. 6, 8, 11, 15, 20, …
10. 0.8, 4, 20, 100, 500, …
11. $1, \frac{1}{2}, \frac{1}{4}, \frac{1}{8}, \frac{1}{16}, \ldots$
12. 5, 5.2, 5.4, 5.6, 5.8, …
13. $a_n = 9n - 5$
14. $a_n = 75(0.4)^{n-1}$
15. $a_n = n^2 + 4$
16. $a_n = n^3 - 1$

For the following exercises, write a formula for each geometric sequence.

17. 7, 28, 112, 448, 1792, …
18. 0.6, 6, 60, 600, 6000, …
19. $1, \frac{4}{5}, \frac{16}{25}, \frac{64}{125}, \ldots$
20. $2, \frac{1}{3}, \frac{1}{18}, \frac{1}{108}, \ldots$
21. 768, 192, 48, 12, 3,…
22. 1250, 250, 50, 10, 2,…
23. 10, 70, 490, 3430, 24010, …
24. 1, 8, 64, 512, 4096, …

For the following exercises, find the indicated term of the geometric sequence. If needed, write the result in scientific notation $N \times 10^k$, with $N$ rounded to three decimal places.

25. Find the 16th term of the sequence 6, 12, 24, 48, …
26. Find the 13th term of the sequence 2, 6, 18, 54, …
27. Find the 15th term of the sequence 1280, 320, 80, 20, …
28. Find the 14th term of the sequence 5625, 1125, 225, 45, …
29. Find the 25th term of the sequence 3.2, 9.6, 28.8, 86.4, …
30. Find the 24th term of the sequence 1.4, 3.5, 8.75, 21.875, …

For the following exercises, the given number is a term in the geometric sequence that follows. Find the term number of that term. Use a graphing calculator table to verify your result.

31. 25,165,824; 6, 24, 96, 384, …
32. 177147; 1, 3, 9, 27, …
33. 2,470,629; 3, 21, 147, 1029, …
34. 109,375; 0.00224, 0.0112, 0.056, 0.28, …
35. 0.01953125; 640, 320, 80, 5, …
36. 0.46875; 80, 240, 120, 60, …
37. 768; 0.046875, 0.09375, 0.1875, 0.375, …
38. 28,697,814; 2, 6, 18, 54, …

### Practice C — Answers

9. 0.0234375
10. 302,330,880
11. $n = 8$

For the following exercises, use the given information to answer the questions.

39. A ball is dropped from a great height, and on the first bounce it reaches a height of 20 feet. Each time it bounces, it returns to $\frac{7}{8}$ of the height of the preceding bounce. Let $a_n$ represent the maximum height of the ball on the $n$th bounce.

   a. Find a formula that describes $a_n$.

   b. What is the maximum height of the ball on the 9th bounce?

40. A ball is dropped from Tim's third floor balcony, and on the first bounce it reaches a height of 8 meters. Each time it bounces, it returns to $\frac{4}{5}$ of the height of the preceding bounce. Let $a_n$ represent the maximum height of the ball on the $n$th bounce.

   a. Find a formula that describes $a_n$.

   b. What is the maximum height of the ball on the 9th bounce?

   c. Will Tim get in trouble for this?

41. You go to work for a company that pays $0.01 the first day, $0.02 the second day, $0.04 the third day, and so on. If the daily wage keeps doubling, to the nearest dollar what will your income be on the 30th day?

42. You go to work for a company that pays $5 the first week, $15 the second week, $45 the third week, and so on. If the weekly pay keeps tripling, what will your pay be for the 12th week?

43. With each cycle, a vacuum pump removes one third of the air in a sealed vessel, and two thirds remains. After one cycle, the sealed glass vessel has a volume of 500 cubic centimeters. Let $a_n$ represent the volume of air remaining in the vessel after $n$ cycles of the vacuum pump.

   a. Find a formula that describes $a_n$.

   b. Find $a_4$ and interpret this value.

   c. How many cycles of the vacuum pump are needed before there is less than 5 cubic centimeters of air in the vessel?

44. A tire swing attached to a tree branch swings with an arc length of 18 feet on its first swing. On each swing following the first, the tire's arc length is $\frac{3}{4}$ of the length of the previous swing. Let $a_n$ represent the arc length on the $n$th swing.

   a. Find a formula that describes $a_n$.

   b. Find $a_6$ and interpret this value.

   c. How many swings does it take before the arc length is less than 1 foot?

45. Aaron and Beatriz both just graduated from college and received job offers. Aaron was offered a starting salary of $40,000 with a 3.75% raise per year. Beatriz was offered a starting salary of $45,000 with a $1000 raise per year. Let $a_n$ be the salary in dollars in the $n$th year. For both Aaron and Beatriz complete the following:

   a. Find a formula that describes $a_n$.

   b. Predict the salary, to the nearest whole dollar, in the 4th year and in the 12th year.

   c. If they each stay with the same company for 30 years, what will their salary be?

   d. In what year does Aaron's salary first become greater than Beatriz's?

# 5.4 Dimensional Analysis

## Overview

In our daily lives we sometimes have the need to convert one unit of measure to another unit of measure. For example, we may wish to convert from pounds to kilograms or from miles to meters. In the sciences, the need to convert units of measure occurs quite often. A chemist may change the density of blood plasma from units of grams per kiloliter to units of kilograms per liter. An engineer may need to convert of a measure of pressure in pounds per square inch to Newtons per square meter. The method of dimensional analysis is a reliable way to calculate these types of conversions.

In this section you will learn how to:

- Cancel units of measure
- Use dimensional analysis to convert single units of measure
- Use dimensional analysis to convert mixed units of measure

## A. Canceling Units of Measure

### Canceling Numerals in Fractions

In mathematics when we talk about *canceling* with respect to fractions, we mean that we divide out like factors. For example, a fraction such as $\frac{6}{14}$ can be reduced to a simpler form by factoring the numerator and denominator and then canceling the factors that are the same.

$$\frac{6}{14} = \frac{2 \cdot 3}{2 \cdot 7} = \frac{\cancel{2} \cdot 3}{\cancel{2} \cdot 7} = \frac{3}{7}$$

We cancel the 2s because 2 divided by 2 equals 1, and 1 times $\frac{3}{7}$ will equal $\frac{3}{7}$.

$$\frac{2}{2} = 1 \quad \text{and} \quad 1 \cdot \frac{3}{7} = \frac{3}{7}$$

In fact, any number divided by itself equals 1. And remember, multiplying or dividing by 1 does not change the value of a number.

$$\frac{a}{a} = 1 \quad 1 \cdot a = a \quad \frac{a}{1} = a$$

We use these facts every time we simplify a fraction or perform dimensional analysis.

We can cancel factors in a fraction whether or not these factors are lined up vertically. As long as the factor occurs in both the numerator and the denominator of the fraction, we can cancel it:

$$\frac{66}{105} = \frac{2 \cdot 3 \cdot 11}{3 \cdot 5 \cdot 7} = \frac{2 \cdot \cancel{3} \cdot 11}{\cancel{3} \cdot 5 \cdot 7} = \frac{22}{35}$$

In fact, we can cancel factors that are in *different* fractions, as long as the factor occurs in both a numerator and a denominator, as below. This process is called *cross-canceling* because we cancel factors across a multiplication symbol:

$$\frac{3}{13} \cdot \frac{7}{2} \cdot \frac{13}{3} \cdot \frac{2}{11} = \frac{3}{\cancel{13}} \cdot \frac{7}{\cancel{2}} \cdot \frac{\cancel{13}}{\cancel{3}} \cdot \frac{\cancel{2}}{11} = \frac{7}{11}$$

## Canceling Units of Measure

Fractions may contain words, often units of measure, which give additional meaning to the fraction in a real word context. For example, $\frac{5 \text{ in}}{10 \text{ sec}}$ can describe the rate at which a small bug crawls: 5 inches in 10 seconds. We can reduce this fraction by canceling factors of 5.

$$\frac{5 \text{ in}}{10 \text{ sec}} = \frac{5 \cdot 1 \text{ in}}{5 \cdot 2 \text{ sec}} = \frac{\cancel{5} \cdot 1 \text{ in}}{\cancel{5} \cdot 2 \text{ sec}} = \frac{1 \text{ in}}{2 \text{ sec}}$$

The simplified fraction tells us that the bug crawls at a rate of $\frac{1}{2}$ inch per second.

A different example of a fraction containing words is $\frac{5 \text{ in}}{10 \text{ in}}$. This fraction might represent the ratio between the length of a small wrench and the length of a larger wrench. In this case, we can not only cancel a factor of 5 from the numerical part of the fraction, but we can also cancel the units that are the same.

$$\frac{5 \text{ in}}{10 \text{ in}} = \frac{5 \cdot 1 \text{ in}}{5 \cdot 2 \text{ in}} = \frac{\cancel{5} \cdot 1 \cancel{\text{ in}}}{\cancel{5} \cdot 2 \cancel{\text{ in}}} = \frac{1}{2}$$

The result means that the smaller wrench is one half as long as the larger wrench. Or put another way, the ratio between their lengths is 1 to 2.

As with purely numerical fractions, we can cancel like units of measure when we are multiplying fractions, as long as the same unit of measure occurs in both the numerator and denominator. The following equation shows how we can cancel units of measure to determine the number of minutes in 450 seconds. We multiply the first fraction by a fraction that represents the equivalency of 60 seconds = 1 minute.

$$\frac{450 \cancel{\text{ sec}}}{1} \cdot \frac{1 \text{ min}}{60 \cancel{\text{ sec}}} = \frac{450}{60} \text{ min} = 7.5 \text{ min}$$

There are 7.5 minutes in 450 seconds. Notice that we did not cancel factors in the numerical part of the fraction this time. Rather, we used a calculator to divide and simplify $\frac{450}{60}$.

## Example 1

Convert 198 inches to feet.

### Solution

We set up the given number of inches as a fraction and multiply by a fraction that represents the equivalency 12 inches = 1 foot. Then we simplify by canceling like units.

$$\frac{198 \text{ in}}{1} \cdot \frac{1 \text{ ft}}{12 \text{ in}} = \frac{198 \cancel{\text{ in}}}{1} \cdot \frac{1 \text{ ft}}{12 \cancel{\text{ in}}} = \frac{198}{12} \text{ ft} = 16.5 \text{ ft}$$

There are 16.5 feet in 198 inches.

## B. Dimensional Analysis with Single Units

**Dimensional analysis** is the procedure by which we multiply a given unit of measure by one or more conversion fractions in such a way that the units cancel, leaving us with the desired units in our answer. This procedure can help you avoid common mistakes that are made when calculating conversions.

In Example 1, we used the equivalency of 12 inches = 1 foot to make a conversion fraction. A **conversion fraction** is a fraction in which the value of the numerator and denominator are equal although they have different units. Any conversion equation can be written as a conversion fraction in two ways. Below are the two ways we can write the conversion fraction that relates inches to feet.

$$\frac{1 \text{ ft}}{12 \text{ in}} \quad \text{or} \quad \frac{12 \text{ in}}{1 \text{ ft}}$$

Here are two more examples of how to write conversion equations as conversion fractions. The equations used are found in the conversion table at the end of this section.

The equation 1 ton = 2000 lb can be written as $\frac{1 \text{ ton}}{2000 \text{ lbs}}$ or $\frac{2000 \text{ lbs}}{1 \text{ ton}}$.

The equation 1 in = 2.54 cm can be written as $\frac{1 \text{ in}}{2.54 \text{ cm}}$ or $\frac{2.54 \text{ cm}}{1 \text{ in}}$.

It's important to know how to choose which form of a conversion fraction to use. We must choose the form that allows us to cross-cancel units to achieve the desired result. However, you're not expected to memorize all of the conversion equations. You just need to know how to find and then use them to perform conversions.

### Example 2

20,000 minutes is equivalent to how many days?

**Solution**

We write the given information as a fraction over 1. We then use the conversion equations:

60 minutes = 1 hour and 24 hours = 1 day

We change the conversion equations to conversion fractions so that the units of minutes and hours occur once each in both the numerator and denominator.

$$\frac{20000 \text{ min}}{1} \cdot \frac{1 \text{ hr}}{60 \text{ min}} \cdot \frac{1 \text{ day}}{24 \text{ hr}}$$

When we cancel like units, we see we are left with days, which is the desired measure.

$$\frac{20000 \text{ min}}{1} \cdot \frac{1 \text{ hr}}{60 \text{ min}} \cdot \frac{1 \text{ day}}{24 \text{ hr}} = \frac{20000}{(60 \cdot 24)} \text{ days} = 13.89 \text{ days}$$

Notice the parentheses around the factors in the denominator in the last step of Example 2. It's important to include these on a calculator in order to obtain the correct value. We don't need to include 1 as a factor for either multiplying or dividing because this won't change the result.

### Example 3

Use the information below to convert units of area.

1. A real estate developer has a plot of land that is 2.34 acres in size. How many square feet ($ft^2$) is this?

2. A carpet-layer is organizing a project that requires 476 square yards ($yd^2$) of carpet. How many square yards is this?

#### Solutions

**1.** We can use the conversion equation: 1 acre = 43,560 $ft^2$. Notice the measurement in feet is already in square units, so we cancel "acres" to leave the desired units.

$$\frac{2.34 \text{ acre}}{1} \cdot \frac{43,560 \text{ ft}^2}{1 \text{ acre}} = \frac{2.34 \; \cancel{\text{acre}}}{1} \cdot \frac{43,560 \text{ ft}^2}{1 \; \cancel{\text{acre}}} = 101,930.4 \text{ ft}^2$$

2.24 acres is equivalent to about 101,930 square feet.

**2.** We can use the conversion equation: 1 yd = 3 ft. Notice these measurements are not in square units, so we need to use the conversion fraction twice to cancel square yards and to leave the desired units of square feet.

$$\frac{476 \text{ yd}^2}{1} \cdot \frac{3 \text{ ft}}{1 \text{ yd}} \cdot \frac{3 \text{ ft}}{1 \text{ yd}} = \frac{476 \; \cancel{\text{yd}^2}}{1} \cdot \frac{3 \text{ ft}}{1 \; \cancel{\text{yd}}} \cdot \frac{3 \text{ ft}}{1 \; \cancel{\text{yd}}} = 4284 \text{ ft}^2$$

We can also square the conversion fraction itself because this is mathematically equivalent to multiplying by the same factor twice. If we work the problem this way, it's important to remember to square all parts of the fraction.

$$\frac{476 \text{ yd}^2}{1} \cdot \left(\frac{3 \text{ ft}}{1 \text{ yd}}\right)^2 = \frac{476 \; \cancel{\text{yd}^2}}{1} \cdot \frac{3^2 \text{ ft}^2}{1^2 \; \cancel{\text{yd}^2}} = 476 \cdot 3^2 \text{ ft}^2 = 4284 \text{ ft}^2$$

▶ 476 square yards of carpet is equivalent to 4284 square feet.

### Example 4

A restaurant is planning to create a concrete patio that is 16 ft × 20 ft × ½ ft. The volume of concrete needed is 160 cubic feet. How many cubic yards is this?

#### Solution

Once again, we use the conversion equation: 1 yd = 3 ft. This time we need to use the conversion fraction three times to cancel cubic feet ($ft^3$) and to leave the desired units, cubic yards ($yd^3$). We can also *cube* the conversion fraction and then all of its factors to simplify the process.

$$\frac{160 \text{ ft}^3}{1} \cdot \left(\frac{1 \text{ yd}}{3 \text{ ft}}\right)^3 = \frac{160 \; \cancel{\text{ft}^3}}{1} \cdot \frac{1^3 \text{ yd}^3}{3^3 \; \cancel{\text{ft}^3}} = \frac{160}{27} \text{ yd}^3 = 5.93 \text{ yd}^3$$

▶ The restaurant will need about 6 cubic yards of concrete to build the patio.

## Practice B

Use dimensional analysis to convert units in the following problems. Remember, you'll find a table of conversion equations near the end of this section. If an answer isn't exact, round to two decimal places. When you're finished, turn the page to check your work.

1. How many minutes are in 4.5 days?
2. Convert 188 kilometers to miles.
3. How cubic feet are in 200 gallons?
4. Convert 1.3 square feet to square centimeters.

## C. Dimensional Analysis with Mixed Units

Many units of measure are combinations of two different units that are either multiplied or divided. We refer to these as mixed units. If the units are divided, they represent either a rate of change or a ratio between units. These rates and ratios can be written in fraction form and used in the process of dimensional analysis. Here are some examples:

The speed limit in a school zone is 20 miles per hour. We write: $\frac{20 \text{ mi}}{1 \text{ hr}}$

Water flow from a fire hose is 350 gallons per minute. We write: $\frac{350 \text{ gal}}{1 \text{ min}}$

The gas mileage of a new car is 38 miles per gallon. We write: $\frac{38 \text{ mi}}{1 \text{ gal}}$

A bicycle tire is inflated to 60 pounds per square inch. We write: $\frac{60 \text{ lb}}{1 \text{ in}^2}$

We can use the method of dimensional analysis to convert from one mixed unit to another mixed unit. To perform a conversion on a mixed unit expressed as a ratio, we often need to multiply by two or more conversion fractions to cancel units in both the numerator and the denominator. Because we can cross-cancel like units, the order in which we use the conversion fractions is not important. We may eliminate the units in either the numerator or the denominator first.

### Example 5

Many stretches of highway have a speed limit of 65 miles per hour. Convert 65 mph to ft/sec (feet per second).

**Solution**

We begin by writing the given rate as a fraction.

$$\frac{65 \text{ mi}}{1 \text{ hr}}$$

It doesn't matter which unit we try to eliminate first, but we'll start with miles in the numerator.

From the conversion chart near the end of this section, we use the equation 1 mi = 5280 ft and write it as a fraction so that we can cross-cancel the miles units.

$$\frac{65 \text{ mi}}{1 \text{ hr}} \cdot \frac{5280 \text{ ft}}{1 \text{ mi}}$$

Next we multiply by the two conversion fractions that will cancel hours and yield seconds.

$$\frac{65 \text{ mi}}{1 \text{ hr}} \cdot \frac{5280 \text{ ft}}{1 \text{ mi}} \cdot \frac{1 \text{ hr}}{60 \text{ min}} \cdot \frac{1 \text{ min}}{60 \text{ sec}}$$

Finally, we cancel like units and use our calculator to find the result.

$$\frac{65 \cancel{\text{mi}}}{1 \cancel{\text{hr}}} \cdot \frac{5280 \text{ ft}}{1 \cancel{\text{mi}}} \cdot \frac{1 \cancel{\text{hr}}}{60 \cancel{\text{min}}} \cdot \frac{1 \cancel{\text{min}}}{60 \text{ sec}} = \frac{65 \cdot 5280 \text{ ft}}{(60 \cdot 60) \text{ sec}} = 95.33 \text{ ft/sec}$$

▸ A speed of 65 miles per hour is equivalent to about 95.33 feet per second.

Notice the use of parentheses in the denominator in the last step of Example 5. We need to use parentheses on a calculator if we want a correct result in one calculation. It's not necessary to put parentheses around the factors in the numerator, but it's essential to put them around the factors in the denominator.

▸ **Example 6**

The water flow from a certain fire hydrant is 500 gallons per minute. Find the equivalent rate in liters per second.

**Solution**

We write the given information as a fraction and check the conversion table for an equation that relates gallons with liters. Look in the section of the table that converts between U.S. customary units and the metric system (SI). Notice there isn't an equation that converts gallons to liters, but we can convert quarts to liters. So we will first need to convert gallons to quarts.

**Practice B — Answers**

1. $\frac{4.5 \cancel{\text{day}}}{1} \cdot \frac{24 \cancel{\text{hr}}}{1 \cancel{\text{day}}} \cdot \frac{60 \text{ min}}{1 \cancel{\text{hr}}} = 4.5(24)(60) \text{ min} = 6480 \text{ min}$

2. $\frac{188 \cancel{\text{km}}}{1} \cdot \frac{1 \text{ mi}}{1.609 \cancel{\text{km}}} = \frac{188}{1.609} \text{ mi} = 116.84 \text{ mi}$

3. $\frac{200 \cancel{\text{gal}}}{1} \cdot \frac{1 \text{ ft}^3}{7.48 \cancel{\text{gal}}} = \frac{200}{7.48} \text{ ft}^3 = 26.7 \text{ ft}^3$ We do *not* need to cube the numbers in the conversion fraction because the units in the numerator are already cubic feet, as desired.

4. $\frac{1.3 \cancel{\text{ft}^2}}{1} \cdot \frac{12^2 \cancel{\text{in}^2}}{1^2 \cancel{\text{ft}^2}} \cdot \frac{2.54^2 \text{ cm}^2}{1^2 \cancel{\text{in}^2}} = 1.3(12^2)(2.54)^2 \text{ cm}^2 = 1207.7 \text{ cm}^2$ We *do* need to square the numbers and the units in *both* conversion fractions so that we can cancel the units given and convert to square centimeters.

$$\frac{500 \text{ gal}}{1 \text{ min}} \cdot \frac{4 \text{ qt}}{1 \text{ gal}} \cdot \frac{1 \text{ L}}{1.057 \text{ qt}}$$

When we multiply these three fractions together, we're left with liters per minute. So next we multiply by a fraction that converts minutes to seconds. Now we can cancel like units and calculate the result.

$$\frac{500 \text{ gal}}{1 \text{ min}} \cdot \frac{4 \text{ qt}}{1 \text{ gal}} \cdot \frac{1 \text{ L}}{1.057 \text{ qt}} \cdot \frac{1 \text{ min}}{60 \text{ sec}} = \frac{500 \cdot 4 \text{ L}}{(1.057 \cdot 60) \text{ sec}} = 31.54 \text{ L/sec}$$

A flow of 500 gallons per minute is equivalent to about 31.54 liters per second.

## Example 7

The density of gold is 19.32 g/cm³ (grams per cubic centimeter). Convert this to lb/in³ (pounds per cubic inch).

### Solution

A conversion fraction is needed to change from grams to pounds and another to change from cubic centimeters to cubic inches. For the latter, since the units are cubed, we will need to cube all of the components of this conversion fraction.

$$\frac{19.32 \text{ g}}{1 \text{ cm}^3} \cdot \frac{1 \text{ lb}}{453.6 \text{ g}} \cdot \left(\frac{2.54 \text{ cm}}{1 \text{ in}}\right)^3$$

Make sure to apply the exponent to all of the numbers and units in the fraction.

$$\frac{19.32 \text{ g}}{1 \text{ cm}^3} \cdot \frac{1 \text{ lb}}{453.6 \text{ g}} \cdot \frac{2.54^3 \text{ cm}^3}{1^3 \text{ in}^3} = \frac{19.32 \cdot 2.54^3 \text{ lb}}{453.6 \text{ in}^3} = 0.70 \text{ lb/in}^3$$

The density of gold is about 0.70 pounds per cubic inch.

Some mixed units are the product of two or more units. In physics, for example, torque and work are measured in foot-pounds (ft-lb) or in Newton-meters (N-m). Sometimes production is measured in man-hours, and sometimes electricity usage is measured in kilowatt-hours. In each of these mixed units, two different units are *multiplied* together. We write this type of mixed unit with a hyphen to indicate that the units are multiplied rather than divided. Notice the word "per" is not used for mixed units that are multiplied.

## Example 8

A project manager for a production team estimates that a certain job will take 580 man-hours. If each person on the 6-person team works 7.5 hours per day, how many work days will it take to complete the project?

### Solution

If we begin with man-hours and know that we have a 6-person team (men or women), we must divide by 6 to determine the number of hours each will work.

$$\frac{580 \text{ man-hours}}{1} \cdot \frac{1}{6 \text{ man}}$$

Next we use the conversion equation, 7.5 hours = 1 day of work, to represent the fact that each team member works 7.5 hours per day. We use this to eliminate hours and obtain the number of days.

$$\frac{580 \text{ man-hr}}{1} \cdot \frac{1}{6 \text{ man}} \cdot \frac{1 \text{ day}}{7.5 \text{ hr}} = \frac{580 \text{ day}}{(6 \cdot 7.5)} = 12.89 \text{ days}$$

It will take the team about 13 days to complete the project.

The method of dimensional analysis can be used to solve many types of problems that involve several conversions steps. The conversion can be done with one equation if we multiple by the correct conversion fractions leaving the desired unit.

## Example 9

The national debt in 2017 was approximately 19.9 trillion dollars ($19,900,000,000,000). If a person could count out $100 per second, without ever taking a break, how many years would it take to count out the national debt?

### Solution

Begin with the given value of the national debt and multiply by a conversion fraction that represents the rate of counting. Write the second fraction to cancel the dollars and obtain the number of seconds it would take to count out the debt.

$$\frac{19,900,000,000,000 \text{ dollars}}{1} \cdot \frac{1 \text{ sec}}{100 \text{ dollars}} = 199,000,000,000 \text{ sec}$$

Now convert seconds into years.

$$\frac{199,000,000,000 \text{ sec}}{1} \cdot \frac{1 \text{ min}}{60 \text{ sec}} \cdot \frac{1 \text{ hr}}{60 \text{ min}} \cdot \frac{1 \text{ day}}{24 \text{ hr}} \cdot \frac{1 \text{ yr}}{365 \text{ day}} = \frac{199,000,000,000 \text{ yr}}{(60 \cdot 60 \cdot 24 \cdot 365)} = 6310.25 \text{ yrs}$$

At the rate of $100 per second, it would take 6310.25 years to count out the national debt.

## Measurement and Conversion Table

### U. S. Customary System
1 yd = 3 ft
1 ft = 12 in
1 mi = 5,280 ft
1 acre = 43,560 ft²
1 lb = 16 oz (dry weight)
1 ton = 2000 lb

16 T = 1 cup
1 cup = 8 fluid oz
1 qt = 4 cups
1 gal = 4 qt
1 gal ≈ 231 in³
1 ft³ ≈ 7.48 gal

### International System of Units (SI) — Metric System
1 m = 1,000,000 microns (μ)
1 m = 1000 mm
1 m = 100 cm
1 km = 1000 m
1 cm = 10 mm
1 hectare (ha) = 10,000 m²

1 kg = 1000 g
1 g = 1000 mg
1 kL = 1000 L
1 L = 1000 mL
1 cm³ = 1 mL
1 m³ = 1000 L

### U. S. Customary and International System Conversions
1 in ≈ 2.54 cm
1 yd ≈ 0.914 m
1 m ≈ 39.37 in
1 m ≈ 3.281 ft
1 mi ≈ 1.609 km
1 lb. ≈ 4.448 N (Newtons)

1 lb ≈ 453.6 g
1 oz ≈ 28.35 g
1 kg ≈ 2.205 lb
1 L ≈ 1.057 qt
1 fluid oz ≈ 29.574 mL
1 ft³ ≈ 28.32 L

## Exercises 5.4

Use dimensional analysis to perform the following single unit conversions. Show the procedure including all of the conversion fractions. Round solutions to two decimal places if needed.

1. 145 oz to lbs
2. 138 in to ft
3. 1.2 mi to in
4. 5 gallons to cups
5. 856 cm to m
6. 9325 mg to g
7. 7.9 L to cm³
8. 3.75 km to cm
9. 450 cm to in
10. 230 in to m
11. 5.9 kg to oz
12. 12 km to ft
13. 200,000 ft² to acres
14. 150 gal to ft³
15. 1542 ft² to yd²
16. 9660 in² to ft²
17. 64,000 cm² to m²
18. 923,000 mm² to cm²
19. 486 in³ to cm³
20. 18 yd³ to m³

Use dimensional analysis to perform the following mixed unit conversions. Show the procedure including all of the conversion fractions. Round solutions to two decimal places if needed. Note: a/b means "a per b"

21. 83.4 m/sec to km/hr
22. 750 ft/min to mi/hr
23. 69 kg/m² to lbs/yd²
24. 522 g/cm² to oz/in²
25. 35 ft/min to cm/sec
26. 134 ft/sec to m/min
27. 256 g/cm³ to lb/in³
28. 9.6 kg/m³ to lb/yd³

29. 28 mi/gal to km/L
30. 36.5 km/L to mi/gal
31. Henry earns $110 per day. Calculate his wage in cents per minute.
32. Hannah earns 4.4 cents per minute. Calculate her wage in dollars per hour.
33. The water flow through a portion of a river dam is 7550 gallons per minute. Calculate the flow rate in cubic feet per second.
34. A public swimming pool is filled at a rate of 57 gallons per minute. Calculate the rate in cubic inches per second.
35. An iceberg is floating south at 8590 cm per hour. Calculate its speed in meters per minute.
36. A motorcycle is traveling 76 km per hour. Calculate its speed in meters per minute.
37. An average human heart pumps blood at a rate of 85 cm3/sec. Calculate this rate in liters per minute.
38. A swimmer's heart is pumping blood at a rate of 6.5 L/min. Calculate this rate in cm3/sec.
39. Water weighs about 8.3 lbs per gallon. Calculate the weight of water in kg per liter.
40. A certain liquid weighs 94 kg per liter. Calculate its weight in lbs per quart.

Use dimensional analysis to solve the following problems. Show the procedure including all of the conversion fractions. Round solutions to two decimal places if needed.

41. Sebastian plans to rent a car when he travels in Europe during his summer break from college. He wants a sense of what some common U.S. speed limits are in km/hr. What will he find for 20 mph, 45 mph, and 55 mph?
42. Sebastian is traveling in Europe and needs to buy gas for his rental car. He sees a sign at a gas station showing the cost is 1.53 euro per liter. If 1 U.S. dollar = .92 euro, what is the cost of the gas in dollars per gallon?
43. In 1983, Carl Lewis was the first man to run a 100 meter race at low altitude in less than 10 seconds. If someone runs 100 meters in 10 seconds, how fast is this in miles per hour?
44. One evening in April 2011, pitcher Aroldis Chapman of the Cincinnati Reds received a standing ovation for a pitch that was a bit high. Why the standing ovation? The stadium radar reading displayed a record breaking velocity of 106 mph. How fast is this in feet per second?
45. While the density of water varies slightly with its temperature, the density of water is very close to one gram per cubic centimeter. Use this fact to estimate the weight in pounds of a gallon of water.
46. The density of gasoline averages about 0.74 grams per cubic centimeter. Convert the density of gasoline to pounds per gallon.
47. A lug nut on the wheel of a particular European car is to be tightened to a torque of 170 N-m (Newton-meters). Convert this torque to ft-lbs (foot-pounds).
48. A tractor pulling a fallen tree off a road does an estimated 34,000 foot-pounds of work. Convert the work done to Newton-meters.
49. Marie commutes from home to work an average of 15,000 miles per year. She spends about $800 per year on gas and estimates that she burns 287 gallons. Estimate what Marie pays for gas in dollars per gallon. What is the mileage (miles per gallon) of her new fuel-efficient car? If Marie commutes to work 180 days per year, what is her daily commute in miles?
50. Suppose a roof is leaking, and during a rainstorm water drips from the ceiling at a rate of 1.5 tablespoons per minute. The homeowner puts a 2 gallon bucket under the leak and goes to bed, worrying that if the storm continues the bucket might overflow. If the homeowner sleeps for six and a half hours, will the bucket overflow?

# SOLUTIONS

# Odd-Numbered Exercises

## Chapter 1: Graphs and Linear Functions

### 1.1 Qualitative Graphs

1. Figure 8d.
3. Figure 8a.
5. Figure 9b.
7. Figure 9c.
9. $p$ is independent, and $t$ is dependent. The vertical and horizontal intercepts are at the origin because if the book has zero pages, it takes no time to read.

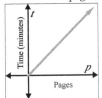

11. $h$ is independent, and $T$ is dependent. The vertical intercept is above zero because the soup starts out hot. There's no horizontal intercept because the temperature of the soup will decrease to room temperature, not 0°C.

13. $d$ is independent, and $h$ is dependent. The vertical and horizontal intercepts are at the origin because when the seed is planted on day zero, the height is 0 cm.

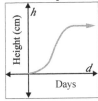

15. The vertical intercept is above zero because Rodrigo began with gas in his car. There is no horizontal intercept unless Roderigo runs out of gas.

17. There is a vertical and horizontal intercept at the origin because before the plane takes off, the altitude is zero feet. There is a second horizontal intercept that represents the altitude is zero feet again when the plane lands in L.A.

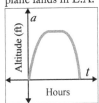

19. The vertical intercept assumes that the volume of the gas is greater than zero cm³ at the beginning. Because the volume is not likely to decrease to zero, there is no horizontal intercept.

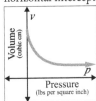

21.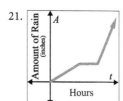

23.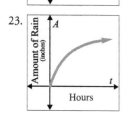

25.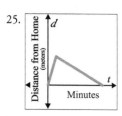

27.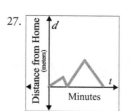

29. The scenarios will vary, but in general, one should describe a dependent variable $y$ that increases as the independent variable $x$ increases. More specifically, the rate of increase slows mid-graph (or mid-scenario) and picks up again after that.

31. The scenarios will vary, but in general, one should describe a dependent variable $y$ that increases then decreases as the independent variable $x$ increases. An example might be the height of a toy rocket, represented by $y$, over a time interval represented by $x$. The rocket is launched, goes up, and then falls to the ground.

## 1.2 Functions

1. A relation is a set of ordered pairs. A function is a special kind of relation in which each input only yields one output.

3. When a vertical line intersects the graph of a relation more than once, that indicates that for that input there is more than one output. At any particular input value, there can be only one output if the relation is to be a function.

5. function

7. not a function

9. function

11. function

13. function

15. function

17. not a function

19. not a function

21. function

23. function

25. not a function

27. function

29. not a function

31. domain: $-2 \le x \le 3$ or $[-2, 3]$; range: $-4 \le y \le 4$ or $[-4, 4]$

33. domain: $-5 \le x < 2$ or $[-5, 2)$
range: $-5 \le y \le 3$ or $[-5, 3]$

35. domain: $x \ge -2$ or $[-2, \infty)$
range: $y \ge -3$ or $[-3, \infty)$

37. domain: all real numbers or $(-\infty, \infty)$
range: $y \le 4$ or $(-\infty, 4]$

39. domain: $1998 \le x \le 2008$ or $[1998, 2008]$
range: $81 \le y \le 139$ or $[81, 139]$

41. domain: $[0, 9.8]$; range: $[0, 225]$

43. domain: $(-12, \infty)$; range: $(-\infty, -7)$

45. a) $y = x^2 + 2$

b) 
| $x$ | -3 | -2 | -1 | 0 | 1 | 2 | 3 |
|---|---|---|---|---|---|---|---|
| $y$ | 11 | 6 | 3 | 2 | 3 | 6 | 11 |

c)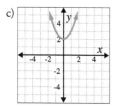

d) domain: $(-\infty, \infty)$; range: $[2, \infty)$

47. a) Multiply the input by $-2$ and add 5 to obtain the output.

b)
| $x$ | -3 | -2 | -1 | 0 | 1 | 2 | 3 |
|---|---|---|---|---|---|---|---|
| $y$ | 11 | 9 | 7 | 5 | 3 | 1 | -1 |

c)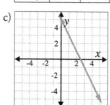

d) domain: $(-\infty, \infty)$; range: $(-\infty, \infty)$

## 1.3 Finding Equations of Linear Functions

1. a) Terry starts at an elevation of 3000 feet and descends 70 feet per second.

   b) 
   | x | 0 | 10 | 20 | 30 | 40 |
   |---|---|---|---|---|---|
   | y | 3000 | 2300 | 1600 | 900 | 200 |

   c)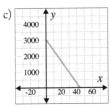

3. a) $y = 8x + 25$

   b)
   | x | 0 | 1 | 2 | 3 | 4 |
   |---|---|---|---|---|---|
   | y | 25 | 33 | 41 | 49 | 57 |

   c)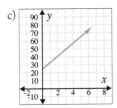

5. increasing
7. decreasing
9. decreasing
11. increasing
13. decreasing
15. $m = 250.25 \approx 250$ The total number of visitors to the museum increases by approximately 250 visitors per month.
17. $m = -1.5$ The scuba diver *descends* at a rate of 1.5 feet per second.
19. $y = 42.50x + 120$ The slope, $m = 42.50$, indicates the plumber charges a rate of 42.50 dollars per hour. The y-intercept (0, 120) indicates the plumber's initial charge is 120 dollars. If the plumber works for $x = 5.5$ hours, then he will charge $353.75.
21. $y = -13x + 4500$ The slope, $m = -13$, indicates that the volume of grain in the silo decreases at a rate of 13 cubic meters per minute. The y-intercept (0, 4500) indicates the initial amount of grain is 4500 cubic meters. After 4 hours, $x = 240$ minutes and there will be 1380 cubic meters of grain remaining in the silo.
23. $y = -3.6x + 100$ The slope, $m = -3.6$, indicates the equipment descends at a rate of 3.6 meters per minute. The y-intercept (0, 100) indicates the initial distance to the bottom of the shaft is 100 meters. After $x = 20$ minutes, the equipment will be 28 meters from the bottom of the shaft.
25. $y = 45x + 180$ The slope, $m = 45$, indicates he farms 45 gold coins per hour. The y-intercept, (0, 180), indicates the number of coins he had when he started farming. After $x = 15$ hours, Marco will have farmed the 855 gold he needs to buy the armor.
27. $y = -12x + 5$
29. $y = 3x - 5$
31. $y = -1.5x + 33$
33. $y = 1.2x + 6$
35. The y-intercept is (0, 120) and indicates that Freddie initially had $120. The x-intercept is (8, 0) and indicates that he played 8 losing rounds and lost all of his money.
37. (a) The slope, $m = -12$, indicates that the water is draining from the trough at a rate of 12 gallons per minute. The y-intercept (0, 350) indicates there were initially 350 gallons of water in the trough.

    b) 182 gallons

    c) When $y = 100$, $x = 20.8\overline{3}$ After approximately 20.8 minutes there will be 100 gallons of water remaining.

    d) When $y = 0$, $x = 29.1\overline{6}$ It will take a little more than 29 minutes before the trough is empty.

## 1.4 Using Linear Functions to Model Data

1. Interpolation is when you make a prediction within the domain and range of the given data.

3. Positive linear relationship means the data presents a nearly linear model with a positive slope (an increasing function). Negative linear relationship means the data presents a nearly linear model with a negative slope (a decreasing function).

5. negative linear relationship

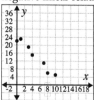

7. positive linear relationship

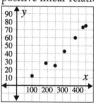

9. Equations will vary slightly depending on points chosen. One possible solution is $y = 112.5x + 11500$. In 2018, $x = 28$, we predict the population of Midgar will be 14,650. This is extrapolation.

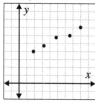

11. Graph 24d

13. Graph 24a

15. $y = 1.7x + 124$ In 2012, $x = 22$: the number of people in the U.S. labor force will be 161.4 million. In 2020, $x = 30$: the number of people in the U.S. labor force will be 175 million.

17. $y = 0.476x + 20.745$; When $y = 35$, $x = 29.947$; In about 30 years (2020) the percentage of persons 25 yrs or older who are college graduates will exceed 35%.

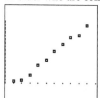

19. $y = -22.803x + 736.727$. In 2016, $x = 26$ and $y = 143.849$. The model predicts there will be about 144 people unemployed in Waldorf in 2016.

21. $y = 1.273x + 40.944$; The model predicts that in 2020, $x = 30$, about 79% of high schoo students will take Algebra 2.

23. $y = 0.066x - 7.149$; A sandwich with 620 calories will have about 33.8 grams of fat.

## 1.5 Function Notation and Making Predictions

1. $-11, -1, 2a + 1$

3. $20, 15, 3a^2 + 20a + 32$

5. $k(2) = 3, t = 4$

7. a) 15
   b) $5, -5$

9. a) 5
   b) $x = 4$

11. a) 1,
    b) 6.5

13. a) $-3$
    b) $x = 0, x = -6$

15. a) 53
    b) $x = 2$

17. $8, 6, 4, 2$

19. $21, 11, 3, -3$

21. $-4, \frac{-3}{2}, \frac{-2}{3}, \frac{-1}{4}$

23. a) $f(40) = 13$, b) A population of 5,000 people produces 2 tons of garbage per week.

25. a) There are 30 ducks in the lake in 1995.
    b) There are 40 ducks in the lake in the year 2000.

27. a) $W = f(t) = 1.5t + 7.5$
    b) $(0, 7.5)$ Arlo weighs 7.5 lbs when he is born.
    c) $(-5, 0)$ This means Arlo would have weighed nothing (0 lbs) 5 months before he was born. This is a breakdown of the model.
    d) $f(6) = 16.5$. Arlo weighs 16.5 lbs when he is 6 months old.
    e) When $W = 24$, $t = 11$. Arlo will weigh 24 lbs in 11 months.

# Chapter 2: Exponential Functions

## 2.1 Properties of Exponents

1. No, the two expressions are not the same. An exponent tells how many times you multiply the base. So $2^3$ is the same as $2 \times 2 \times 2$, which is 8. $3^2$ is the same as $3 \times 3$, which is 9.
3. Scientific notation is a method of writing very small and very large numbers in the form of $a \times 10^n$.
5. 81
7. 243
9. $\frac{1}{16}$
11. $\frac{1}{11}$
13. 1
15. $4^9$
17. $12^{40}$
19. $\frac{1}{7^9}$
21. $a^4$
23. $b^6 c^8$
25. $ab^2 d^3$
27. $m^4$
29. $\frac{q^5}{p^6}$
31. $\frac{y^{21}}{x^{14}}$
33. 25
35. $72a^2$
37. $\frac{c^3}{b^9}$
39. $\frac{a^{14}}{1296}$
41. $\frac{n}{a^9 c}$
43. $\frac{1}{a^6 b^6 c^6}$
45. $3.14 \times 10^{-5}$
47. 16,000,000,000
49. 0.000022 m
51. $1.0995 \times 10^{12}$
53. 0.00000000003397 in.
55. 602,214,130,000,000,000,000,000
57. $\approx 4.8312 \times 10^{25}$
59. $\approx 1.3555 \times 10^{19}$
61. $\approx 1.3876 \times 10^{14}$
63. $\approx 5.2 \times 10^{-8}$
65. $\approx 9.4182 \times 10^{20}$
67. $\approx 1.1360 \times 10^{-8}$

## 2.2 Rational Exponents

1. 2
3. 3
5. 16
7. 4
9. $\frac{1}{3}$
11. $\frac{1}{8}$
13. −4
15. −4
17. $3^{6/5}$
19. 320
21. 2
23. $\frac{1}{2}$
25. $a^{3/5}$
27. $\frac{1}{b^{1/2}}$
29. $5a^3$
31. $30ab^2$
33. $a^{1/2}$
35. $\frac{7}{b^{1/11}}$
37. $\frac{1}{a}$
39. $\frac{2b^{6/5}}{3c^{4/5}}$

## 2.3 Exponential Functions

1. For an exponential function, if the value of the base is greater than 1, $b > 1$, then the function represents exponential growth. If the value of the base is between 0 and 1, $0 < b < 1$, then the function represents exponential decay.
3. The $y$-intercept occurs when $x = 0$. For $f(x) = ab^x$, $f(0) = ab^0 = a$, so the $y$-intercept is at $(0, a)$.
5. Exponential. The base is a constant, and the independent variable $x$ is the exponent.
7. Exponential. The base is a constant, and the independent variable $t$ is the exponent.
9. Not exponential. The base is the independent variable $x$, and there is no exponent.
11. 1944
13. $\frac{8}{9} \approx .889$

15. 18
17. ≈ 101.136
19.
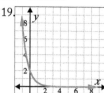
domain: $(-\infty, \infty)$ and range: $(0, \infty)$

21.
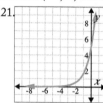
domain: $(-\infty, \infty)$ and range: $(0, \infty)$

23. exponential decay; (0, 300)
25. exponential growth; (0, 0.5)
27. exponential decay; (0, 3)
29. exponential growth; (0, 1)
31. B
33. A
35. E
37. D
39. C
41. The forest represented by the function $B(t) = 82(1.029)^t$
43. After $t = 20$ years, forest A will have 43 more trees than forest B.
45. a) $2700; b) $4175.18
47. a) shrinking; b) 3970 people; c) about 216 people

## 2.4 Finding Equations of Exponential Functions

1. $a = 8, b = 2, f(x) = 8(2)^x$
3. $m = -7, b = 11, f(x) = -7x + 11$
5. $a = 243, b = \frac{1}{3}, f(x) = 243\left(\frac{1}{3}\right)^x$
7. $b = \pm 7$
9. $b = 3$
11. $b = -0.5$
13. $b \approx \pm 2.667$
15. There are no real number solutions.
17. $b \approx 0.839$
19. $a = 6, b = 5, y = 6(5)^x$
21. $a = 450, b \approx 0.447, y = 450(0.447)^x$
23. $a = 4.5, b \approx 1.200, y = 4.5(1.2)^x$
25. $a = 11.77, b \approx 0.750, y = 11.77(0.75)^x$
27. Let $f(t)$ be the population of chicken turtles $t$ years after the initial release.
   a) (0, 300) and (7, 550)
   b) $f(t) = 300(1.09)^t$
   c) $f(12) = 300(1.09)^{12} = 844$ Twelve years after the initial release, the chicken turtle population will be about 844.
29. Let $f(t)$ be the value of the Prius $t$ years after 2007.
   a) (0, 19500) and (9, 5600)
   b) $f(t) = 19500(0.871)^t$
   c) $f(13) = 19500(0.871)^{13} = 3237.95$ In 2020, the value of the car will only be about $3238.

## 2.5 Using Exponential Functions to Model Data

1. exponential growth, 19% increase per unit of time
3. exponential decay, 2% decrease per unit of time
5. exponential growth, 2.85% increase per unit of time
7. exponential decay, 50% decrease every 12 units of time (or approximately 5.61% decrease per unit of time)
9. $f(t) = 42{,}000(1.0254)^t$
11. $f(t) = 250\left(\frac{1}{2}\right)^{t/28}$ or $f(t) = 250(.9755)^t$
13. $f(t) = 1500(2)^{t/12}$ or $f(t) = 1500(1.0595)^t$
15. $f(t) = 2340(0.905)^t$
17. a) $f(t) = 2700(1.0325)^t$
   b) In 5 years, about $3168.21, and in 10 years, about $3717.61.
19. a) $f(t) = 1000(1.1041)^t$
   b) In 25 years, about $11,887.95.
21. a) $P = f(t) = 9740(0.928)^t$
   b) About 712 people in 2015.

23. a) $V = f(t) = 32000 (0.76)^t$
b) In 4 years, about $10,676.
25. a) $f(t) = 40 \left(\frac{1}{2}\right)^{t/1620}$
b) After 2000 years, 17 mg.
27. $200 \left(\frac{1}{2}\right)^{120/20} = 3.125$ mg
29. a) $y = 84.638 (1.012)^x$
b)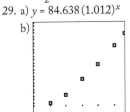
c) 1.2% increase per year.
d) In 2025, about 375.1 million people, but the answers may vary slightly due to rounding.

31. a) $y = 0.690 (1.084)^x$
b)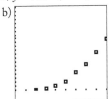

c) Since $a \approx 0.690$, we estimate that the revenue collect by the IRS in the initial year 1900 (in this problem) was 0.690 billion dollars or about $690,000,000. Since $b \approx 1.084$, the percent change is 8.4% increase per year.
d) In 2018, the IRS will collect about 9382.2 billion dollars or about $9,382,000,000,000.

33. a) $y = 402.388 (2.283)^x$
b)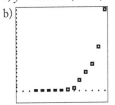

c) In 2000, there were approximately 500,317,295 computers connected to the Internet, but answers may vary due to rounding.

# Chapter 3: Logarithmic Functions

## 3.1 Logarithmic Functions

1. 3
3. 5
5. 2
7. 4
9. 7
11. 0
13. −3
15. −2
17. −5
19. 0
21. 1
23. ½
25. ⅓
27. ½
29. ¼
31. 5
33. 3.831
35. 1.462
37. −0.604
39. 4
41. −2
43. 3
45. $\frac{1}{2}$
47. 2.5; 7.9
49. 2; 5.8

## 3.2 Properties of Logarithms

1. $4^3 = 64$
3. $5^4 = 625$
5. $2^{-3} = \frac{1}{8}$
7. $10^0 = 1$
9. $w^r = m$
11. $\log_2(256) = 8$
13. $\log_{49}(7) = \frac{1}{2}$
15. $\log(10{,}000) = 4$
17. $\log_4(97) = x$
19. 144
21. 1
23. $\frac{1}{1000} = 0.001$
25. 4
27. 25
29. 2
31. 7
33. 81
35. 2.1129
37. 4.2009
39. 3.2161
41. 7.2380
43. 2.7030
45. 0.9783
47. 9.2788
49. 6.8480
51. $x = -25, x = 1.1946$
53. 1.5150, 4.9230
55. no real-numbered solutions
57. a) $f(t) = 2200(1.047)^t$
    b) $3817.53
    c) 17.9 years
59. in 26.1 years
61. a) $f(t) = 15{,}000(0.96)^t$
    b) 9972 whales
    c) in 15.4 years
63. in 4.7 years
65. a) 180 mg
    b) After 12 hrs there are 45 mg of caffeine remaining.
    c) in 25 hrs.
67. a) The population is increasing by 1.3% per year.
    b) In 2008, the population was 8.649 thousand people.
    c) in 2019
69. a) about 680.3 million airline travelers
    b) in 2032
71. in 2021
73. 9 years
75. 11.9 years

## 3.3 Using Logarithms to Make Predictions with Exponential Models

1. 2.3823
3. 8.4338
5. $e^7 = 1096.6$
7. $e^{-0.3} = 0.74$
9. $e^{6.2} = 493$
11. $\ln(126) \approx 4.836$
13. $\ln(403.43) \approx 6$
15. $\ln(0.0067) \approx -5$
17. $x = 54.5982$
19. $x = 11.3891$
21. $x = 24.7355$
23. $x = 4.8866$
25. $x = 5.7323$
27. $x = 7.8396$
29. $x = 5.1358$
31. a) There were initially 500 bacteria.
    b) about 725 bacteria
    c) 14.1 hours
33. a) The initial investment was $15,000.
    b) $23,290.61
    c) 22.3 years
35. 3.7 minutes
37. 12.9 years
39. 47.8 years

# Chapter 4: Quadratic Functions

## 4.1 Expanding and Factoring Polynomials

1. $48x^7y^8$
3. $-14a^5b^7$
5. $4x^3 + 36x^2 - 20x$
7. $-21x^4 + 33x^3 - 6x^2$
9. $-18a^3b^2 + 15a^2b^2 - 12ab^2$
11. $2x^4y + 18x^3y^2 - 2x^2y^3$
13. $x^2 - 2x - 8$
15. $x^2 + 19x + 88$
17. $x^2 - 15x + 54$
19. $x^3 + 8x^2 - 4x - 32$
21. $8a^4 - 4a^3 + 24a^2 - 12a$
23. $20x^5 + 4x^4 - 15x^3 - 3x^2$
25. $12a^2 - 17ab - 5b^2$
27. $x^2 + 18x + 81$
29. $a^2 - 14a + 49$
31. $4m^2 - 12m + 9$
33. $25x^2 + 120x + 144$
35. $x^2 - 4$
37. $b^2 - 36$
39. $16x^2 - 1$
41. $4a^2 - 49b^2$
43. $x^3 - 9x^2 - 31x - 18$
45. $6x^3 - 11x^2 - 19x + 28$
47. $2x(7 + 2y - 9y^2)$
49. $15xy(2x^2 - 3xy + 9y^2)$
51. $18a^2b^2(2a^2 - ab + 3b^2)$
53. $16x^3(5x^4 - y)$
55. $(a+11)(a-2)$
57. $(x+5)(x+7)$
59. $(x-8)(x-1)$
61. prime polynomial
63. $(x-6)(x+5)$
65. $(t-12)(t+3)$
67. $(7x-1)(x+7)$
69. $(2b-3)(b+8)$
71. $(5t-4)(t-3)$
73. $-5(x+3)(x+7)$
75. $-4x(x+6)(x-1)$
77. $(x+1)(x-1)$
79. $(2m+3)(2m-3)$
81. $(5x+14)(5x-14)$
83. $(12a+7)(12a-7)$
85. $x = 0, x = -8$
87. $a = 0, a = 7$
89. $x = 11, x = -11$
91. $m = 10$
93. $x = \frac{5}{2}, x = -\frac{5}{2}$
95. $x = 6, x = 3$
97. $x = -7, x = 5$
99. $x = -\frac{2}{3}, x = -\frac{8}{3}$
101. $x = \frac{1}{2}, x = -5$
103. $x = -\frac{8}{3}, x = -10$
105. $x = -\frac{2}{7}, x = 3$
107. $x = \frac{5}{2}, x = \frac{4}{3}$
109. $x = 0, x = \frac{5}{4}$
111. $x = 3, x = -2$
113. $x = 6, x = -2$
115. $x = -5, x = 5$
117. $x = 0, x = -1, x = -4$
119. $x = 0, x = -4, x = -11$

## 4.2 Quadratic Functions in Standard Form

1. a) opens up

b)
| $x$ | $-3$ | $-2$ | $-1$ | $0$ | $1$ | $2$ | $3$ | $4$ |
|---|---|---|---|---|---|---|---|---|
| $f(x)$ | $4$ | $-5$ | $-8$ | $-5$ | $4$ | $19$ | $40$ | $67$ |

c) Not all points from the table are on the graph.

d) $(-1, -8)$
e) $(0, -5)$

3. a) opens down

b)
| $x$ | $-3$ | $-2$ | $-1$ | $0$ | $1$ | $2$ | $3$ | $4$ |
|---|---|---|---|---|---|---|---|---|
| $f(x)$ | $-18$ | $-9$ | $-2$ | $3$ | $6$ | $7$ | $6$ | $3$ |

c) Not all points from the table are on the graph.

d) $(2, 7)$
e) $(0, 3)$

5. a) opens up
   b) (0, −4)
   c) (0, −4)
7. a) opens up
   b) (0, 13)
   c) (−4, −3)
9. a) opens down
   b) (0, 5)
   c) (1, 6)
11. a) opens up
    b) $(0, \frac{5}{4})$
    c) $(\frac{1}{2}, 1)$
13. vertex: (−1.5, 7)
    domain: all real numbers
    range: $y \leq 7$
15. vertex: (−1, 5.6)
    domain: all real numbers
    range: $y \leq 5.6$
17. vertex: (3, −9.9)
    domain: all real numbers
    range: $y \geq -9.9$
19. vertex: (−2, 4)
    domain: all real numbers
    range: $y \geq 4$
21. a) $A = -2W^2 + 130W$
    b) $W = 32.5$ ft.
    c) The maximum area is 2112.5 sq. ft.
23. The manufacturer should produce 20 garden hoses per day for a minimum production cost of $700.
25. 16 feet
27. $x = 30$; Producing 30 units per day will yield a maximum profit of $1000.
29. a) 2 meters
    b) $f(7) = 447.9$ After 7 seconds the rocket is 447.9 meters high.
    c) 10 seconds
    d) 492 meters
31. a) (0, 120) The stone is 120 ft above the beach when it is first tossed.
    b) (1.25, 145); The stone reaches a maximum height of 145 feet after 1.25 seconds.
33. $P = -20n^2 + 200n + 97{,}500$
    $n = 5$; Planting 5 more trees will yield a maximum crop of 98,000 apples.

## 4.3 The Square Root Property

1. 14
3. $2\sqrt{7}$
5. $6\sqrt{5}$
7. $\frac{4}{9}$
9. $\frac{\sqrt{17}}{12}$
11. $\frac{6\sqrt{5}}{5}$
13. $\frac{\sqrt{7}}{7}$
15. $\frac{\sqrt{21}}{7}$
17. ±6
19. $\pm\sqrt{14}$
21. $\pm 2\sqrt{10}$
23. $\pm\frac{\sqrt{2}}{3}$
25. ±2
27. $\frac{1}{3}, 5$
29. $-6 \pm \sqrt{7}$
31. $-9 \pm 2\sqrt{3}$
33. −8, 10
35. $3i$
37. $5i\sqrt{2}$
39. $\frac{2}{5}i$
41. $8i\sqrt{3}$
43. $\pm 2i$
45. $\pm 3i\sqrt{5}$
47. $\pm i\sqrt{2}$
49. $9 \pm 10i$
51. $-1 \pm 2i\sqrt{2}$
53. $\frac{7}{4} \pm \frac{3}{4}i$

## 4.4 The Quadratic Formula

1. $4, -\frac{1}{2}$
3. $\frac{3 \pm 2\sqrt{6}}{5}$
5. $\frac{1 \pm i\sqrt{11}}{6}$
7. $-\frac{4}{3}$
9. (−1, 0), (−1.5, 0), and (0, 3)
11. (−3.225, 0), (−0.775, 0), and (0, −5)
13. (4.449, 0), (−0.449, 0), and (0, −2)
15. (0.425, 0), (−1.175, 0)
17. (0.563, 0), (9.237, 0)
19. $x \approx 1.653$, $x \approx -6.653$
21. $t \approx 4 \rightarrow$ April 1999, $t \approx 24 \rightarrow$ Dec. 2000
23. 2.728 secs.
25. discriminant = 73; two real-number solutions; two x-intercepts
27. discriminant = −92; two imaginary-number solutions; no x-intercepts
29. discriminant = 0; one real-number solution; one x-intercept
31. discriminant = 1; two real-number solutions; two x-intercepts

33. $5 \pm 6i$
35. $2, -9$
37. $-0.208, 3.208$

39. $1, -\frac{1}{4}$
41. $-3 \pm i\sqrt{3}$
43. $1.589, -1.743$

45. $\pm i\sqrt{7}$
47. $-6, 5$
49. $-2, 5$

## 4.5 Modeling with Quadratic Functions

1. $R = -0.0125 x^2 + 45x$
   1800 thousand phones (or 1,800,000 phones)
   40500 thousand dollars (or $40,500,000)
3. 2.449 secs; 37.388 meters
5. 2100 rpm; 850 horsepower
7. a) At t = 0 sec, the softball is thrown from a height of 6.5 ft, and at 2 secs, the height is 32.5 ft
   b) It takes 1.406 sec to reach a maximum height of 38.14 ft
   c) 2.95 sec
   d) 0.5 and 2.313 seconds
9. a) $y = 48.81 x^2 + 25x - 181$
   b) about 11,179 kg

11. a) $y = .05 x^2 + 4.55x - 32.15$
    b) $45.5 \times 100 = 4550$ units sold.
    c) $x \approx 7.72$ and $x \approx 83.28$. These are the break–even values. The company will make a profit if it sells between 772 and 8328 units.
13. a) linear: $y = 1.753x + 14.673$,
    quadratic: $y = 0.097 x^2 + 0.788x + 16.120$
    b) The quadratic model appears to be the better fit because the data points are closer to the curve when we graph the equation and the scatter plot in the same window.
    c) linear: $60.251 billion; quadratic: $102.18 billion

# Chapter 5: Further Topics in Algebra

## 5.1 Variation

1. $y = 20x$
3. $p = 3q^3$
5. $V = 6t^4$
7. $y = \frac{8}{x}$
9. $h = \frac{1.28}{d^2}$
11. $r = \frac{15}{\sqrt{A}}$
13. $y = 34$
15. $t = 3$
17. $y = 8$
19. $t = 2$
21. $Q$ increases
23. $T$ increases
25. less resistance
27. 3 seconds
29. 48 inches
31. 49.75 pounds
33. $C = 33.33$ amperes

## 5.2 Arithmetic Sequences

1. A sequence where each successive term of the sequence increases (or decreases) by a constant value.
3. We find whether the difference between all consecutive terms is the same. This is the same as saying that the sequence has a common difference.
5. 17, 29, 41, 53, 65; arithmetic with $d = 12$
7. $\frac{1}{2}, \frac{1}{3}, \frac{1}{4}, \frac{1}{5}, \frac{1}{6}$; not arithmetic
9. −2, 13, 28, 43, 58; arithmetic with $d = 15$
11. 99, 96, 91, 84, 75; not arithmetic
13. arithmetic; $d = \frac{1}{2}$
15. The sequence is not arithmetic because $16 − 4 \neq 64 − 16$.
17. arithmetic; $d = −21$
19. arithmetic; $d = −1.9$
21. $a_n = 5n + 3$; $a_{25} = 128$
23. $a_n = −7n + 34$; $a_{25} = −141$
25. $a_n = −25n + 13$; $a_{25} = −612$
27. $a_n = 1.75n + 21.75$; $a_{25} = 65.5$
29. $n = 20$
31. $n = 238$
33. $n = 82$
35. $n = 35$
37. No; $n \approx 25.3$ which is not a positive whole number.
39. a) $a_n = 3n + 18$
    b) $a_{12} = 54$, There are 54 seats in the 12th row.
    c) 40 rows
41. a) $a_n = 950n + 31{,}550$
    b) $a_5 = 36{,}300$, She will make \$36,300 in the 5th year of employment.
    c) in the 20th year
43. a) $a_n = −8n + 217$
    b) $a_{14} = 105$, After 2 weeks (14 days) there are 105 food units left.
    c) 27 days

## 5.3 Geometric Sequences

1. A sequence in which the ratio between any two consecutive terms is constant.
3. Both geometric sequences and exponential functions have a constant ratio. However, their domains are not the same. Exponential functions are defined for all real numbers, and geometric sequences are defined only for positive integers.
5. geometric; $r = 3$
7. arithmetic; $d = −20$
9. neither
11. geometric; $r = \frac{1}{2}$.
13. arithmetic; $d = 9$
15. neither
17. $a_n = 7(4)^{n-1}$
19. $a_n = \left(\frac{4}{5}\right)^{n-1}$
21. $a_n = 768\left(\frac{1}{4}\right)^{n-1}$
23. $a_n = 10(7)^{n-1}$
25. 196,608
27. $4.768 \times 10^{-6}$
29. $9.038 \times 10^{11}$
31. $n = 12$
33. $n = 8$
35. $n = 16$
37. $n = 15$
39. a) $a_n = 20\left(\frac{7}{8}\right)^{n-1}$
    b) 6.872 ft.
41. \$5,368,709.12
43. a) $a_n = 500\left(\frac{2}{3}\right)^{n-1}$
    b) $a_4 = 148.148$; About 148 cubic cm of air remain after 4 cycles.
    c) 13 cycles
45. a) Aaron: $a_n = 40000(1.0375)^{n-1}$
    Beatriz: $a_n = 1000n + 44000$
    b) Aaron: $a_4 = \$44{,}671$, $a_{12} = \$59{,}969$
    Beatriz: $a_4 = \$48{,}000$, $a_{12} = \$56{,}000$
    c) Aaron: $a_{30} = \$116{,}336$
    Beatriz: $a_{30} = \$74{,}000$
    d) year 9

## 5.4 Dimensional Analysis

1. 9.06 lbs
3. 76,032 in
5. 8.56 m
7. 7900 cm³
9. 177.17 in
11. 208.15 oz
13. 4.59 acres
15. 171.33 yd²
17. 6.4 m²
19. 7964.11 cm³
21. 300.24 km/hr
23. 127.10 lb/yd²
25. 17.78 cm/sec
27. 9.25 lb/in³
29. 11.90 km/L
31. 7.64 cents/min
33. 16.82 ft³/sec
35. 1.43 m/min
37. 5.1 L/min
39. 1.00 kg/L
41. 32.18 km/hr, 72.41 km/hr, 88.50 km/hr
43. 22.37 mph
45. A gallon of water weighs about 8.34 pounds.
47. 125 ft-lb
49. $2.78 per gallon, 52.26 miles per gallon, 83.33 miles per day

CPSIA information can be obtained
at www.ICGtesting.com
Printed in the USA
LVHW010905111121
703022LV00001B/5